ART AND SYMMETRY IN EXPERIMENTAL PHYSICS

Related Titles from AIP Conference Proceedings

589 New Developments in Fundamental Interaction Theories: 37th Karpacz Winter School of Theoretical Physics
Edited by Jerzy Lukierski and Jakub Rembieliński, October 2001, 0-7354-0029-6

570 SPIN 2000; 14th International Spin Physics Symposium
Edited by Kichiji Hatanaka, Takashi Nakano, Kenichi Imai, and Hiroyasu Ejiri, June 2001, 0-7354-0008-3

551 Atomic Physics 17: XVII International Conference on Atomic Physics; ICAP 2000
Edited by Ennio Arimondo, Paolo De Natale, and Massimo Inguscio, February 2001, 1-56396-982-3

549 Intersections of Particle and Nuclear Physics: 7th Conference, CIPANP2000
Edited by Zohreh Parsa and William J. Marciano, December 2000, 1-56396-978-5

545 Spin Statistics Connection and Commutation Relations: Experimental Tests and Theoretical Implications
Edited by Robert C. Hilborn and Guglielmo M. Tino, November 2000, 1-56396-974-2

543 Atomic and Molecular Data and Their Applications: ICAMDATA—Second International Conference
Edited by Keith A. Berrington and Kenneth L. Bell, November 2000, 1-56396-971-8

541 Theoretical High Energy Physics: MRST 2000
Edited by C. R. Hagen, November 2000, 1-56396-966-1

539 Symmetries in Subatomic Physics: 3rd International Symposium
Edited by X.-H. Guo, A. W. Thomas, and A. G. Williams, October 2000, 1-56396-964-5

506 X-Ray and Inner Shell Processes: 18th International Conference
Edited by R. W. Dunford, D. S. Gemmell, E. P. Kanter, B. Krässig, S. H. Southworth, and L. Young, February 2000, 1-56396-713-8

270 Time Reversal– The Arthur Rich Memorial Symposium
Edited by Mark Skalsey, Philip H. Bucksbaum, Ralph S. Conti, and David W. Gidley, 1993, 1-56396-105-9

To learn more about these titles, or the AIP Conference Proceedings Series, please visit the webpage **http://proceedings.aip.org**

ART AND SYMMETRY IN EXPERIMENTAL PHYSICS

Festschrift for Eugene D. Commins

Berkeley, California   20–21 May 2001

EDITORS
Dmitry Budker
University of California, Berkeley
Philip H. Bucksbaum
University of Michigan, Ann Arbor
Stuart J. Freedman
University of California, Berkeley

Melville, New York, 2001
AIP CONFERENCE PROCEEDINGS ■ VOLUME 596

Editors:

Dmitry Budker
Department of Physics
University of California, Berkeley
Berkeley, CA 94720-7300
USA

E-mail: Budker@socrates.berkeley.edu

Philip H. Bucksbaum
Randall Laboratory
University of Michigan
Ann Arbor, MI 48109-1120
USA

E-mail: phb@umich.edu

Stuart J. Freedman
Department of Physics
University of California, Berkeley
Berkeley, CA 94720-7300
USA

E-mail: SJFreedman@lbl.gov

The article on pp. 253-274 was authored by a U.S. Government employee and is not covered by the below mentioned copyright.

Authorization to photocopy items for internal or personal use, beyond the free copying permitted under the 1978 U.S. Copyright Law (see statement below), is granted by the American Institute of Physics for users registered with the Copyright Clearance Center (CCC) Transactional Reporting Service, provided that the base fee of $18.00 per copy is paid directly to CCC, 222 Rosewood Drive, Danvers, MA 01923. For those organizations that have been granted a photocopy license by CCC, a separate system of payment has been arranged. The fee code for users of the Transactional Reporting Service is: 0-7354-0040-7/01/$18.00.

© 2001 American Institute of Physics

Individual readers of this volume and nonprofit libraries, acting for them, are permitted to make fair use of the material in it, such as copying an article for use in teaching or research. Permission is granted to quote from this volume in scientific work with the customary acknowledgment of the source. To reprint a figure, table, or other excerpt requires the consent of one of the original authors and notification to AIP. Republication or systematic or multiple reproduction of any material in this volume is permitted only under license from AIP. Address inquiries to Office of Rights and Permissions, Suite 1NO1, 2 Huntington Quadrangle, Melville, N.Y. 11747-4502; phone: 516-576-2268; fax: 516-576-2450; e-mail: rights@aip.org.

L.C. Catalog Card No. 2001096132
ISBN 0-7354-0040-7
ISSN 0094-243X
Printed in the United States of America

CONTENTS

Preface vii
Group Photo ix

SYMPOSIUM OVERVIEW, HISTORY, AND RECOLLECTIONS

Gleanings from the ComminsFest Symposium 3
    J. D. Jackson
Weak Interactions of Leptons and Quarks Using Radioactive Atomic Beams 13
    P. A. Vetter
A Random Walk in Science 21
    S. Chu

SEARCHES FOR P,T VIOLATIONS IN ATOMS AND MOLECULES

The Search for a Permanent Electric Dipole Moment—Still Active, Still Important 39
    E. N. Fortson
New Search for a Permanent Electric Dipole Moment of ^{199}Hg 47
    M. V. Romalis, W. C. Griffith, J. P. Jacobs, and E. N. Fortson
Measuring the Electron Electric Dipole Moment in Ybf 62
    B. E. Sauer, J. J. Hudson, M. R. Tarbutt, and E. A. Hinds
Search for the Electric Dipole Moment of the Electron Using Metastable PbO 72
    D. DeMille, F. Bay, S. Bickman, D. Kawall, L. Hunter, D. Krause, Jr.,
    S. Maxwell, and K. Ulmer
Progress towards Fundamental Symmetry Tests with Nonlinear Optical Rotation 84
    D. F. Kimball, D. Budker, D. S. English, C.-H. Li, A.-T. Nguyen,
    S. M. Rochester, A. O. Sushkov, V. V. Yashchuk, and M. Zolotorev
Atomic Tests of Discrete Symmetries at Berkeley 108
    D. S. English, D. F. Kimball, C.-H. Li, A.-T. Nguyen, S. M. Rochester,
    J. E. Stalnaker, V. V. Yashchuk, D. Budker, S. J. Freedman,
    and M. Zolotorev

SEARCHES FOR NEW PARTICLES AND INTERACTIONS

Hunting the Fifth Force on the Snake River 123
    W. R. Bennett, Jr.
Remarks on Search Methods for Stable, Massive, Elementary Particles 156
    M. L. Perl

MORE ADVENTURES IN ATOMIC, MOLECULAR, AND NUCLEAR PHYSICS

From Helium-6 to Krypton-81 171
    Z.-T. Lu
Quantum Control 176
    P. H. Bucksbaum
The Creation and Measurement of Chiral Coherences 186
    R. A. Harris and J. D. Walls
CASCADE: A New Efficient and Position Sensitive Detector for Thermal Neutrons on Large Areas 193
    M. Klein, H. Abele, D. Fiolka, and C. J. Schmidt

THEORETICAL PERSPECTIVES

Symmetries and the Connection between Spin and Statistics in Rigorous Quantum Field Theory 201
    E. H. Wichmann
Electric Field Distribution in Nuclei Produced by the P,T-Odd Nuclear Schiff Moment 232
    V. V. Flambaum and J. S. M. Ginges
Night Thoughts on Consciousness and Time Reversal 246
    A. Zee

ASTROPHYSICS AND COSMOLOGY

Supernovae, Dark Energy, and the Accelerating Universe—What Next? 253
    S. Perlmutter
Mid-Infrared Stellar Interferometry and Diameters of Cold Stars 275
    C. H. Townes

Author Index 285

Preface

On May 20-21, 2001, colleagues, friends, former and present students, gathered at Berkeley to honor Professor Eugene David Commins on the occasion of his retirement and 69th birthday. This volume contains articles based on talks given at the CommminsFest Symposium that was held at the Physics Department, as well as several additional contributions by authors from different parts of the World. The diversity of topics, as well as the balance between theory and experiment are dictated by Prof. Commins' past and present scientific interests, spanning a very broad range -- from experimental nuclear, atomic, and molecular physics, and elementary particle physics, to theoretical astrophysics.

In addition to science, those who attended the Symposium also experienced much art and music, both life-long passions of Prof. Commins. We hope the readers will enjoy reprints of Prof. Commins' drawings made over the years that can be found in this volume.

Finally, we wish to acknowledge Professors William Chinowsky, Roger Falcone, and Ms. Carol Dudley, who have co-organized the CommminsFest Symposium with us, and the staff and students of the Berkeley Physics Department, who helped make this event a success.

                                      D. Budker, P. H. Bucksbaum, and S. J. Freedman
                                      August 2001, Berkeley and Ann Arbor

CommninsFest Symposium participants. Color version of this photograph, as well as additional Symposium photographs, Symposium information, and Prof. Commins' art are available at http://socrates.berkeley.edu/~genesymp.

SYMPOSIUM OVERVIEW, HISTORY, AND RECOLLECTIONS

Gleanings from the CommninsFest Symposium

J. D. Jackson

Department of Physics, University of California, Berkeley and Lawrence Berkeley National Laboratory
Berkeley, CA 94720

Abstract. Notes taken on the sixteen talks of the day-long Symposium have been worked into an informal narrative of what I heard and learned, perhaps imperfectly, from the lectures.

MORNING SESSION 1

I arrived a little late. Stuart Freedman was just completing his opening remarks with a transparency listing Gene Commins's 28 Ph.D. students. I sat down near the front and began taking notes on my yellow checked pad. The taking of notes was a resuscitation of an old habit, invoked today to stave off involuntary nods on this important occasion. By the end of the day, I had six pages of handwritten notes from which I reconstruct this informal summary. Readers are warned that alleged remarks and results of the speakers found in these notes should not be quoted or trusted, but should be verified by consulting their contributions.

Bill Bennett of Yale, by historical chance Gene's brother-in-law, began at 9 a.m. sharp, with Roger Falcone in the Chair. Bennett is someone I was glad finally to see in person. He and I had interacted by mail and telephone some years ago on the dangers or lack thereof of EMFs from the power lines. Some recollections of Gene at Columbia as Polykarp Kusch's student preceded Bennett's main topic, his search for the "fifth force," a once hot topic triggered by the re-analysis 60 years later by Fishback et al of Eötvös's 1922 data on the equivalence of gravitational and inertial masses. Experiments in the mid-1980s gave conflicting results (East Coast, yes; West Coast, yes and no). Bill devised a clever experiment using lead, copper, and the Snake River.

As he began to describe things thusly, I thought, "This sounds as if it is going to be a shaggy dog story. I bet he went fishing on the Snake River with copper lures weighted with lead sinkers!" But, no, it turned out to be a real experiment with a disk made of two halves separated by a diameter, one of copper and the other of lead. The disk was suspended by a VERY fine fiber to form a torsion balance to be attracted by a known large mass, namely, the water (1.7×10^5 metric tonnes) in a lock on the Snake River. With 12 minutes to fill or empty, numerous "lock-in" (and-out") repetitions were possible (although barges or other craft in the lock could cause systematic errors). Other experiments had perhaps larger nearby masses (e.g., mountains), but lacked the desirable control of keeping the apparatus in place and changing the attracting mass. He had applied to the NSF for funding, but never received even an acknowledgment, so he scraped together some personal funds and started anyway. The whole apparatus was assembled in New Haven, loaded into a truck, and driven by Bill and his wife to the Snake. Results of this first experiment in late 1988 were good enough to rule out the West Coast "yes" experiment.

After publication of his results, Bennett received a call from the NSF. They had found his proposal! And they wanted to fund it! So in 1990, he mounted a second, greatly improved experiment, the details of which I did not totally comprehend, except that there were lots of remote video cameras to monitor the lock, etc. The latest result is null, with an accuracy comparable to the Adelberger (West Coast) experiments.

Nice story, and told in 35 minutes.[Typical of the day, the Chairs were relaxed, to say the least, in enforcing time limits. But things worked out just fine.]

At 9:35, **Saul Perlmutter** of LBNL reviewed the supernovae searches, the findings on dark matter and dark energy, and projected the future for cosmological studies. The data from the supernovae searches (which show evidence of acceleration in the expansion of the universe and so the presence of a vacuum energy - Einstein's cosmological constant or equivalent), the cosmic microwave background (CMB), and galactic clusters can be summarized in a plot of vacuum energy versus mass density. The bands overlap with $\Omega_\Lambda \approx 0.7$ and $\Omega_M \approx 0.3$, with large errors. The vacuum energy is a big puzzle. What can it be? The naive idea that it might be just the sum of the zero-point energies of all the quantum fields in the universe fails by an astronomically large number of orders of magnitude (too large!). Another puzzle is why the mass density and the vacuum energy are within a factor of two of each other at the present time.

With the present large errors, new experiments are needed. One experiment, SNAP, now in the pre-proposal stage, is a satellite with a 2-meter telescope, a 10^9 pixel mosaic that scans 1 square degree of the sky, and a 3-arm spectrograph. The device will find and measure supernovae at red shifts of $z = 0.2 \rightarrow 1.5$-2.0 with a great improvement in precision. [Perlmutter finished within his allotted 20 minutes.]

Next, **Norval Fortson** of the University of Washington spoke on searches for static electric dipole moments (EDMs) of particles. He reviewed the fundamentals - the invariances of parity (P) and time reversal (T) must be violated to have non-vanishing EDMs; "large" values would indicate physics beyond the Standard Model.

The first observations on the neutron set limits in the range of 10^{-20} e cm, now improved to 10^{-25} e cm. The Standard Model with 6 quarks (and so a CKM matrix with a CP-violating phase) is consistent with CP violation in K^0 decay and gives EDMs of the order of 10^{-30} e cm. Supersymmetric models (SUSY) can yield values in the range 10^{-25} - 10^{-27} e cm, conveniently (?) just below the current limit for the neutron. Forston made the point that, for the electron bound in atoms or molecules, the Sandars effect (a relativistic effect largest in heavy atoms) can cause an enhancement of order 500 in the effects caused by the electronic EDM on the energy levels and transitions. This is the effect that makes the experiments of Commins and his students competitive.

Fresh from his graduation ceremony a few days earlier, **Chris Regan**, Commins's newly minted Ph.D student, spoke on his search for an EDM of the electron in thallium. The enhancement factor is $d_{Tl}/d_e \approx -585$. After his praise of Gene as mentor, Chris pointed out that the previous version of his experiment, done by others, had been completed in 1994. His improved apparatus is the expected tour-de-force with multiple lasers, precision machining and alignment, etc. Four atomic beams of thallium (up & down, east & west) pass through magnetic and electric fields, with sodium used for control of systematics (negligible EDM expected, but a magnetic moment). Extreme magnetic shielding is essential. Beside the apparatus, but outside the shielding, milligauss fluctuations are seen, displayed for us in a dramatic transparency. The causes? The Birge Hall elevator is the main culprit, but BART trains, too. Data had to be taken in the wee small hours, after BART shut down and before it began again - 4 hours of quiet during the week, 6 hours on the weekends! Such is the life of an experimenter.

Chris gave a clear and detailed discussion of the physics, of the method of data analysis, and of the systematic errors, all within his allotted 20 minutes. His result is $|d_e| < 1.5 \times 10^{-27}$ e cm (90% C.L.)

Mike Romalis, from the University of Washington, spoke next, on the EDM of ^{199}Hg atoms. His experiment utilized "stationary" polarized atoms within two cells (10^{14} atoms per cell) in parallel electric and magnetic fields. Modulation of conditions produces a signal at 15 hertz, with a run lasting 200 seconds. The electric field direction is reversed after each run. He said that he had taken data over approximately 5×10^4 reversals. His preliminary result is $d_{Hg} = -(1.06 \pm 0.49 \pm 0.4) \times 10^{-29}$ e cm, with the upper limit, $|d_{Hg}| < 2.1 \times 10^{-28}$ e cm.

Improvements will be four cells instead of two, with factors of 2 to 4 in precision hoped. There will also be a new experiment with ^{129}Xe in the liquid phase, a radically new technique.

11:00 - 11:25 A break for coffee and bagels. [Just 20 minutes behind schedule]

MORNING SESSION 2

After the enjoyable break, when I introduced myself to Bill Bennett and we noted that EMF concerns never go away, **Larry Hunter** of Amherst College, one of Gene's former students, spoke of an experiment that he and several undergraduates have performed over the past two or three years. It concerns the measurement of the electric potential difference between inner and outer surfaces of a hollow rotating dielectric cylinder that is also magnetically permeable. [The results appear in a paper in the July 2001 issue of the American Journal of Physics (Vol. 69): "Measurement of the relativistic potential difference across a rotating magnetic dielectric cylinder," by J. B. Hertzberg, S. R. Bickman, M. T. Hummon, D. Krause, Jr., S. K. Peck, and L. R. Hunter, pp. 648-654.]

Interest in this topic has resurfaced after 85 years because of questions about (a) electromagnetism in rotating coordinate systems, and (b) whether a 1913 experiment of the Wilsons actually demonstrated what was claimed. At the time (and until recently) their experiment had been viewed as a vindication of electrodynamics within the framework of special relativity, particularly the 1908 prediction of Einstein and Laub. Pellegrini and Swift (1995) triggered a series of papers in Am. J. Phys. by claiming that the results for rotating systems obtained by local replacement with inertial frames moving rectilinearly with the same instantaneous velocity were incorrect; the significance of the agreement of the Wilson's data with a well-known formula (See problem 11.31 of the third edition of my *Classical Electrodynamics*.) was cast in doubt. By 1999, the question had shifted to challenging the Wilsons use of an artificial magnetic dielectric made of small steel balls embedded throughout wax as a substitute for a medium with both magnetic and electrical polarizability. Analysis showed that the conductivity of the spheres caused the competing theories to give the same result.

Larry and his undergraduates set out with modern techniques to repeat the experiment. Their paper discusses the theory and experiment in detail. Suffice it to say that they chose yttrium-iron garnet (YIG) as the real magnetic dielectric, created an artificial dielectric akin to the Wilsons, and had a non-magnetic dielectric (nylon) as a control. The apparatus is small (length, inner and outer radii of the cylinders are approximately 9.0, 1.27, and 1.80 cm, respectively) with rotation frequencies up to 100 Hz. The voltages at 100 Hz are typically 8 mV. The results show unambiguously that the original formula derived with local inertial frames is correct, while the Pellegrini-Swift modification is not. The artificial magnetic dielectric results agreed with the YIG data, but could not test "local inertial" versus "rotating" electrodynamics.

Larry and Amherst College are to be commended for showing that interesting physics experiments can be done by undergraduates at an institution without a graduate program.

We were now running 30 minutes late as **David DeMille** of Yale began his talk at 11:45 on the search for an EDM in metastable lead oxide (PbO). [Eulogy of Gene as supervisor] Why a molecule instead of an atom? Because molecules can be 100% polarized, there can be very large enhancement factors. Unfortunately, one loses signal because of the distribution of excitations over rotational states. It is necessary to use free radicals (unpaired electron), with the consequence of high temperatures and difficult chemistry. Now, PbO is not a free radical in its ground (singlet) state. A laser is used to convert the singlet state to a triplet (2 unpaired electrons). DeMille's cell holds roughly 10^{16} molecules, compared to 10^8 atoms in the Tl beam experiment of Chris Regan. The technique exploits two triplet states, one $J = 1^-$, the other $J = 1^+$. In an applied electric field the $m = \pm 1$ states mix and split apart; the $m = 0$ state is not affected. The energy level diagram appears as a capital X with the four m = ±1 states at the ends of the arms and the m = 0 state at the center. The laser is used to populate the m = 0 level. Larry Hunter has collaborated with DeMille in measuring relevant spectroscopic properties.

My notes indicate that DeMille's promising technique is still under development.

Sharp at noon, as the Campanile chimed, **Tony Zee** began his talk on "Theoretical implication of the electric dipole moment of the electron." He began by saying he was going to do something "crazy" (I'm not sure that is the word he used, but it conveys the sense). He did not wish to speak about the boring computation with Barr of EDMs [Phys. Rev. Lett. **65**, 21-24 (1990)]. Instead, he would speak about time reversal invariance (TRI) or the lack thereof. He said he did not understand TRI and then showed a transparency of a couple sitting across from each other. The woman is saying, "We can pause, Stu; we can even try fast-forwarding, but we can never rewind." That turned out to be the theme of Zee's talk. He pointed out that, although Einstein unified space and time, time goes only one way. Are consciousness and TRI connected, perhaps in opposition? Mention of consciousness led to an apparent digression about Schrödinger's statements in his book, "What is life?", where Schrödinger discusses the universal consciousness of all of mankind and extrapolates to the conclusion that "We have become God," an inference with which Zee cannot agree.

Back on TR, Tony stressed that for human beings time moves inexorably forward, with a minimum time of perception that must be of the order of 0.1 seconds. He also stressed that for physicists reality is empty space. [By which I guess he means that we draw our conclusions about Nature with small numbers of particles interacting in controlled conditions. Pace, condensed matter folk!]

Zee's parting shot was that the passage of time for humans may be related to violation of TRI and the existence of electronic EDMs! He left us with a question: How long and how many electrons' precessions of their EDMs before the human brain recognizes the passage of time?

Listeners questioned whether Zee was confusing the macroscopic (thermodynamic) violation of TRI with the microscopic violation associated with EDMs.

I have a question of my own: Given the current upper limit on the electron's EDM, how strong an electric field in the brain is necessary for the dipolar energy to exceed kT?

It's 12:15 and we still have four talks to go before lunch!

The next one is by **Paul Vetter** of LBNL, who happens to be Polykarp Kusch's grandson [What does that make him with respect to Gene Commins?]. Vetter spoke on the history of weak interaction experiments with radioactive beams. The first such experiment was done by Jerrold Zacharias in 1942 with ^{40}K beams - with a half life of 10^9 years, ^{40}K barely qualifies as radioactive! [But it's pretty rare!] The interest was in spin determinations and also magnetic moments of unstable nuclei. Luther Davis (1949) studied 2.6 yr ^{22}Na (J = 3); K. F. Smith (1951) measured 15 hr ^{24}Na (J = 4); Commins (1959) did 0.8 s ^6He (J = 0). The ^6He beta decay featured importantly and misleadingly in the determination of the mixture of invariants in the nuclear four-fermion weak interaction (another story).

By 1963 the studies had moved on to the details of the weak interactions. Commins and David Dobson (present with his wife at the symposium) used polarized 17 s ^{19}Ne to measure the asymmetry of the beta emission with respect to the nuclear spin direction (the so-called A parameter). At present, the so-called D parameter of the neutron (the coefficient of $\mathbf{s}_n \cdot (\mathbf{v}_e \times \mathbf{v}_\nu)$, non-vanishing if TRI is violated in beta decay) is being measured by Stuart Freedman's group at Gaithersburg.

Derek Kimball of UC Berkeley described work in progress on tests of fundamental symmetries with nonlinear optics. He first indicated that linear Faraday rotation as a means of measuring EDMs was limited to $|d_e| > 10^{-19}$ e cm and so was not useful today. In nonlinear Faraday rotation, the index of refraction depends on light intensity and gives possibility of greater sensitivity. With paraffin-coated cells from Russia the relaxation rate for Cs atoms can be reduced to below 1 Hz. Thus, EDMs at the level of $|d_e| \approx 2 \times 10^{-26}/\sqrt{Hz}$ e cm seems feasible. While calculations agree with observations on the cells, puzzling effects (number of atoms in the cells, desorption from the paraffin walls, etc.) are still to be explained and/or eliminated.

Zheng-Tian Lu of Argonne National Laboratory, formerly a Freedman student, described his work on optical spectroscopy of rare atoms. One example is ^{81}Kr, a radioactive atom with a half life of 2.1×10^5 yr. Its relative abundance is 10^{-12}, produced by cosmic ray bombardment in the atmosphere. Lu obtains this isotope from Greenland ice cores at the rate of one atom every 10 minutes! Another Kr isotope with a different origin is ^{85}Kr, a fission product with a half life of 10.7 yr, now present in the atmosphere, mostly from nuclear fuel reprocessing. Such isotopes have applications in environmental studies and in the history of climate.

An important isotope for Lu is ^{6}He. Precise study of the hyperfine structure of excited states will permit determination of the nuclear charge radius for comparison with theoretical calculations.

Shortly after 1:00 p.m. **Phil Bucksbaum** from U. Michigan (former Commins student and co-author with Gene of a well-received book on weak interactions) spoke on the topic "Quantum Control." He warmed us up by boldly writing the time-dependent Schrödinger equation on the blackboard ($H\psi = i\hbar\partial\psi/\partial t$) and observing that we can control the time evolution of the wave function by controlling the Hamiltonian. A striking transparency of a Japanese print of a gigantic foaming, breaking wave towering above a small boat, with Mt. Fuji in the distant background, was a lesson that the sense of motion was intuitive. The quantum analogy (lost on me a bit) is the phases associated with components of the wave function. Wave packets are coherent superpositions of different waves, in Phil's case different colors of light. With optical pulse shapers that modify the spectral make-up of a pulse as it develops, the coherence can be controlled. Phil had nice transparencies of various examples, but they went by so fast that I have no notes to help me explain the intricacies.

He put forth the idea that the systems could contain learning algorithms with learning feedback. He foresaw exciting applications.

And at 1:20 we broke for **lunch**, with the afternoon sessions to begin at 3 p.m. An excellent barbecue lunch was served on the "patio" tucked into the space between LeConte and Birge Halls, with the guys from the departmental machine shop officiating at the grills. While we ate, Martin Perl questioned me about how one could estimate limits on the mass of very massive charged particles before one could shake them out of matter with mechanical accelerations of, say, a hundred times g. Little did I know (because I hadn't looked at the program) that -

AFTERNOON SESSION 1

At 3:05 p.m. the first afternoon session began with "The search for massive stable fundamental particles" by **Martin Perl** of SLAC. Marty was a fellow graduate student of Gene Commins at Columbia, Rabi's student. He remarked that two characteristics of Columbia physics students were contrariness and arrogance. He discussed the mass composition of extraterrestrial particle fluxes and various negative searches. For example, there have been searches for "heavy water," i.e., HXO, where X^+ is very massive. For $Mc^2 < 10^8$ GeV, the upper limit of abundance is extremely small. Slowly moving neutral massive particles are being sought through their collisions in matter to produce low speed recoils (Sadoulet).

Perl then summarized the searches for fractionally charged particles, e.g., quarks, with "Millikan oil drop" experiments. Drops totalling 17 mg of silicone oil have been studied; nothing unusual has been seen. Similar studies of Allende meteorite material are also negative.

Without tipping his hand, Perl indicated he was still exploring new ways of looking for very massive stable particles.

The one truly theoretical talk of the symposium was given by **Eyvind Wichmann** of UC Berkeley. His title was "Symmetries and Quantum Field Theories." He first reminded us that classical physical quantities have well defined transformation properties under translations, rotations, inversions, etc. In quantum theory the operators of observables have correspondingly certain group theoretical transformations. Experimentally, parity (P), charge conjugation (C), and the combined PC symmetries are known to be broken. The breaking of time reversal (T) invariance is inferred from the so-called PCT theorem.

Eyvind then asked What is the basis of the PCT theorem? And the spin-statistic theorem? He chose to discuss these theorems in the framework of the Wightman version of rigorous quantum field theory (Postulates include finite-dimensional spinor and tensor fields, locality, causality.). [A moment of confusion and delay occurred when Eyvind eschewed the transparency projector in favor of chalk on the blackboard. And not the big stubby soft chalk used by Bucksbaum to write the Schrödinger equation, but serious, thinner, hard chalk, more difficult to read from the back of the room.] For spin and statistics he just jotted down a few formulas and mentioned that Nick Burgoyne had been the first with a rigorous proof, with slightly less complete proofs by Lüders and Zumino. He said that he could be brief because a clear discussion appears in the book by Streater and Wightman ("PCT, spin and statistics, and all that," by R.F. Streater, A.S. Wightman. Princeton, N.J. : Princeton University Press, 2000, paperback reprinting.).

On the PCT theorem, Wichmann sketched a few formulas of the proof by Res Jost, which required the exploitation of the complex Lorentz group, also needed by Burgoyne. I think both he and the audience appreciated that the intricacies of the proof could not be conveyed in a 20-minute talk (that actually stretched to 35). He cited two more theorems of interest - a recent one by J. Mund that shows that the PCT theorem is restricted to theories with particle content (!?!) and another, that local internal symmetry operators must commute with the Poincaré group and the PCT transformation.

A change of pace occurred from 4:10 to 4:35 as **Dima Budker** presided over the presentation of some gifts to Gene Commins. One was a book from Gene's Australian colleagues and friends. The other was a very handsome, hand-made, leather-bound volume of letters of appreciation and photographs on the occasion of Gene's retirement from teaching. The book was the creation of an artist friend of Budker's. Its cover and clasps had intricate tooling of the leather to show various physics symbols related to Gene.

Then there was a short coffee break and photo op for a group picture on the west steps of LeConte Hall.

AFTERNOON SESSION 2

The final session of the day - the "Nobel" session (although we had heard from one of the three Nobel Laureates already) - began with **Charles Townes** of UC Berkeley, who discussed "Stellar sizes, what do we know?" Charlie remarked that measurement of stellar sizes had a long history. Michelson used interferometry to measure the diameter of Betelgeuse (α Orionis, a red giant) as 48 milliarcsecond, with a $\pm 10\%$ error, and also noted limb darkening. At the Keck telescope a mask with holes in it implements the interference aspect. Measurements of angular diameters versus wavelength yield puzzling results - some stars show constancy, others show a sizable increase with wavelength. What is going on?

Infrared (approximately 10 micron) spectroscopy and interferometry is Townes's tool. His apparatus consists of two trailer-based telescopes on Mt. Wilson, chosen for its excellent "seeing." He showed us a block diagram of the data-taking system and described how Fourier analysis of the data gives a diffraction pattern with rapid little wiggles at scales or the order of 1 milliarcsecond from dust scattering. Data for Betelgeuse gave a (uniform disk) radius of $R = 27.37 \pm 0.17$ milliarcseconds, larger than in the visible. The variable star, o Ceti, has a radius of $R = 23.91 \pm 0.25$ milliarcseconds. Townes said that some theorists were bothered by this result. They want a much smaller radius (16) to fit their models of the oscillations, which have a period of one year. Other theorists invoke a nonlinear rearrangement of material, with 30-40% change in size, to get rough agreement with Townes's radius.

Another issue is whether all stars are spherical. Are the oscillations breathing modes or changes in shape? At present we do not know. We need two-dimensional interferometry, and must measure at different wavelengths, on and off spectral lines, etc. Preliminary data of the Townes group disagrees with those of a British group. More work needed! Townes says perhaps two more years.

The final talk of the Symposium, and the only *Power Point* presentation, was given by Gene's former student, **Steve Chu** of Stanford. Advertised as "Random walk in science (in 20 minutes)," it ran a bit longer. Chu summarized his own career, with much praise of Commins, his philosophy, and his standards along the way.

The first part of the story began in January 1972 and ended in September 1978. He began with Commins on a trial basis. Frank Calaprice had left; Mel Simmons and Stuart Freedman were about to depart. Steve said it was nice to be Gene's only student. He worked initially on a theoretical problem, the theory of proto-star formation (Gene must have been teaching an astrophysics course.). After three months, Steve decided he wanted to be an experimenter. Thus he gave himself an "Incomplete." [For non-Berkeley folk, an Incomplete is an official grade that can be given to a student who has completed a substantial part of the required work, but not all, and can be expected to complete the work in a prompt fashion in the coming months. It is believed that many grad students' records are littered with "Incompletes," never removed.]

Steve's next "Incomplete" (all of these are his own private grades) was obtained from a year's work on the energy dependence of the up-down asymmetry (that A parameter) in beta decay. I am not sure why that didn't work out. Another year and another "Incomplete" came from research on the Lamb shift in high Z atoms with Dick Marrus. Then came his long stint (2.5 yrs as a grad student and 2 yrs as a post-doc) measuring parity violation in atomic transitions. This was his thesis work, for which he gave himself a "C^-". His highest self-awarded grade ("B^+") was for a one-week experiment in which he showed that the human ear (of some students) can violate the relation $\Delta v \Delta t \geq 1$ by a factor of 10. The idea was to ask observers to identify musical tones heard for shorter and shorter times.

Chu said that he carried away from his six and a half years of association with Gene the lessons: (1) Only do important experiments, even if they are nearly impossible, and (2) Do not be ashamed to abandon an experiment if it *is* impossible.

Steve's next interlude (1978 - 1987) was at Bell Labs where he worked on Anderson localization; pulse propagation in absorbing media ($v_g > c$, even $v_g > \infty$); positronium, with Alan Mills; almost, but not, the quantum Hall effect with Dan Tsui; and laser cooling of atoms. The latter, termed "optical molasses," is the best known of Steve's work at Bell Labs. In 1985 they achieved cooling of Na atoms to 240 µK. He said he didn't deserve an "A" for that because three years later Phillips pushed the temperature down by a factor of six, to 40 µK. Another lesson from Gene: Physics is precision measurements.

Steve's story now moved (in 1987) to Stanford, where he first continued the area of research begun at Bell Labs. The "atomic fountain" in which atoms are tossed up at walking speeds, rise, and then fall under gravity, led to precision measurements of the fine structure constant α (one part per billion). He hopes for at least a factor of two improvement on that. His personal activities at Stanford and his group have bifurcated, with half continuing in precision atomic physics and half working in biophysics. Study of the "folding" of biological molecules is one current area of research.

Steve ended his presentation with some photos of Gene Commins, Gene and Steve's mother, and Gene and his wife Ulla. His final comment was that Gene's students could make a first class physics department, in the top 20 in the country or perhaps better.

I forgot to mention that Steve stumbled in starting his *Power Point* presentation - I give him a "B^{--}" for that - but by the end had brought his final grade up to an "A^{-}."

The Symposium ended about 5:40 p.m. with **Gene Commins** thanking the organizers and speakers and indeed all the attendees for a most enjoyable and informative day. A reception and dinner at the Faculty Club followed, but my notes do not extend that far.

ACKNOWLEDGMENTS

I thank Eugene D. Commins, a scholar, a gentleman, and an esteemed colleague for more than thirty years, for allowing the ComminsFest Symposium, May 21, 2001 to honor him and his wonderful group of Ph.D. students. Thanks go to the organizers Phil Bucksbaum, Dmitry Budker, Willi Chinowsky, Roger Falcone, and Stuart Freedman for arranging this happy event where all the participants, speakers and audience alike, could celebrate Gene and learn and enjoy.

Weak Interactions of Leptons and Quarks using Radioactive Atomic Beams

P.A. Vetter

Lawrence Berkeley National Laboratory, Berkeley, California 94720

Abstract. A major theme in Eugene Commins' career has been the use of atomic beam machines on radioactive isotopes, predominantly to study the Weak Interaction.

On the occasion of Professor Commins' retirement, it's interesting to reflect on accomplishments of his career in research, particularly some of the experiments performed some time ago. Commins hasn't worked directly in the subfield I'm going to discuss for quite some time, but his influence is still felt among the current researchers in similar experiments. My goal is to delineate Eugene Commins' application of atomic beam techniques to the study of the Weak Interaction. At a time when radioactive atomic beams were in their infancy, he made influential contributions to this subfield and recognized the potential of bringing atomic beam techniques to accelerator-based experiments on the Weak Interaction. It's fair to say that he paved the way for the continued productivity of this field today.

It's probably not necessary to point out the influence on twentieth-century physics of the atomic beam magnetic resonance technique, so I'll forego a general historical introduction to the topic, and begin with the earliest applications of atomic beams to radioactive isotopes. Generally, the early experimental investigations of Weak processes concentrated on beta decay, which necessitates working with artificial radioactive species. But the first example of an application of atomic beam techniques (reference [1] – an experiment performed by Zacharias at the Columbia Rad Lab), used the naturally ocurring radioisotope ^{40}K, which has a natural abundance of about one part in 10^4. As was typical for the early magnetic resonance beam research at the time, the interest was in measuring the spin and magnetic moment as fundamental quantities for atomic and nuclear physics. With a half-life of more than one billion years, however, this isotope presented little opportunity for study of its decay properties, and is interesting nowadays as a geochronometer rather than a Weak Interactions laboratory. (The application of radioactive atomic beam techniques to "natural" radioactivity has been recognized again as an interesting new problem, but we leap ahead.) But that experiment was in some sense a genesis of lots of work done today, and it was recognized as a pivotal experiment by the early researchers. Zacharias did suggest that had the experiment not found the expected large integer-valued nuclear spin (the spin is $4\hbar$), the case of ^{40}K would have presented a substantial anomaly for the Fermi theory of beta decay. But the spin measurement turned out well for the theory. The fundamental problem of a very small abundance of the desired isotope in the presence of an enormous background was solved by Zacharias by using a "flop-in" magnetic focus rather than a "flop-out" focus (as all the other Columbia machines to that time had been) – i.e. tiny subtractive signal on huge background is traded for tiny additive signal on a much smaller background. Substitute the action of a resonant laser on the radioactive species of interest for the radiofrequency state selection and you have the isotopic selective, Zeeman-slowed MOT capture apparatuses of today. But this leaps ahead again...

The next approach to radioactive beams was by Luther Davis at MIT[2], with a measurement of the spin and magnetic moment of the 2.6 year half-life ^{22}Na. The beam oven in this experiment used a tiny sample of ^{22}NaN$_3$ diluted in 10^4 parts of stable NaN$_3$. The activity was created by deuteron bombardment of a magnesium target and then chemical separation of the activity. Davis notes that "[t]he beam intensity of ^{22}Na was naturally small, a few thousand atoms/sec., and the radiofrequency transition intensities were less than 100 atoms/sec." To make a measurement with such tiny signals, a much more sensitive detection scheme was necessary, and this experiment used a mass spectrometer magnet to separate the ^{22}Na from the ^{23}Na, and an electron multiplier to detect the individual ^{22}Na atoms. The atomic beam is ionized by a hot filament at the end of the beam line, and the ions are extracted electrically and run through the mass-spectrometer magnet. Davis' laboratory went on to study several other long-lived radioisotopes in reference [3]. The first paragraph of the paper discusses the motivation for the work: "Attempts to compare the various forms of the Fermi theory of beta-ray disintegration with experiments usually meet with the difficulty that those characteristics of the decay which make it possible to observe the nuclear angular momentum directly are just those characteristics which make the observation of the energetics of the decay more difficult and less precise. ... It is therefore desirable to develop a method for the observation of nuclear moments which can be applied to more energetic substances with shorter half-lives, since it is for such materials that decay schemes and spectra can be well-studied."[3] In other words, in order to compare to known data on endpoint energies and beta spectra, one should work with activities with nice high energies. But high endpoint energies imply short half-lives, since loosely speaking, the beta decay rate is proportional to the fifth power of the endpoint energy. And a short half-life presents significant challenges to the best technique for measuring nuclear moments – atomic beam magnetic resonance. The half-life of ^{22}Na is still 2.6 years, and the technique developed by Davis is limited to isotopes which can be prepared by irradiation, chemical separation, and insertion into the source oven.

Returning to Columbia again briefly, an experiment by Nafe and Nelson, [4] measured the hyperfine splitting of tritium. While not motivated by the radioactive decay properties of tritium or Weak Interaction studies, the experiment is technically interesting in the continuing development of the field. This experiment would be viewed today as a rare isotope atomic beam rather than a radioactive atomic beam, but such political distinctions need not concern us. Because of the extreme natural scarcity of tritium, the sample in the experiment was obtained from the Atomic Energy Commission (1949), and Nafe and Nelson realized that the usual gas consumption of an rf discharge source to produce atomic hydrogen would be unsuitable for their limited stock of tritium. Their solution was a recapture technique (to which the story will return later...). The atomic beam chamber was pumped by a three-stage mercury diffusion pump which compressed the throughput gas. The exhaust of the mercury pump was then vented to the discharge tube, so that the remaining molecular tritium or unused beam was recirculated and used again, and recovered after the experiment. The result of the measurement and later studies with a more efficient discharge tube by Prodell and Kusch in 1956 [5] found no hyperfine anomaly in hydrogen, a disappointing but fascinating result not relevant to the main flow of the narrative.

The next development in radioactive atomic beam work came with the experiments by Smith and Bellamy [6] at the Cavendish Laboratories in Cambridge. Here, the minute atomic beams were detected by counting the radioactivity deposited on a target which could be removed from the end of the beamline. Smith and Bellamy measured the spin and moments of ^{24}Na ($t_{1/2} = 15$ h) and ^{42}K ($t_{1/2} = 12.4$ h). This is the first real combination of radioactivity and atomic beam work. Yet the removal of the condensed activity and target from the beamline was cumbersome and time consuming, and the half-lives amenable to this technique required long atomic beam bombardments (which was typical of high-precision atomic beam work).

At last, we come to the appearance of Eugene Commins, whose paper on the spin of ^{6}He was even more succinct than his predecessors in this nascent field. Commins did his graduate work at Columbia University studying the hyperfine splitting of ^{3}He as a test of quantum electrodynamics and a search for nuclear structure effects, but he performed an experiment on the side that proved to be the inspiration for much of his later work on beta decays using radioactive beams, and indeed for his long involvement with Weak Interactions.

Commins' experiment on ^{6}He was a timely strike to test an exotic hypothesis which was designed to save the budding (V-A) Weak theory from embarassment. Conflicting experimental results left a situation which suggested either unusual and unexpected messy physics, or else a serious experimental error. At that time, 1957-8, the structure of the Weak coupling between the leptons and nucleons in beta decay was still uncertain. The allowed Lorentz-invariant tensor structure of the coupling could be either a combination of vector and axial vector couplings (VA) between lepton and nucleon currents, or a combination of scalar and tensor couplings (ST). Other possible combinations (e.g. VS) were disfavored by spectral shape factor measurements and spin selection rules, while pseudoscalar coupling is disfavored by the argument that a P term would vanish in the limit of nonrelativistic nucleons. The result of an experiment at Brookhaven Laboratory by Ruby and Rustad [7] suggested that ST was the correct coupling. This experiment measured the beta-neutrino correlation of ^{6}He by a coincident daughter ion detection technique. The ionized, recoiling daughter nucleus (^{6}Li$^+$) was detected using an electron multiplier, and the direction and energy of the associated $\beta-$ was detected using a movable scintillation detector. The correlated distribution of the betas and recoil ions as a function of beta energy was measured and compared to calculations with different fundamental couplings. The decay of ^{6}He is purely Gamow-Teller: ^{6}He(0^+) $\to ^6$Li(1^+) and therefore sensitive to only axial vector and tensor couplings. Their result for the correlation between the direction of the beta and the recoil ion strongly favored the kinematics given by tensor coupling. This result disagreed with the new (V-A) theory of the weak interaction by Feynman and Gell-Mann, [8]. At the time, either the experimental result was erroneous, or the theory was incorrect. But it was possible that the interpretation of the experiment was incorrect: if the ground state of ^{6}He were not spin

0, but instead spin 1, then the transition would not be purely Gamow-Teller, and the beta-neutrino correlation would be sensitive to scalar and vector couplings instead. The beta-neutrino correlation coefficient **a** (which appears in the seminal paper by Jackson, Treiman, and Wyld [9]) can be written in terms of the coupling constants for the various tensor terms:

$$a = \frac{\left(|C_V|^2 + |C_V'|^2 - |C_S|^2 - |C_S'|^2\right) + \frac{|\langle GT \rangle|^2}{3|\langle F \rangle|^2}\left(|C_T|^2 + |C_T'|^2 - |C_A|^2 - |C_A'|^2\right)}{\left(|C_V|^2 + |C_V'|^2 + |C_S|^2 + |C_S'|^2\right) + \frac{|\langle GT \rangle|^2}{|\langle F \rangle|^2}\left(|C_T|^2 + |C_T'|^2 + |C_A|^2 + |C_A'|^2\right)} \quad (1)$$

Here the primed coefficients reflect an ambiguity of how to place parity-violation in the Weak Lagrangian coupling lepton and nucleon currents. In the Standard Model today, the couplings are $C_V = C_V' = 1$ and $C_A = C_A' \approx 1.25$.

Commins' experiment was a test of this latter hypothesis. To demonstrate that the nucleus was spin 0, one could search for a magnetic moment of ^{6}He, and attempt to set as small a limit as practical. Commins and Kusch applied an atomic beam technique in order to search for a nuclear magnetic moment. Using the same gas production and handling system as used by Ruby and Rustad, the ^{6}He enters a Stern-Gerlach type atomic beam system to search for a deflection of the nuclear magnetic moment. In this respect, there was nothing particularly novel or challenging about the atomic beam aspect of the experiment. The experiment was "flop out," so that the ^{6}He beam, if its magnetic moment were deflected by the inhomogeneous magnet, would miss the detector and a decrease in beam intensity would be observed. In this case, "flop in" was not necessary to supress a large stable component of the beam, because there was no stable component to the beam (ideally) other than background gas. In the publication of the results, Commins suggested, "[t]he only novel feature of the apparatus is the method of detection."[10] But as it turned out, the detection technique was more important than the result of the measurement.

In a brief experimental run, Commins and Kusch measured the magnetic moment of ^{6}He to be consistent with zero nuclear magnetons, with an uncertainty of perhaps as much as 0.15 μ_N. They observed that within error, there was no decrease in beam intensity when the deflecting magnets were turned on. With such a small magnetic moment, the spin was very unlikely to be anything other than $0\hbar$. This strongly suggested that the ^{6}He beta decay was Gamow-Teller. The result by Ruby and Rustad [7] also disagreed with later measurements in ^{6}He by a group at the University of Illinois, [11]. That result, from measuring the spectral shape of the emitted recoil ions from the decay, was that the beta-neutrino correlation was $-\frac{1}{3}$, consistent with axial vector coupling, rather than the $+\frac{1}{3}$ measured in [7]. Furthermore, the Ruby and Rustad experiment conflicted with the result of the now famous experiment by Goldhaber, Grodzins, and Sunyar [12], using the Gamow-Teller decay of ^{152m}Eu to measure the helicity of the emitted neutrino. That experiment was consistent with axial vector coupling (A), although the authors cautiously noted that it could be possible in principle that the couplings were different for β+ and β− decays (a possibility acknowledged in [8] in the discussion of possible Lorentz covariant terms). The final consensus is that the Ruby and Rustad result was erroneous. Possible causes of systematic error include contamination by long-lived activities like ^{16}N, a kinematic shift in the accepted data caused by the fast forward motion of the ^{6}He gas through the counting chamber, or treatment of the geometric acceptance function of the beta detector.

The detection technique in Commins and Kusch was the first *in situ* measurement of the radioactivity of an atomic beam. Previous radioactive atomic beams either did not detect the radioactivity of the beam at all, or detected the accumulated activity deposited on a catch foil over a long period by removing the target and counting elsewhere. In the work of Commins and Kusch at Brookhaven, at the end of the atomic beam line was a cavity with a long, narrow entrance channel (effectively a beam oven at the wrong end). Atoms entering this cavity are trapped by the low geometric probability of re-emergence. Plastic scintillator was placed outside the thin-walled cavity, and beta hits in the scintillator were counted with a photomultiplier tube. This technique allows measurements on isotopes with very short half-lives. It could also be said to foreshadow the development of traps for radioactive atoms. Commins saw the appeal of using the atomic beam technique to provide spin-prepared nuclei for detection, and this ^{6}He experiment was a springboard for very fruitful work for him over the next two decades.

Commins' next use of radioactive atomic beams, in Berkeley, built on the *in situ* detection technique. Commins first began work on ^{19}Ne at the 60-inch cyclotron in Berkeley, and then moved the apparatus to finish initial studies of ^{19}Ne and ^{35}Ar at the 90-inch cyclotron at Livermore. The first ^{19}Ne experiment [13], measured the parity-violating beta-asymmetry coefficient, **A**, of the beta decay of polarized ^{19}Ne. An atomic beam of ^{19}Ne was prepared by bombarding a high-pressure gas target of SF_6 with 13 MeV protons. The target gas was continuously flowed to the atomic beam source, and the SF_6 was condensed out using a liquid nitrogen cold trap. The ^{19}Ne was polarized using an inhomogeneous magnetic field (the "A" magnet in atomic beamer's notation) and movable collimator to select a given nuclear polarization state. The beam impinged on a detector cell with a very narrow entrance channel. The pyrex cell bulb was teflon coated to prevent spin relaxation as the trapped ^{19}Ne bounced around before its decay. A homogeneous magnetic field maintained the nuclear polarization, and one Geiger detector on each end of the cell measured the up-down beta decay asymmetry of the polarized nuclei. The same apparatus was also used to measure the asymmetry in ^{35}Ar, reported in [14].

The ^{19}Ne experiment increased in complexity when it was moved to the newly completed 88" cyclotron at Lawrence Berkeley Lab. The next phase of the experiment confronted the time-reversal violating **D** coefficient of beta decay. In a general V-A description of the weak interaction, in principle, the coupling constants C_V and C_A could be relatively imaginary, which would result in an interference term in a beta decay rate proportional to the difference $i(C_V C_A^* - C_A C_V^*)$. This imaginary term is explicitly T-odd, and would manifest its presence in the beta decay rate as a correlation term between the initial nuclear spin direction and the momenta of the leptons: $d\Gamma \propto \vec{\sigma}_N \cdot (\vec{p}_e \times \vec{p}_\nu)$. Again, seeking such a correlation in the decay rate requires polarized activity – the beam apparatus provided polarization of approximately 90%, and the up/down detector for measuring the **A** coefficient was used as a polarization monitor. In the case of the **D** coefficienct, however, both the β+ and the recoil daughter nucleus, ^{19}F must be detected. The **D** detectors were a pair of very complex beta-ion detectors with octagonal symmetry. The **D** detectors went through two generations from [15] to [16], beginning as in-flight detectors using scintillator for the positrons and fluorine ions, and ending as a cell or trap volume detector with semiconductor Si(Li) detectors.

The experiment to measure the T-odd **D** term was a null measurement, and as such, statistics were crucial. In order to increase the beam flux, recirculating diffusion pumps were used at the beam source end to capture and reuse ^{19}Ne which did not pass through the defining collimator slits. The longer paper [17] includes a schematic diagram of the apparatus indicating ten diffusion pumps. One of the recirculating pumps (closest to the beam source) was a mercury diff pump, as the usual silicone oil was reported to suffer from radiation damage. Fortunately, in the later version of the experiment [16], the number of diffusion pumps seems to have been trimmed to only seven. The result of the recapture technique was a factor of 25 enhancement in beam flux, but more importantly, it demonstrated a sensitivity to the issue common to radioactive atomic beam experiments – the scarcity of the activity. In such experiments, the radioisotope of interest is produced and purified with great effort. Yet atomic beam sources can be profligate with their charge (as Commins is well aware with the thallium electric dipole moment beam experiment). The simplest atomic beam oven – an ideal box with a hole through a thin wall, effuses its charge with a $\cos\theta$ distribution, and the desired beam geometry through various slits, magnets, and channels subtends a very small forward angle. Using naturally occuring isotopes in a beam oven, one usually effuses large amounts of material and selects only $\theta = 0 \pm \varepsilon$ forward components of the beam. Recapture of atoms not in the forward direction is in a sense equivalent to increasing the production rate of the isotope of interest. In some applications, "wicking" ovens are uses to conserve the charge, but in applications with natural isotopes, this is really a labor saving device to avoid refilling the oven quite so often.

The ^{19}Ne apparatus subsequently moved to Princeton University and Calaprice's group, where it continued to provide improved tests of **D** and other correlation parameters to still higher precision. A clever new version of the recoil ion detector and capture cell ultimately achieved a measurement of $\mathbf{D} = 0.0004 \pm 0.0008$[18], consistent with no T violation, but greater than the expected final state effects.

The impact of Commins' work using radioactive atomic beams is felt in a new generation of experiments using radioactive atomic beams. Our group's work at the 88" Cyclotron using laser-trapped radioactive ^{21}Na uses a radioactive atomic beam of sodium to measure the beta-neutrino correlation coefficient, **a**. In our case, the counting cell with a narrow entrance channel and long holding time has been replaced with a magneto optic trap. The MOT, rather than a narrow geometric acceptance, derives its long holding time from the cooling lasers: initially high-velocity atoms are slowed as they are loaded into the trap and cooled to velocities much lower than the escape velocity of a MOT. The state selector magnets in Commins' original beam work are parallelled by our Zeeman slower, which acts on the magnetic dipole moment of the atomic beam and increases the available beam flux to load into the MOT.

The atomic beam loaded magneto-optic trap is a convenient technique for beta decay correlation studies since it combines an efficient capture of the atoms with a relatively low-background environment for counting the activity. These advantages were discussed in the first reported demonstration of magneto-optic trapped radioactive atoms in [19]. The concept of counting a radioactive atomic beam *in situ* is now very mature, and with our apparatus, we reconstruct the kinematics of the decay by detecting the momentum of the beta, and the recoil nucleus (using an electric-field guided time-of-flight technique similar in principle to the later ^{19}Ne **D** coefficient detector).

The 88" Cyclotron at Lawrence Berkeley Lab provides a 2 μA beam of protons at 25 MeV for the reaction ^{24}Mg(^{1}H,^{4}He)^{21}Na. Our target consists of nine disks of compressed MgO powder, held at normal incidence to the beam by a "comb" of tantalum. The target disks are contained in a sealed ceramic crucible which is heated to roughly 1200 C to enhance the diffusion of the ^{21}Na from the refractory MgO. The ^{21}Na evolves from the target as a neutral atomic beam through narrow collimator tubes in the side of the target crucible. The forward flux of the atomic beam is increased by the use of cooling lasers immediately after leaving the oven nozzles. The atomic beam is slowed from its initial thermal velocity by a counterpropagating (slowdown) laser beam. The tapered magnetic field of our Zeeman slower contiuously adjusts the atomic resonance frequency to compensate for the changing Doppler shift of the atoms as they are slowed. Reference [20] describes the current state of the apparatus installed at the 88" Cyclotron.

To measure the beta neutrino correlation, we detect the ionized recoil daughter nucleus, ^{21}Ne. The initial momentum of the ^{21}Ne after the beta decay determines its total time of flight in the electric fields: ^{21}Ne initially moving towards the beta detector have a longer flight time than those moving towards a microchannel plate detector. The time-of-flight for the ion events encodes the initial kinematics of the recoil nuclei. The kinematics of the decay produce a known time-of-flight distribution for recoil ions. The detected ion TOF spectrum can be calculated by Monte-Carlo simulations of the beta decay and ion trajectories in the electric field of the trapping chamber. These simulations generate template timing curves. The data are fit to a sum of contributions from the isotropic and beta-neutrino correlated time-of-flight curves. This fit generates our measured value of **a**.

In total, our data set has a statistical uncertainty for **a** of about 1.1%, and a combined systematic uncertainty estimated at 6%. The systematic uncertainty is dominated by our inability to account for backscattered events in the tail of the ^{21}Ne$^+$ peak. This effect is being addressed with a redesigned beta detector and collimators. Based on our current data set, we believe that laser-trapped radioactive atoms are capable of providing new and interesting measurements of beta decay correlation parameters. Our study of the β-ν correlation provides another datum in the search for exotic scalar, tensor, or second class currents. We believe that there are no insurmountable obstacles for $< 1\%$ uncertainty measurments.

Trapping other atomic species of interest for electroweak and nuclear physics is also possible – the MOT technique has been applied to alkalis, alkaline earth, noble gases, and a few select transition elements, and radioactive atoms have been trapped by several other groups for such work. In particular, highly efficient atomic beam-loaded MOT techniques have been applied by Z.T. Lu at Argonne for isotopic analysis and geochronometry[21].

To conclude, Commins guided radioactive atomic beams from a demonstration or "trick" field to productivity. His experiments were capable of sensitive, serious tests of the Weak Interaction. He accomplished this by recognizing the key challenges of the field, including the *in situ* radiation detection of the beam, pioneering the "cell-trap" end station to enhance count rate, and attacking the problems of radioisotope scarcity and source purity.

REFERENCES

1. Zacharias, J.R., *Phys. Rev.* **61**, 270 (1942).
2. Davis, L., *Phys. Rev.* **74**, 1193 (1948).
3. Davis, L., Nagle, D.E., and Zacharias, J.R., *Phys. Rev.* **76**, 1068 (1949).
4. Nelson, E.B., and Nafe, J.E., *Phys. Rev.* **75**, 1194 (1949).
5. Prodell, A.G., and Kusch, P., *Phys. Rev.* **106**, 87 (1957).
6. Smith, K.F., *Nature* **167**, 942 (1951). Bellamy, E.H. and Smith, K.F., *Phil. Mag.* **44**, 33 (1953).
7. Rustad, R.M., and Ruby, S.L., *Phys. Rev.* **97**, 991 (1955).
8. Feynman, R.P., and Gell-Mann, M., *Phys. Rev.* **109**, 193 (1958).
9. Jackson, J.D., Treiman, S.B., and Wyld, H.W., *Phys. Rev.* **106**, 517 (1957).
10. Commins, E.D., and Kusch, P., *Phys. Rev. Lett.* **1**, 208 (1958).
11. Hermannsfeldt, W.B. *et al.*, *Phys. Rev. Lett.* **1**, 61 (1958),
12. Goldhaber, M., Grodzins, L., and Sunyar, A.W., *Phys. Rev.* **109**, 1015 (1958).
13. Commins, E.D., and Dobson, D.A., *Phys. Rev. Lett.* **10**, 347 (1963).
14. Calaprice, F.P., Commins, E.D., and Dobson, D.A., *Phys. Rev.* **137**, B1453 (1965).
15. Calaprice, F.P., Commins, E.D., Gibbs, H.M., Wick, G.L., and Dobson, D.A., *Phys. Rev. Lett.* **18**, 918 (1967).
16. Calaprice, F.P., Commins, E.D., and Girvin, D.C., *Phys. Rev. D* **9**, 519 (1974).
17. Calaprice, F.P., Commins, E.D., Gibbs, H.M., Wick, G.L., and Dobson, D.A., *Phys. Rev.* **184**, 1117 (1969).
18. Hallin, A.L. *et al*, *Phys. Rev. Lett.* **52**, 337 (1984).
19. Lu, Z-T., Bowers, C.J., Freedman, S.J., Fujikawa, B.K. Mortara, Shang, S-Q., Coulter, K.P., and Young, L., *Phys. Rev. Lett.* **72**, 3791 (1994).
20. Rowe, M.A., Freedman, S.J., Fujikawa, B.K., Gwinner, G., Shang, S.Q., and Vetter, P.A., *Phys. Rev. A* **59**, 1869 (1999).
21. Chen, C.Y. *et al.*, *Science* **286**, 1139 (1999). See also Lu's paper in this volume.

A Random Walk in Science

Steven Chu [†]

Physics Department, Stanford University, Stanford, CA 94305-4060

Abstract. The following memoir is a brief summary of a random walk in science, beginning with the eight years I spent as a student and postdoc with Eugene Commins.

MY YEARS AT BERKELEY

I first met Gene Commins in the fall of 1970, my first quarter at Berkeley. I did not have a statistical mechanics course at the University of Rochester and needed remedial undergraduate training as preparation for the qualifying exam. While thermodynamics, with its mysterious definition of entropy and jumble of Maxwell derivatives, left me utterly confused, statistical mechanics, as taught by Gene seemed to be breathtakingly elegant. I later learned that all of the other courses taught by Gene which included weak interactions, stellar evolution, astrophysics, and quantum mechanics, also shared a similar beauty.

After passing the qualifying exam the following year, I was recruited by Gene just before I was about to seek him out as a potential advisor. I admired him for his breadth of knowledge and teaching ability but didn't know about his future research plans. Instead of choosing a field of study, I decided to choose an advisor and remain open as to what I would do as a graduate student.

Like many entering graduate students who learned physics almost exclusively in the classroom, I wanted to be a theoretical physicist. My heroes were Galileo, Newton, Maxwell, Einstein, and also the contemporary giants such as Feynman, Gellman, Yang and Lee. At that time, the number of available jobs in physics was shrinking and prospects were especially difficult for budding young theorists. I recall the theoretical faculty admonishing us about the perils of theoretical physics: unless we were as good as Feynman, we would be better off in experimental physics. To the best of my knowledge, this warning had no effect on either me or my fellow students.

Gene had just ended a series of beta decay experiments and was getting interested in astrophysics. I had an interest in astrophysics, and spent the summer before entering Berkeley at the National Radio Astronomy Observatory modeling high redshift radio source galaxies in an effort to determine the deceleration of the universe. I emerged after two months with my first non-result: the red shift study could not extract a meaningful deceleration parameter because of likely galactic evolution.

[†] The author has taken the liberty of quoting himself without attribution.

Perhaps in deference to my desire to do theory, Gene asked me to think about proto-star formation of a closely coupled binary pair. However, in the next two months, instead of working on the theoretical problem he gave me, I played in the lab.

My most successful play-experiment was motivated by my interest in classical music. I noticed that one could hear out-of-tune notes played in a very fast run by a violinist. A simple estimate suggested that the frequency accuracy, $\Delta\nu$ times the duration of the note, Δt did not seem to satisfy the uncertainty relationship $\Delta\nu\Delta t \geq 1$. In order to test the frequency sensitivity of the ear, I connected an audio oscillator to a linear gate so that a pure tone of varying duration could be generated. I then asked my fellow graduate students to match the frequency of an arbitrarily chosen tone by adjusting the knob of another audio oscillator until the notes sounded the same. Students with the best musical ears could identify the center frequency contained in a short tone burst (that sounded like a "click") with an accuracy of $\Delta\nu\Delta t \sim 0.1$. In the language of spectroscopists, our ears were able to split the center of the line by a factor of 10.

By this time, it was becoming obvious (even to me) that I would be much happier as an experimentalist and I told Gene. He agreed and started me on a beta-decay experiment looking for so-called "second-class currents." His idea was to use the polarized proton beam at the 88-inch cyclotron to create partially polarized nuclei and then measure the energy dependence of the beta decay asymmetry with respect to the nuclear polarization. By then, Gene's most recent students, Stuart Freedman and Mel Simmons, had graduated and I became an only child. After a year of building up the apparatus, the experiment was going nowhere. I remember a time when we were finishing a late night run at the 88" cyclotron. (We were mostly assigned to the owl shift since Gene was Chair of the Department and rarely came to beam scheduling meetings and I was just a kid.) The experiment failed to produce any data, but rather than showing any frustration he might have been feeling, he looked at me and said, "This experiment is not good enough for you. You deserve better!" Despite this failure, Gene let me know how much he believed in me.

After that experimental run, we abandoned the beta decay experiment to measure the Lamb Shift in high-Z hydrogen-like ions. Dick Marrus had pioneered the use of the HILAC accelerator and beam-foil technology to create helium and hydrogen-like ions and study their forbidden transitions. Gene and Dick suggested that I consider using a CO_2 laser to excite the $2S_{1/2}$ to $2P_{1/2}$ Lamb Shift transition of a highly stripped ion. One could then tune the laser across the resonance by using the Doppler shift of the light relative to the accelerated ions. On the other hand, I was becoming fascinated by the rapidly developing technology of tunable dye lasers and was looking for an excuse to play with these new toys. I convinced Gene that I could build a powerful (one joule per pulse) Rhodamine-6G dye laser that could excite the $2S_{1/2}$ to $2P_{3/2}$ interval in hydrogen-like sulfur ion. Gene agreed to let me try and then gave me a free hand in designing and constructing an appropriate dye laser for this experiment. Happy as a clam, I set about to become a "laser jock."

In 1974, Claude and Marie-Anne Bouchiat published their proposal to look for parity non-conserving effects in atomic transitions. The unified theory of weak and electromagnetic interactions suggested by Weinberg, Salam, and Glashow postulated a neutral mediator of the weak force in addition to the known charged interactions. Such an interaction would show up as a very slight asymmetry in the absorption of left and right circularly polarized light in a magnetic dipole transition. Gene and I were drawn to this fundamental problem and were excited by the prospect that a table-top experiment could say something decisive about high energy physics. The opportunity was too good to pass up and I abandoned yet another experiment.

The parity experiment needed a state-of-the-art, reliable laser with high average power and narrow bandwidth. I brashly told Gene not to worry; the experience gained building the first dye laser would allow me to build the new laser in no time. I underestimated the difficulty of the project, but after two iterations, succeeded in constructing the required laser [1].

By this time, Ralph Conti had joined the group and was also enlisted into the experiment. Gene also took on another graduate student, David Neuffer. Dave and I were close friends and roommates at Rochester and Berkeley, and since I was a classmate with him in most of my undergraduate and graduate courses, I could vouch for his abilities as a physicist. He was extremely shy, somewhat narcoleptic, and was having a rough time fitting in with a high energy experimental group. Gene assigned Dave the task of refining the estimate of the size of the parity asymmetry first estimated in the Bouchiat and Bouchiat paper. He did an admirable job and went on to become a Wilson Fellow at Fermi Lab and a well-respected accelerator physicist [2].

Gene hated to write proposals and carried out most of his work on a shoestring budget. Luckily, all three of his students had NSF fellowships. The experiment required only relatively modest funding that was supplied through Gene affiliation with Lawrence Berkeley Laboratory. Gene and I were well matched: he didn't have any money and I loved the idea of scrounging around in the surplus yards and junk piles of LBL and Lawrence Livermore Lab for equipment. The best thing about scientific prospecting is the feeling you get when you strike the Mother Lode. Perhaps my best hit occurred when I found 8 brand new hydrogen thryatrons, still in their original shipping boxes, but sitting out in the open in a Livermore junk yard.

The goal of our experiment was to measure a 01.% asymmetry in an extremely forbidden magnetic dipole transition, and progress was slower than expected. After two years, I told Gene that I had suffered enough as a graduate student and would like to continue on the experiment as a postdoc. Besides, my NSF fellowship had run out, and if I could get a NSF postdoctoral fellowship, he wouldn't have to pay me as a postdoc. He could then use his limited resources to fund other graduate students such as Ralph, whose NSF fellowship had also run out. Gene agreed to downgrade my thesis to the observation of the forbidden dipole transition in thallium rather the observation of parity non-conservation [3].

Our thallium experiment work was beginning to attract many students, with Phil Bucksbaum, Larry Hunter and Persis Drell joining the group within a few years. We were also being noticed by the outside world. The high energy measurements of neutral currents was beginning to sort itself out (no more alternating currents), but two atomic physics experiments were reporting a null result significantly below the prediction of the minimal $SU(2) \times U(1)$ model of Weinberg and Salam. Steven

Weinberg would call up Gene every few months, hoping to get preliminary news of a parity violating effect with the promise that he would not tell anybody else except his wife. Dave Jackson and I would sometimes meet at the university swimming pool. During several of these encounters, he squinted at me and tersely asked, "Got a number yet?" The unspoken message was, "How dare you swim when there is important work to be done!"

During much of the last three years at Berkeley, I spent a lot of time teaching Phil and later Persis Drell about lasers. I loved playing the role of the older, wiser graduate student taking care of impressionable young charges. Persis would sometimes tease me that what I really wanted was a puppy dog to follow me around Birge and LeConte Halls. In addition to teaching Phil and Persis about lasers and experimental physics, I maintain that I also taught Persis how to use a stick shift and drink bourbon.

In the fall of 1978, we published our first results showing a parity violating effect. [5] Unfortunately, we were scooped: a few months earlier, a beautiful high energy experiment at the Stanford Linear Collider had seen much more convincing evidence

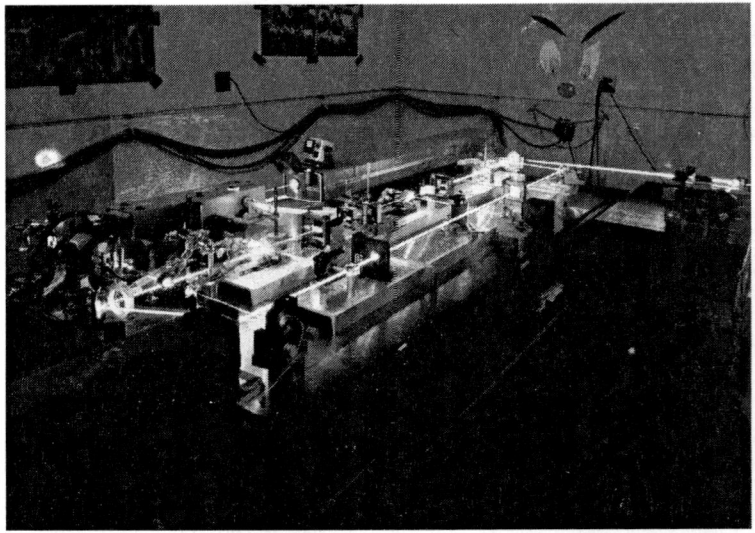

Fig. 1. Photo of the dye laser built for Persis Drell's Ph. D. thesis. Light from a cw dye laser and amplified by three YAG-pumped traveling wave amplifiers [4] was directed though a hole in the wall into the adjacent lab. Gene's contributions included the artwork surrounding the hole.

of

neutral weak interactions between electrons and quarks. Combined with the fact that earlier atomic physics experiments produced conflicting results (two reported no parity violating effect, and one reported an effect consistent with the Weinberg-Salam model), our result was largely dismissed by the high energy physics community. Thus, our work at Berkeley had little impact on the larger scientific community.

I worked side-by-side Gene for most of my seven years with him, often times more than 10 hours a day. What did I learn from him during all this time? Gene once told me that the only thing he taught me was how to solder. This is not true: he *tried* to teach me to solder, but in the end, I had to figure it out for myself. What Gene taught me was that I should only work on fundamental questions, that it is sometime OK to abandon an experiment that isn't going well, and that it is sometimes important to stick doggedly with other experiments. Gene also stressed the importance of writing clearly. When I presented him with the first draft of my thesis, a two week long core dump of the thallium work, he showed a rare moment of irritation. I can still remember hearing him lecturing me, "Be more precise. Let the reader know exactly what the ideas are and what was done." I began to appreciate that clear writing was synomomous with clear thinking, and I try to instill the same appreciation of good writing in my own students.

One of the joys of working side by side with Gene was that I was regaled with wonderful stories of his childhood in Princeton, rubbing elbows with immortal figures. His father, Saxe Commins, was a famous editor for Random house and his mother, Dorothy, a professional pianist [6]. Eugene O'Neill was his godfather and Albert Einstein played duets with his mother regularly. William Faulkner would sometimes stay at their home, writing in the attic bedroom at night while lubricated with a fifth of bourbon. In the morning, a few pages could be found on the kitchen table for Saxe's critical comments before continuing in the succeeding night with a fresh bottle. When Faulkner won the Nobel Prize, he appeared at the Commins' doorstep, again seeking a caretaker to see him through the ordeals of the Nobel ceremony.

Table 1: How I spent my time with Gene, 1972-1978

Topic	Time spent on project	Letter grade
Theory of protostar formation	~ 3 months	Incomplete
Violation of $\Delta v \Delta t \geq 0.1$	~ 1 week	Success, but not for credit
Beta decay asymmetry	~ 1 year	Incomplete
Lamb Shift of High-Z ions	~ 1 year	Incomplete
Parity violation in atomic transitions	2.5 years + 2 years postdoc	C -

Despite my less than stellar performance outlined in Table 1, I was offered a job as Assistant Professor at Berkeley. However, I spent all of my graduate and postdoctoral days at Berkeley, and the faculty was concerned about inbreeding. As a solution, they hired me and gave me the choice of either starting my own group in the fall, or taking a leave of absence for two years. I accepted the job, and realizing that I had a narrow view of science, took the opportunity to broaden myself, and went to Bell Labs on leave.

LIFE AT BELL LABS

I joined Bell Laboratories in the fall of 1978. I was a member of a group of brash, young scientists that included future Nobel Prize winners, Doug Osheroff, Bob Laughlin, Horst Stormer, and Dan Tsui. We felt like the "Chosen Ones," with no obligation to do anything except the research we loved best. The joy and excitement of doing science permeated the halls. The cramped labs and office cubicles forced us to interact with each other and follow each other's progress. The animated discussions were common during and after seminars and at lunch and continued on the tennis courts and at parties.

Bell Labs management supplied us with funding, shielded us from extraneous bureaucracy, and urged us not to be satisfied with doing merely "good science." The atmosphere was too electric to abandon, and I never returned to Berkeley. To this day, I feel guilty about this decision and hope the faculty at Berkeley have forgiven me.

With the hope that I would get interested in condensed matter physics, my Department Head, Peter Eisenberger, suggested that I spend my first six months in the library and talk to people before deciding what to do. During this time, he chided me not be content with anything less than "starting a new field." I responded that I would be more than happy to do that, but needed a hint as to *what* new field he had in mind.

I spent the first year at Bell writing an internal memo, reviewing the current status of x-ray microscopy. I also started an experiment on energy transfer in ruby with Hyatt Gibbs and Sam McCall, former graduate students of Gene and Erwin Hahn.

The goal of our experiment was modest: we wanted to confirm a measurement of Stan Geschwind, also at Bell Labs. The previous year, Phil Anderson had announced that the *only* clear example of a "mobility edge" separating localized states from extended states in a disordered solid was Stan's measurement of the mobility of excitons in ruby. The excitation was thought to diffuse rapidly in the crystal through short range resonant exchange coupling before being localized in trapping states formed with near-neighbor Cr^{3+} ion pairs. By tuning a laser over an inhomgeneously broadened absorption line of ruby, Stan realized that one could excite different densities of Cr^{3+} impurity sites. With low densities (weak coupling), the excitons were believed to be localized, while strong coupling should lead to de-localized excitons. Instead of confirming Stan's work, we ended up with a disappointing conclusion. There was no sharp transition between localized and extended states. A separate measurement showed that the excitons were coupled via long range dipole-dipole forces so that Anderson's theory did not even apply to this system [7].

During those early years, I also began planning the experiment on the optical spectroscopy of positronium. Positronium, an atom made up of an electron and its anti-particle, was considered the most fundamental of all atoms, and a precise measurement of its energy levels was a long standing goal ever since the atom was discovered in 1950. The problem was that the atoms would annihilate into gamma rays after only 140 nanoseconds, and it was impossible to produce enough of them at any given time. At the time we started the experiment, there were 12 published attempts to observe the optical fluorescence of the atom. People only publish failures if they have spent enough time and money so their funding agencies demand something in return.

My management thought I was ruining my career by trying an impossible experiment, and after two years of no results, they strongly suggested that I abandon my quest. But I was stubborn and also had a secret weapon: Allen Mills. Our strengths complimented each other beautifully, but in the end, he helped me solve the laser and metrology problems while I helped him with his positrons. We finally managed to observe a signal working with only ~4 atoms per laser pulse. Two years later and with 20 atoms per pulse, we refined our methods and obtained one of the most accurate measurements of quantum electrodynamics corrections to an atomic system [8]. The measurement made a big, but short lived splash as one of the most accurate test of a bound state QED system. However, after the excitement died down, it can be described as a "tour-de-force experiment," a polite translation of "impressive technically, but why did you do it?"

During the time I was working on positronium, my technician, Steve Wong, and I were also beginning to build up an apparatus that was to use picosecond transient grating spectroscopy to search for systems where Anderson Localization would be applicable. I also wanted to use this apparatus to study other phase transitions such as the Mott metal-insulator transition. While I was setting up this experiment, I began to put the apparatus through simple tests such as the measurement of the speed of the propagation of a pulse through our GaP:N samples. Much to our surprise, the pulse seemed to emerge faster from the absorbing sample as compared to the absence of a sample as we tuned the laser to the center of the resonance. A more careful measurement revealed that we could get the pulse velocity to go to infinity and become strongly negative. A negative pulse velocity means that the peak of the pulse leaves the sample even before the peak enters the sample.

The reaction of this finding followed a not-too-uncommon sequence of reactions from fellow scientists: 1) the work is wrong, 2) the work is trivial, and finally, 3) you were not the first person to discover this. True to form, the first set of colleagues I told thought the measurement was wrong. Next, Michael Sturge pointed out that Garret and McCumber treated this case theoretically more than a decade ago. Their paper was followed by initial observations that the speed of a pulse can be made to go slightly faster than 3×10^{10} cm/sec. I happened to stumble onto a system that made the effect too large to miss. The basic physics behind this effect *was* trivial: one could show in a few lines that a Gaussian pulse would propagate in an absorbing medium with the group velocity $v_g = c/[n_0 + \omega (dn/d\omega)]$ provided the Taylor series expansion of $n(\omega)$ remained valid. The effect is due to pulse shaping where the initial part of the pulse is attenuated much less than the trailing part of the pulse. The fun part of the effect is that a Gaussian pulse of a given width would emerge as a Gaussian pulse with the same width.

The pulses traveled "faster than a speeding bullet" and I realized how the popular press might jump on this work. Despite the spectacular effect, the basic physics could be explained by classical E&M straight out of Jackson. After thinking about how Gene might handle the situation, I opted for the low key approach and chose the boring title "Linear Pulse Propagation in an Absorbing Medium". [9] This effect has been since discovered several times after my own rediscovery, and because the attention given to "super-luminal" pulse propagation in the popular press, this work was propelled to notoriety two decades later.

During my years at Bell Labs, I continued to follow my graduate student habit of amassing a number "incompletes." In one attempt, I tested the feasibility of a new neutrino end-point experiment using molecular or atomic hydrogen. The idea was that complicated final state interactions of the tritium decay could mask a signature of the most energetic electrons of the β-decay spectrum, and these interactions in an atomic or molecular system were easier to calculate. The experiment was doomed from the start, given that the object was to make 200Å thick films that would not to leak hydrogen. I was also planning to enter surface science by constructing a novel electron spectrometer that could potentially increase the energy resolution of traditional electron energy loss spectroscopy by more than an order of magnitude. My idea was based on threshold ionization of atoms using a high repetition picosecond laser. If atoms could be ionized, one atom per pulse because of space charge effects, one could achieve the surface sensitivity of electron probes with the resolution of optical methods. Furthermore, a lot of science could be found with that increase of resolution.

In fall of 1983, I became head of the Quantum Electronics Research Department and moved to the Holmdel branch of Bell Labs. While designing the electron spectrometer, I began talking informally with Art Ashkin, a colleague at Holmdel. Art had a dream to trap atoms with light, and found in me an increasingly attentive listener. That fall I was also joined by my new post-doc, Leo Hollberg, who was going to help construct an electron energy loss spectrometer. However, Leo was trained as an atomic physicist and was also developing an interest in the possibility of manipulating atoms with light.

Leo and I decided, on the spur of the moment, to drive to Massachusetts to attend a workshop on the trapping of ions and atoms organized by Dave Pritchard at MIT. I was ignorant of the subject and lacked the primitive intuition that is essential to add something new to a field. As an example of my profound lack of understanding, I found myself wondering why the dipole force would change sign when the light was tuned above and below the atomic resonance. Looking back on these early fumblings, I am embarrassed by how long it took me to recognize that the effect can be explained by freshman physics. On the other hand, I was not alone in my lack of intuition. When I asked a Bell Labs colleague about this effect, he answered, "Only Jim Gordon really understands the dipole force!"

The dipole force was demonstrated by the Bell Labs team of John Bjorkholm, Richard Freeman and Art Ashkin in 1978. However, they had reached an impasse, and their management stopped the work shortly afterwards. After arriving at Holmdel, I began to realize that the way to hold onto atoms with light was to first get them very cold. Laser cooling was going to make possible Art Ashkin's dreams, plus a lot more. I promptly dropped most of my other experiments and with Leo Hollberg and my new technician, Alex Cable, we began our laser cooling experiments.

In my excitement, I told my director, Chuck Shank that my plan was so simple it would be like falling off a log. With the previous experimental efforts still fresh in his mind, he was more sanguine. During the year-end performance review session with the department heads of our laboratory in the fall of 1983, he declared that I "earned the right" to pursue whatever I wanted, but did not want other principal investigators such as John Bjorkholm to work on the project.

By the winter of 1984, we were making good progress, and I invited John to join our effort. During the following 2 years, our team made good progress, first with the demonstration of optical molasses [10], the first optical trap for atoms [11] and the magneto-optic trap [12]. Meanwhile, Art Ashkin had demonstrated the optical tweezers trap for micron sized particles in his lab [13] and then his later discovery that he could hold onto live cells with his tweezers trap.

During this time, we were beginning to see signs that optical molasses was more than it was originally cracked up to be. In 1986, we reported that we could align the laser beams so that the atom storage time would increase from roughly 0.5 seconds to over 30 seconds. About the same time, Bill Phillips' NIST group reported molasses lifetimes with a very different frequency dependence from the one predicted by the simple diffusion formula we published in our first molasses paper. This group also found that the trap was more stable to beam imbalance than had been expected. In our collective euphoria over cooling and trapping atoms, the research community had not performed the basic tests to measure the properties of optical molasses, and I was the most guilty.

STANFORD

In 1987, I decided to leave the cozy ivory tower that was Bell Labs. Ted Hänsch had left Stanford to become co-director of the Max Planck Institute for Quantum Optics, and I was recruited by Art Schawlow to replace him. Within a few months, I also received offers from Berkeley and Harvard, and I thought now was the time to sell while my stock was temporarily high. At last I could begin a career in the mode of Gene and spawn scientific progeny.

My first three graduate students, Mark Kasevich, David Weiss and Michale Fee, were outstanding and I continue to be blessed with superb students and postdocs. However, I soon realized that it took a very special person like Gene to get in the trenches and work day after day alongside of his students. Also, I could not contain my scientific appetite, and within a decade, my group expanded to the point that I now struggle to keep up with my students, both past and present.

My original plan was to explore variants of the magneto optic trap as a bright source of atoms and then use this source to do fundamental physics. In another chamber, I wanted to study the quantum reflection of atoms from cold surfaces, a problem I got interested in while at Bell Labs. The problem can be simply posed as follows: consider an atom with a long de Broglie wavelength λ incident on an idealized, short-range, attractive potential. In general there is a transmitted wave and reflected wave, but in the limit where λ is much greater than the length scale of the potential, the probability of reflection goes to unity. An atom near a surface will experience an attractive van der Waals force with a $1/z^3$ potential and further away, the attractive potential would become $1/z^4$ due to "retarded potential" effects first discussed by Casimir. It has been predicted that the probability of reflection, R would go to unity as $\lambda \rightarrow \infty$ for $n > 3$. There are subtleties to the problem when inelastic scattering channels are included, and theorists became polarized between two outcomes, $R = 0$ and $R = 1$.

My plan of research was soon altered by a discovery that sent shock waves through the laser cooling community. In the spring of 1988, Bill Phillips and co-workers reported that sodium atoms in optical molasses could be cooled to temperatures eight times colder than the limit predicted by theory. Within a few months, groups led by Carl Wieman, Claude Cohen-Tannoudji and myself verified that sodium and cesium atoms in optical molasses could be cooled well below the Doppler limit. I didn't pick up on the several clues dancing before us and simply blew the opportunity.

At the end of June of 1988, Claude and I attended a conference on spin polarized quantum systems in Torino, Italy. In a self-flagellating talk, I gave a summary of the new surprises in laser cooling known at that time. After our talks, Claude and I had lunch and compared the findings in our labs. The theory that predicted the minimum temperature for two-level atoms was beyond reproach. We felt the lower temperatures must be due to the fact that the atoms we were playing with were (real) atoms with Zeeman sub-levels and hyperfine splittings. Our hunch was that the cooling mechanism probably had something to do with the Zeeman sub-level structure and not the hyperfine structure since $v_{rms} \sim 5 v_{recoil}$ for both cesium and sodium despite the fact that their hyperfine structures are 9.2 GHz and 1.77 GHz, respectively. From Phillips' work, we also knew that the magnetic field had to be reduced to below 0.05 Gauss to achieve the best cooling. All of these clues suggested that the cooling must somehow involve optical pumping.

After the conference, Claude returned to Paris while I was committed to remain in Europe, trapped in "viewgraph space." My next stop was Munich, where I told Ted Hänsch that I thought it had to be an optical pumping effect. My knowledge of optical pumping was rudimentary, so I spent half a day in the physics department library reading about the subject. As I read about optical pumping, Claude's name kept on popping up and it began to dawn on me that Claude and Jean Dalibard were better positioned to figure out this puzzle.

The "eureka" moment occurred when I realized that laser cooling was due to a combination of optical pumping, light shifts, and the fact that the polarization in optical molasses changed at different points in space. The International Conference on Atomic Physics met the following week, and on the Sunday before the conference, Jean and I met and compared notes. It was immediately obvious that the cooling models Jean and Claude and I concocted were the same. Jean, already scheduled to give a talk at the conference, gave a summary of their model. I was generously given a "post-post-deadline" slot in order to give my account of the new cooling mechanism [14].

Both groups followed up the work with a more detailed development of optical molasses in multi-level atoms. Claude and Jean approached the problem with an elegant treatment of a model system consisting of a spin 1/2 ground state and a spin 3/2 excited state [13]. We took a more brute force approach and generalized the Optical Bloch equations to include the Zeeman sublevels of a real sodium atom [14].

While this working was going on, Mark Kasevich and a postdoc, Erling Riis, were given the task of producing an atomic fountain as a first step in the quantum reflection experiment, which was to be Mark's thesis. The idea was to launch the atoms upwards in an atomic fountain with a slight horizontal velocity. When the atoms reached their zenith, they would strike a vertically oriented surface with extremely low velocity.

As they were setting up this experiment, I asked them to do the first of a number of "quickie experiments." I was able to convince my students to embark on a series of fast diversions that would only take a few weeks. The first detour was to use the atomic fountain to do some precision spectroscopy.

While I was a graduate student with Gene, I learned about an attempt by Zacharias to make an atomic fountain of atoms by directing an atomic beam upwards. Although most of the atoms would crash into the top of the vacuum chamber, the very slowest atoms in the Maxwell distribution were expected to follow a ballistic trajectory and return due to gravity. Zacharias wanted to use the fountain to make an ultra-precise atomic clock that could measure the gravitational red shift. Unfortunately, the experiment failed. The slowest atoms in the Maxwell-Boltzman distribution were scattered by faster atoms overtaking them from behind and never returned to the microwave cavity.

With our source of laser-cooled atoms, it was a simple matter for us to construct an atomic fountain [17]. After the theory of polarization gradient cooling was developed, we realized that there was a much better way to launch the atoms. By pushing with a single laser beam from below, we would heat up the atoms due to the random recoil kicks from the scattered photons. However, by changing the frequency of the molasses beams so that the polarization gradients would be in a frame of reference moving relative to the laboratory frame, the atoms would cool to polarization gradient temperatures in the moving frame. The atoms could be launched with precise velocities and with no increase in temperature [18]. This work was quickly followed by the first cesium atomic fountain and the first cesium frequency standard [19]. Today, the world time standard is based on atomic fountain clocks, and with improved cooling methods, it is now possible to launch cesium atoms in a fountain with a temperature of less than 200 nanoKelvin [20].

The next "quickie" to follow the atomic fountain was the demonstration of normal incidence reflection of atoms from an evanescent sheet of light. Balykin and collaborators deflected an atomic beam through a small angle with an evanescent sheet of light extending out from a glass prism [21]. We wanted to demonstrate an "atomic trampoline" trap by bouncing atoms from a curved surface of light created by internally reflecting a laser beam from a plano-concave lens. This demonstration would also be a prelude to our quantum reflection experiment. Unfortunately, the lens we used produced a considerable amount of scattered light, and the haze of light "levitated" the atoms and prevented us from seeing bouncing atoms. Mark ordered a good quality lens and we settled for bouncing atoms from the dove-prism surface with the intent of completing the work when the lens arrived [22].

While waiting for the delivery of our lens, I began to think about the next stage of the quantum reflection experiment. The velocity spread of the atoms in the horizontal direction would be determined by a collimating slit. The quantum reflection experiment would require exquisitely cold atoms with a velocity spread corresponding to an effective temperature in the range of tens to hundreds of picoKelvin and very narrow collimating slits would reduce the flux of atoms to distressingly low levels. Ultimately, the collimating slits would cause the atoms to diffract.

Instead of using collimating slits, we realized the velocity selection could be made spectroscopically using the Doppler effect. Usually, the Doppler sensitivity is limited by the linewidth of the optical transition. However, if we induced a two-photon transition between two ground states with laser beams at two different frequencies, the linewidth would be determined by the time it took to induce the two-photon transition, and with an atomic fountain, we would have lots of time. The second frequency could be generated by an electro-optic or acousto-optic modulator so that the frequency jitter of the excitation laser would not enter the transition.

Despite the fact that the resonance would depend on the frequency difference of the two beams, the Doppler sensitivity would depend on the frequency sum if the two laser beams were counter-propagating. This idea of using a two-photon Raman transition would allow us to achieve Doppler sensitivity equivalent to an ultra-violet transition, but with the frequency control of the microwave domain. In a proof of principle experiment, we created an ensemble of atoms with a velocity spread of 270µm/sec corresponding to an effective 1-dimensional temperature of 24 picoKelvin [23].

By 1990, we were aware of several groups trying to construct atom interferometers based on the diffraction of atoms by mechanical slits or diffraction gratings, and their efforts stimulated us to think about different approaches to atom interferometry. We realized that our velocity selection method could be used to put an atom into a superposition of two atomic states with different velocities. This effect allowed us to design a new type of atom interferometer based on optical pulses of light, and because the atoms would be in an atomic fountain, we would automatically obtain very high precision. In our first atom interferometer paper, we demonstrated a resolution in g, the acceleration due to gravity, of $\Delta g/g = 10^{-6}$, followed with improved experiment that achieved a resolution of $\Delta g/g = 3 \times 10^{-8}$ [24]. With a number of refinements, including the use of an actively stabilized vibration isolation system, we have been able to achieve a precision of 0.6 parts in 10^{10} and an absolute accuracy of 3 parts per billion [25].

Our ability to measure small velocity changes with stimulated Raman transitions suggested another application of atom interferometry. If an atom absorbs a photon of momentum $p_\gamma = h\nu/c$, it will receive an impulse $\Delta p = M\Delta v$. Thus, $h/M = c\Delta v/\nu$, and since Δv can be measured as a frequency shift, we could make a precision measurement of h/M. An accurate value of h/M would be a better determination of the fine structure constant, α, since α can be expressed in terms of quantities in the above relation that can be measured precisely in terms of frequencies or frequency shifts

$$\alpha^2 = (2R_\infty/c)(m_p/m_e)(M_{atom}/m_p)(h/M_{atom}).$$

After this revelation, Dave Weiss' thesis project was shifted to a measurement of h/M. The interferometer geometry he chose was previously discussed as an extension of the Ramsey technique into the optical domain, and with Brent Young, he obtained a resolution of roughly a part in 10^7 in h/M$_{Cs}$ [26].

In a second generation version of this experiment, we used an adiabatic transfer method demonstrated by Klaus Bergmann and collaborators [27]. The beauty of an adiabatic transfer method is that it is insensitive to the small changes in experimental parameters and other potentially troublesome systematic effects. By varying the intensity of our two laser beams with acousto-optic modulators, we were able to construct an atom interferometer using the adiabatic transfer method [28]. After many refinements and considerable effort, we are in the final stages a measurement h/M$_{Cs}$ with a quoted uncertainty in α of ~3 or 4 parts in a billion [29]. Curiously, the most accurate methods of determining α are direct applications of three Nobel Prizes: the Josephson effect (56 ppb), the quantized Hall effect (24 ppb), the magnetic moment of the electron with the QED calculation (3.8 ppb).

BIOPHYSICS

In 1986, while the world was excited about atom trapping, Art Ashkin began to play with optical tweezers trapping of micron-sized particles [13]. While experimenting with colloidal tobacco mosaic viruses, he noticed tiny, translucent objects in his sample. Rushing into my lab, he excitedly proclaimed that he had "discovered life!" What Art had discovered were bugs in his apparatus: bacteria that had eventually grown in his sample beads and water. He followed his discovery by showing that infrared light focused to megawatts/cm^2 could be used to trap live e-coli bacteria and yeast for hours without damage [30]. Steve Block and Howard Berg soon adapted the optical tweezers technique to study the mechanical properties of the flagella motor [31].

By this time, I had gotten to Stanford and wanted to be part of the fun. However, rather than manipulate single cells, I wanted to hold on to individual molecules by attaching micron-sized polystyrene spheres to the ends of the molecule and grabbing onto the plastic handles with laser beams focused by the microscope objective. Steve Kron, an M.D./Ph.D. student in the medical school, introduced me to molecular biology in the evenings, and by 1990, we could stretch a single, fluorescently labeled DNA molecule immersed in water [32].

My original goal in developing methods to manipulate DNA was to study the motion of enzymes moving on the molecule. However, once we began to play with the molecules, we noticed that a stretched molecule of DNA would spring back like a rubber band when released from the tweezers. Using our new ability to visualize and manipulate individual molecules of DNA, my group began to answer polymer dynamics questions that have persisted for decades [33]. Even more thrilling, we discovered something completely new: identical molecules in the same initial state will choose several distinct pathways to a new equilibrium state [34]. This "molecular individualism" was never anticipated in previous polymer dynamics theories and has fundamentally altered our thinking of polymer dynamics.

Following this discovery, we asked ourselves whether biochemical processes occurred as the set of well-ordered events depicted in textbooks, or whether processes such as DNA replication, RNA transcription, protein translation, and macromolecule folding have a much richer behavior. We have recently been applying fluorescence resonance energy transfer to study the behavior of individual biomolecules. Donor and acceptor dyes attached to two sites of a biological molecule can be used to measure the distance between the two dyes: donor fluorescence emission is strongly quenched in a distance and orientation dependent manner by the acceptor, while the acceptor emission increases due to the energy transfer. Thus, a measurement of a change in fluorescence from the two dyes can be used as an indicator of a change in the conformation of the host molecule. Fluorescence resonant energy transfer (FRET) was first seen at the single molecule level by Shimon Weiss and collaborators [35].

With this technique, we have been studying a variety of biological problems including the observation of enzymatic activity and RNA folding [36]. We have also begun to study more complex interacting systems such as neurotransmission at the synapse and the manufacture of proteins by the ribosome.

Our work in biology and polymer dynamics is a departure from the problems in fundamental physics that was the focus of Gene's work. I began to choose research areas by sometime sniffing out areas where a novel technology was beginning to emerge and simply followed my nose without the clear vision of the future. Sometimes, I felt that I was disappointing Gene with my wayward habits of working on things that were not manifestly important. Gene and I are similar in many ways, but in this respect we are different, and one of the best things about being mentored by Gene was that he allowed me to be different from him.

In my random walk into areas that I knew nothing about, I was armed with a self-confidence that I would not have had if I were not Gene's student. I always felt that Gene treated me as special. Gradually, I realized that Gene treated all of his students this way. He believed in us, and as a consequence, we began to believe in ourselves. In this extraordinarily nurturing atmosphere, we developed the self-confidence needed to sustain us through the failures that we would encounter in our future research. It is no small wonder that a disproportionate number of his former students have gone on to become distinguished professors and researchers.

I am not sure how things would have turned out for me had I worked for another advisor. What I am certain about is that I would have achieved far less in my own career, and I

The prodigal son returns from a random walk.

owe a great deal to Gene for believing in me, for shaping me, and for remaining my model of what a scientist and mentor should be.

REFERENCES

1. S. Chu and E. Commins, Appl. Optics **16**, 2619 (1977); S. Chu and R.W. Smith, Opt. Comm. **28**, 221 (1979).
2. D. Neuffer and E.D. Commins, Phys. Rev. **A 16**, 844 (1977); Phys. Rev. **A 16**, 1760, (1977).
3. S. Chu, E. Commins, and R. Conti, Phys. Lett. **60A**, 96 (1977).
4. P. Drell and S. Chu, Opt. Communications **28**, 343 (1979).
5. R. Conti, P. Bucksbaum, S. Chu, E. Commins, and L. Hunter, Phys. Rev. Lett. **42**, 343 (1979).
6. Dorothy Commins, What is an Editor? Saxe Commins at Work (Univ. of Chicago Press, 1978).
7. S. Chu, H.M. Gibbs, S.L. McCall, and A. Passner, Phys. Rev. Lett. **45**, 1715 (1980); S. Chu, H.M. Gibbs, and S.L. McCall, Phys. Rev. **B24**, 7162 (1981).
8. S. Chu and A.P. Mills, Phys. Rev. Lett. **48**, 1333 (1982); S. Chu, A.P. Mills, Jr. and J.L. Hall, Phys. Rev. Lett. **52**, 1689 (1984).
9. S. Chu and S. Wong, Phys. Rev. Lett. **48** 738 (1982); Also, see Comments, Phys. Rev. Lett. **49**, 1293 (1982).
10. S. Chu, L. Hollberg, J.E. Bjorkholm, A. Cable, and A. Ashkin, Phys. Rev. Lett. **55**, 48 (1985).
11. S. Chu, J.E. Bjorkholm, A. Ashkin, and A. Cable, Phys. Rev. Lett. **57**, 314 (1986).
12. E.L. Raab, M. Prentiss, A.E. Cable, S. Chu, and D.E. Pritchard, Phys. Rev. Lett. **59**, 2631 (1987).
13. A. Ashkin, J.M. Dziedzic, J.E. Bjorkholm, and S. Chu, Optics Lett. **11**, 288 (1986).
14. J. Dalibard, C. Solomon, A. Aspect, E. Arimondo, R. Kasier, N. Vansteenkiste, and C. Cohen-Tannoudji, in *Atomic Physics 11*, eds. S. Haroche, J.C. Gay and G. Grynberg, (World Scientific, Singapore, 1989). pp. 199-214; S. Chu, D. Weiss, Y. Shevy and P. Ungar, op cit. pp 636-638.
15. J. Dalibard and C. Cohen-Tannoudji, J. Opt. Soc. Am. B **6**, 2023 (1989).
16. P.J. Ungar, D.S. Weiss, E. Riis, and S. Chu, J. Opt. Soc. Am. B **6**, 2058 (1989).
17. M.A. Kasevich, E. Riis, S. Chu, and R.G. DeVoe, Phys. Rev. Lett. **63**, 612 (1989).
18. D.S. Weiss, E. Riis, M. Kasevich, K.A. Moler, and S. Chu, in *Light Induced Kinetic Effects on Atoms, Molecules, and Molecules*, eds. L. Moi, S. Gozzini, C. Gabbanini, E. Arimondo, F. Strumia, (ETS Editrice, 1991) pp. 35-44.
19. A. Clarion, C. Salomon, S. Guellati and W.D. Phillips, Europhys. Lett. **16**, 165 (1991); K. Gibble and S. Chu, Metrologia **29**, 201, (1992).
20. P. Treutlein, K.-Y. Chung and S. Chu, Phys. Rev. **A63**, 51401 (2001).
21. V.I. Balykin, V.S. Letokhov, Yu.B. Ovchinnikov and A.I. Sidorov, Phys. Rev. Lett. **60**, 2137 (1988).
22. M.A. Kasevich, D.S. Weiss, and S. Chu, *Opt. Lett.*, **15**, 667 (1990).
23. M. Kasevich, D. Weiss, E. Riis, K. Moler, S. Kasapi and S. Chu, Phys. Rev. Lett. **66**, 2297 (1991).
24. M. Kasevich and S. Chu, Phys. Rev. Lett. **67**, 181 (1991); M. Kasevich and S. Chu, Applied Physics B **54**, 321 (1992).
25. A. Peters, K.-Y. Chung and S. Chu, Nature **400**, 849-852 (1999); A. Peters, K.Y. Chung and S. Chu, Metrol., **38** (2001); Philipp Treutlein, Keng Yeow Chung and Steven Chu, Phys. Rev. **A63**, 51401 (2001).
26. D.S. Weiss, B.C. Young, and S. Chu, Phys. Rev. Lett. **70**, 2706 (1993); D.S. Weiss, B.C. Young and S. Chu, Applied Physics B **59**, 217-256 (1994).
27. U. Gaubatz, P. Rudecker, M. Becker, S. Schiemann, M. Kültz, and K. Bergmann, Chem. Phys. Lett. **149**, 463 (1988).
28. M. Weitz, B.C. Young, and S. Chu, Phys. Rev. Lett. **73**, 2563 (1994).
29. J. Hensley, A. Wicht, B. Young and S. Chu, to be published (2001).
30. A. Ashkin and J.M. Dziedzic, Science **235**, 1517 (1987); A. Ashkin, J.M. Dziedzic and T. Yamane, Nature **330**, 769 (1987).

31. S. Block, D.F. Blair and H.C. Berg, Nature **338**, 514 (1989).
32. S. Chu and S. Kron, Int. Quantum Electronics Conf. Tech. Digest, (Optical Soc. of Am., Washington DC, 1990) p 202; M. Kasevich, K. Moler, E. Riis, E. Sunderman, D. Weiss, and S. Chu, *Atomic Physics 12*, eds. J.C. Zorn and R.R. Lewis, (Am. Inst. of Physics, 1990), pp. 47-57.
33. See, for example, T. Perkins, D.E. Smith and S. Chu, Science **64**, 819 (1994); T.T. Perkins, S.R. Quake, D.E. Smith and S. Chu, Science **264**, 822 (1994); T.T. Perkins, D.E. Smith, R.G. Larson, and S. Chu, Science **268**, 83 (1994); D.E. Smith, T.T. Perkins and S. Chu, Phys. Rev. Lett. **75**, 4146 (1995); S.R. Quake and S. Chu, Nature, **388**, 151 (1997); D.E. Smith, H.P. Babcock and S. Chu, Science **283**, 1724 (1999); H. Babcock, D.E. Smith, J. Hur, E. Shaqfeh and S. Chu, Phys. Rev. Lett. **85**, 2018-2021 (2000).
34. T.T. Perkins, D.E. Smith and S. Chu, Science **276**, 2016-2021 (1997); D.E. Smith and S. Chu, Science **281**, 1335 (1998).
35. T. Ha, et al., *Proc. Nat'l. Acad. Sci. USA*, **93**, 6264-6268 (1996).
36. T. Ha, X. Zhuang, H.D. Kim, J.W. Orr, J.R. Williamson and S. Chu, Proc. Natl. Acad. Sci. USA **96**, 9077-9083 (1999); X. Zhuang, T. Ha, H.D. Kim, T. Centner, S. Labeit and S. Chu, PNAS **97**, 14241-14244 (2000); X. Zhuang, L.E. Bartley, H.P. Babcock, R. Russell, T. Ha, D. Herschlag, and S. Chu, Science **288**, 2048-2051 (2000).

SEARCHES FOR P,T VIOLATIONS IN ATOMS AND MOLECULES

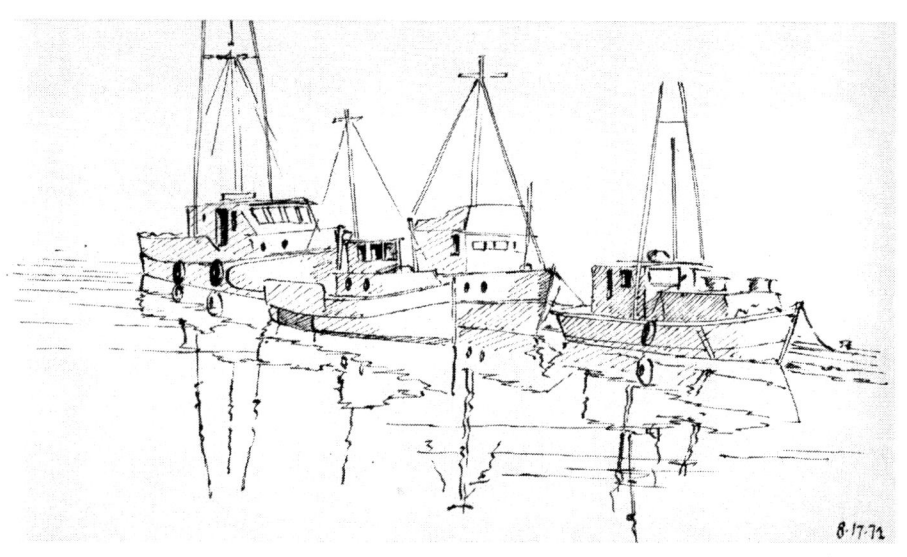

The Search for a Permanent Electric Dipole Moment - Still Active, Still Important

E. Norval Fortson

Department of Physics, Box 351560
University of Washington, Seatttle, WA 98195

Abstract. There has been exciting progress in recent years in the search for a permanent electric dipole moment of atoms, molecules, and the neutron. An EDM can exist only if time reversal symmetry is violated. Although such a dipole has not yet been detected, theories of possible new physics, such as Supersymmetry, predict the existence of EDMs within reach of modern experiments. In this brief introduction and survey, I discuss current and planned EDM experiments, and summarize what recent results imply about the size of possible T violating (and hence CP violating) interactions among elementary particles.

INTRODUCTION

It is a pleasure to contribute to these proceedings honoring Eugene Commins, a long time friend and physics colleague. My topic is particularly appropriate since through the Berkeley thallium experiment Gene has played such a crucial role in the ongoing search for a permanent electric dipole moment (EDM) of some simple system such as an atom, a molecule or the neutron.

This short report is intended to supply only a brief introduction to EDM experiments and their great significance for elementary particle theory. Individual experiments are discussed fully by M. V. Romalis, C. B. Regan, and D. P. DeMille elsewhere in the proceedings.

Half a century ago, a precise search for an EDM of the neutron was undertaken by Smith, Purcell, and Ramsey [1,2]. They set what seemed at the time a remarkably small upper limit, $d(n) < 5 \times 10^{-20}$ e-cm. Since then, there have been many further searches for an EDM of the neutron, with ever increasing precision. Likewise there have been continually improved searches for an EDM of an atom or a molecule - including experiments sensitive to an intrinsic EDM of the electron. Thus far, all EDM experiments have yielded a null result. Nevertheless, most elementary particle theories that attempt to go beyond the Standard Model [3] predict that EDMs should exist and be large enough to detect by experiments now underway or soon to begin.

The existence of an EDM of any non-degenerate quantum system would imply a breakdown of time-reversal symmetry (T), and through the CPT theorem, a violation of CP symmetry as well [4]. (C is charge conjugation, or particle/antiparticle symmetry, and P is parity, or space-inversion symmetry.) The only known example of CP violation in nature was discovered over 35 years ago in the decay of the K_0 meson [5]. For many years after this discovery, the search for a neutron EDM provided an exacting test of theories put forward to account for the K_0, and ruled out most of them as the experimental upper limit on the neutron moment steadily decreased to its current value [6,7]. Atomic and molecular EDM experiments made equally striking advances as well, starting in the 1960s [8,9], and emerging again in the 1980s with results on xenon [10], thallium fluoride [11,12], cesium [13], thallium [14,15], and mercury [16,17]. Recent work includes just completed measurements on thallium [18] and mercury [19], and a host of new experiments as discussed below. These experiments should provide an exacting test of Supersymmetry and other theories of possible new physics.

UNDERLYING THEORY

It is now generally accepted that a satisfactory explanation of the observed CP violation (or equivalently, T violation) in the K_0 system is given by the Standard Model [3], in which CP violation occurs as a phase factor (the KM phase) in the interaction of quarks with W bosons. This model by itself predicts EDMs too small to be observed in current or contemplated experiments. Thus, if an EDM is found, it will be compelling evidence for the existence of some sort of physics beyond the Standard Model.

There is no shortage of theories of such new physics, but by far the most cherished among particle theorists is Supersymmetry (SUSY) [20]. SUSY incorporates quantum gravity consistently, and also solves the *gauge hierarchy* problem, i.e. it protects the huge energy gap between grand-unification/quantum-gravity at 10^{16} - 10^{19} GeV and the electroweak scale at 100GeV. A feature of SUSY that is of great importance for EDMs is the existence of scalar particles with T-odd phase angles that have no natural reason to be small, just as the KM phase angle is about 45^0 in the Standard Model. Such scalar particles automatically generate EDMs of observable size if their mass scale M is in the 100 - 1000 GeV range required for SUSY to protect the gauge heirarchy [3,20]. A number of authors have pointed out that EDM searches therefore have a good chance of being the first experiments to discover SUSY [21]. Specific limits on SUSY set by recent EDM experiments have now been calculated [22], and are presented later in this report.

There are also arguments from proposals of Cosmological baryogenesis at the electroweak energy scale that there must be some additional source of CP violation, as perhaps in SUSY, since the CP violation in the Standard Model alone seems unable to account quantitatively for the observed matter/antimatter asymmetry in the Universe [23].

The way such T violation at the fundamental elementary particle level would generate an observable EDM depends upon the system under study. The neutron is sensitive almost exclusively to T violation in the quark sector, while atoms and molecules have bound electrons and are therefore sensitive to T violation in the lepton sector as well as the quark sector.

In atoms and molecules there are actually a number of ways that T violating interactions at the particle level could give rise to an EDM, and *all are enhanced considerably in heavy atoms* [4,8,24,25]. Calculations have been made of the atomic

EDM due to an EDM distribution in the nucleus, to a T violating force between electrons and nucleons, and to an intrinsic EDM of the electron itself, corresponding respectively to hadronic (quark-quark), semi-leptonic (electron-quark), and purely leptonic interactions as the chief source of T violation.

Which of the possible effects will predominate in a given atom or molecule depends upon the net electronic angular momentum J. In systems with J = 0 (i.e. systems with only closed electronic shells, such as Hg, Xe, and TlF), the EDM vector points along the nuclear spin **I**, and the greatest sensitivity is to purely hadronic T-violation inside the nucleus. In this case, the important quantity is the nuclear *Schiff Moment* [3,4], which measures the part of the nuclear EDM that is not completely shielded from the outside world by the atomic electrons. Although shielding does reduce the size of EDMs in closed shell atoms, it turns out that this loss can be more than compensated by the extra experimental EDM sensitivity attained in these atoms. Another source of an EDM along **I** could in principle be a tensor-pseudotensor form of electron-nucleon T-violation [24].

In systems with non-zero J (i.e. paramagnetic systems such as Cs, Tl or open-shell molecules) the EDM has a component parallel to **J**, and the greatest sensitivity is to an intrinsic electron EDM, or to a scalar-pseudoscalar [24] form of electron-nucleon T-violation. The great atomic theory discovery here, made in the 1960s by Sandars [8], is that the effect of an electron EDM is actually *enhanced* in a heavy atom, by over a factor of 100 in cesium and considerably more in thallium and other heavier atoms.

EXPERIMENTS

General Concept

All experiments are based on what should happen when a spinning elementary particle, atom or molecule having an EDM is placed in the electric field that exists between two oppositely charged parallel plates. In the manner of a spiining top, the spin will precess about the electric field axis due to the electric torque on the dipole. The longer the spin remains in the electric field without being otherwise disturbed, i.e. the longer the spin relaxation time T_2, the larger will be its angle of precession due to an EDM and the more sensitive will be the experiment. When the electric field direction is reversed by reversing the sign of the voltage between the plates, the sense of spin precession about the field axis also reverses. This behavior helps distinguish the precession due to an EDM from that due to other torques.

More formally, when the system under study is placed in external electric and magnetic fields **E** and **B**, the Hamiltonian may be written in terms of the electric and magnetic dipole moments d and μ as:

$$H = -(d\mathbf{E} + \mu\mathbf{B})\cdot\mathbf{F}/F \qquad (1)$$

where **F** is the total angular momentum, which for atoms is **I** + **J**. In all experiments, the search for an EDM consists of looking for a change in this energy (or equivalently, a change in the torque on **F**) when the direction of **E** is reversed. For example, if **E** and **B** are parallel to each other, an EDM would cause the frequency of Larmor precession about the field axis to change when **E** is reversed relative to **B**.

Eq. (1) demonstrates the well known fact mentioned above that an EDM violates T (and also P). Under time reversal, $\mathbf{B} \to -\mathbf{B}$ and $\mathbf{E} \to \mathbf{E}$. Thus $H_T \neq H$, and the system is clearly asymmetric under T. Under parity, $\mathbf{B} \to \mathbf{B}$ and $\mathbf{E} \to -\mathbf{E}$. Therefore $H_P \neq H$, and the system also violates P.

EDM measurements have been carried out thus far either on a beam of particles or on a sample of them held in a cell. Beams are usually the method of choice for such atoms and molecules as Tl and YbF which undergo rapid spin relaxation in a cell. Cells are usually chosen for those systems which have a long spin relaxation time, such as Cs, Xe, Hg, or the neutron. The long T_2 obtainable in a cell means a sharp resonance line and high precision. (For somewhat different reasons the proposed PbO experiment discussed below will also be carried out in a cell.) A cell also has the advantage of averaging the $\mathbf{v} \times \mathbf{E}$ motional magnetic field nearly to zero. However, cells cannot support as large an electric field as beams, and cells usually have larger high voltage leakage currents.

There are proposals for using laser cooled and trapped alkali atoms for EDM experiments [26]; also for using cold atoms having closed shells [27]. Such methods may combine the advantages of cells and beams [28].

Tables 1 and 2 below list some current and planned EDM experiments and the method used for each.

TABLE 1. Electron spin EDM experiments.

System	Method	Location
^{205}Tl	Beam	Berkeley
PbO	Cell	Yale
YbF	Beam	Sussex
^{133}Cs	Cell	Amherst
^{133}Cs	Cold atom trap	Penn St., Stanford, Texas

TABLE 2. Nuclear spin EDM experiments.

System	Method	Location
Neutron	Cell	Los Alamos, Grenoble....
^{199}Hg	Cell	Seattle
^{129}Xe	Cell	Michigan, Harvard
^{129}Xe	Cell (liquid Xe)	Princeton

Some Recent and Current Experiments

Neutron

Recent neutron experiments have been carried out at ultra-cold neutron (UCN) reactor facilities as in Grenoble, France. Magnetic mirrors are used to line up the neutron spins and monitor the spin direction. Very slow neutrons, with speeds less than 6 meters/sec, enter a containment cell in which there is a large electric field of about 15,000 volts/cm. Because such slow neutrons readily bounce off the inner walls of the cell, each neutron remains inside for about a minute, during which time a neutron EDM would cause the spin to precess in the electric field. The failure to see any such precession in the most recent experiment yields the upper limit size $d(n) < 10^{-25}$ e-cm [6,7], almost six orders of magnitude smaller than the original 1951 limit. Some further improvement is expected by improving the use of magnetometers to subtract magnetic field perturbations.

A radically new idea [29] now under development at Los Alamos is to produce and store UCN in a superfluid ^{4}He bath that contains a small concentration of polarized ^{3}He. The polarized ^{3}He serves as a neutron polarizer and spin precession analyzer, as well as a magnetometer. In principle, this new experiment could improve the neutron EDM accuracy by a factor of well over 100 for several reasons: 1. a larger electric field due to the dielectric strength of ^{4}He, 2. a larger number of stored neutrons, 3. a longer storage time due to the lower temperature of the walls, and 4. the elimination of magnetic field systematics through the use of the ^{3}He co-magnetometer.

Mercury

A precise search for an EDM of the mercury atom is being carried out in our laboratory at the University of Washington in Seattle. In mercury the occupied electronic shells are all filled and there is no net electronic spin. The ^{199}Hg isotope has a net nuclear spin, with $I = 1/2$, and as discussed above could have a net EDM due to the same sorts of T-violating interactions of quarks as could give rise to a neutron EDM. In the experiment, circularly polarized light is beamed through transparent cells filled with mercury vapor in which electric fields of 10,000 volts/cm are applied. This light transfers its angular momentum to the atomic nuclei by optical pumping, and also reveals the instantaneous direction of the precessing nuclear spins by the amount of light absorbed by the vapor. Precession frequency shifts as small as one nanohertz (the equivalent of only one complete rotation in 30 years) can be measured. The latest experiment sets a new upper limit on the EDM size given by $d(^{199}\text{Hg}) < 2.1 \times 10^{-28}$ e-cm [19], thus far the smallest EDM limit set on any system.

We have developed a new version of the mercury experiment in Seattle, adding two more cells to make a stack of four: a middle pair of Hg cells with opposing electric fields for measuring an EDM as before, sandwiched between an outer pair having no applied electric fields. The outer cells serve as magnetometers that are sensitive to magnetic field gradients (due to leakage currents, shield magnetization, etc.) that could mimic an EDM signal in the middle pair of cells. Both this version and the previous version of the experiment are discussed by Michael Romalis elsewhere in these proceedings.

Xenon

Xenon, like mercury, is a closed shell atom. The ^{129}Xe isotope has nuclear spin 1/2, and is sensitive to the same sorts of T-violating interactions as ^{199}Hg above, although the magnitude of the EDM would be reduced an order of magnitude relative to Hg because of the lower atomic number Z. Still, because of the longer T_2 with xenon, and possibility of utilizing a larger atomic density compared to mercury, there is the potential for greater EDM sensitivity in xenon in the long run. The best limit on the EDM of ^{129}Xe thus far was obtained recently using ^{3}He in the same cell as a co-magnetometer, and operating both atomic species as nuclear spin masers [30].

A new idea is being tested by Michael Romalis [31]. Liquid ^{129}Xe can be spin polarized and liquified in bulk quantities, and the spin precession detected by a SQUID magnetometer. The dielectric strength of liquid xenon permits much stronger electric fields than in current cell experiments with atoms. Thus, the ultimate shot noise limited EDM sensitivity is likely to be much better than in ^{199}Hg. The challenge will be to control magnetic perturbations down to the desired level.

Thallium and Cesium

As pointed out above, atoms with an unfilled electronic shell can have net electronic spin, and can reveal the existence of an intrinsic EDM of the electron. During the past 15 years, measurements have been carried out at Amherst College using cesium vapor cells, and at Berkeley using a beam of thallium atoms in a very long apparatus to prolong the flight time of the beam atoms in the region of large electric field. Both experiments employ a variant of optical pumping to align the atomic electron spins and monitor their precession. The most sensitive previously published upper bound on the electron EDM came from thallium, with the magnitude $d(e) < 4 \times 10^{-27}$ e-cm [15]. This value has been improved in a major new thallium experiment at Berkeley, employing multiple Tl and Na beams to cancel out $\mathbf{v} \times \mathbf{E}$ motional fields and other magnetic effects, yielding a new upper limit of magnitude: $d(e) < 1.6 \times 10^{-27}$ e-cm [18].

Molecules

Another kind of experiment, using ideas originally developed at Oxford [32], takes advantage of the large internal electric fields in a polar molecule to measure the EDM associated with the nucleus or an electron inside the molecule. Earlier work centered on TlF [11,12], a closed shell molecule that enhances the effect of quark-quark interactions associated with the Tl nuclear spin. More recently, the first experiment with a paramagnetic molecule, YbF, was undertaken at Sussex [33]. A major advantage of heavy paramagnetic molecules is that they not only take advantage of the large internal molecular electric field, but also exhibit the heavy atom enhancement of the effect of the electron EDM already discussed.

One problem with paramagnetic molecules (known as free-radicals in Chemistry) is that they are difficult to concentrate at high density because of their strong reactivity. This problem may be solved in a new experiment now being tested at Yale [34]. The stable (closed shell) ground state of PbO can be selectively excited by laser light into a metastable, spin-polarized paramagnetic state. This excited state is the one to be probed for the EDM by application of an electric field. Just a few volts/cm are needed to completely polarize this excited state of the molecule. The ground state PbO vapor, because it is not particularly reactive, can be contained in the cell at high density.

IMPLICATIONS OF RECENT RESULTS

A look at Table 3 below reveals the kind of constraints that are now being placed on CP violating phases of various gauge theories by the recent mercury and thallium EDM limits (thallium supplying the most sensitive limit on the electron EDM). The neutron EDM sets similar, but slightly larger, bounds [3,22] compared to mercury. The limits shown in the table give only a general guide; specific theories usually have 2 or more phases, and a given experiment places bounds on only a linear combination of the phases. Nevertheless, since the natural size of these phases would be of order unity, the table does indicate that EDM experiments already require that the mass scale at which CP violation in these theories would have such a natural size must be M = 1000 GeV or higher. For more information about the specific implications for supersymmetry, see reference [22] and the article by Michael Romalis in these proceedings.

TABLE 3. Upper limts on CP violating interactions in gauge models set by the most recent Hg and Tl experimental results. The quantity R is the ratio $(M/100GeV)^2$, where M is the mass scale (in GeV) at which the gauge symmetry is broken. For more details about the parameters, see reference [3].

CP violating Parameters	Limit from ^{199}Hg experiment $d(^{199}Hg) < 2.1 \times 10^{-28}$ e-cm	Limit from ^{205}Tl experiment $d(e) < 1.6 \times 10^{-27}$ e-cm
Supersymmetry (ε^{susy})	$< 2 \times 10^{-3}$ R	$< 2 \times 10^{-2}$ R
Multi-Higgs (ε^{Higgs})	$< (0.4/\tan\beta)$ R	$< (0.3/\tan\beta)$ R
Lift-Right Symmetric (x^{LR})	$< 1 \times 10^{-3}$ R	$< 2 \times 10^{-2}$ R
QCD phase ($\bar{\theta}_{qcd}$)	$< 2 \times 10^{-10}$	--

In conclusion, it continues to be of paramount interest to push the precision of EDM experiments further, in order to test for the existence of supersymmetry or other new physics and, as mentioned in the introduction, to help address the problem of the observed matter/antimatter asymmetry in the Universe. It could well turn out that EDM experiments will provide the first glimpse of what physics lies beyond the Standard Model.

ACKNOWLEDGMENTS

I wish to thank Michael Romalis, Clark Griffith, Matthew Swallows, and my other colleagues on the mercury EDM experiment. This work was supported by NSF Grant PHY 9732513.

REFERENCES

1. E. M. Purcell and N. F. Ramsey, Phys. Rev. **78**, 807 (1950).

2. J. H. Smith, E. M. Purcell, and N. F. Ramsey, Phys. Rev. **108**, 120 (1957).
3. S. M. Barr, Int. J. Mod. Phys. **A8**, 209 (1993).
4. I. B. Khriplovich and S. K. Lamoreaux, *CP Violation without Strangness: electric dipole moments of particles, atoms, and molecules*, Springer, New York (1997).
5. J. Christenson, J. Cronin, V. Fitch, and R. Turlay, Phys. Rev. Lett. **13**, 138 (1964).
6. K. F. Smith *et al.*, Phys. Lett. B **234**, 191 (1990).
7. P. G. Harris, *et al.*, Phys. Rev. Lett. **82**, 904 (1999).
8. P. G. H. Sandars, Phys. Lett. **14**, 194 (1965); Phys. Lett. **22**, 290 (1966).
9. M. C. Weisskopf et al, Phys. Rev. Lett. **21**, 1645 (1968); M. A. Player and P. G. H. Sandars, J. Phys. **B3**, 1620 (1970).
10. T. G. Vold, F. Raab, B. Heckel, and E. N. Fortson, Phys. Rev. Lett. **52**, 2229 (1984).
11. E. A. Hinds and P. G. H. Sandars, Phys. Rev. **A21**, 471 and 480 (1980); D. A. Wilkening, N. F. Ramsey, and D. J. Larson, Phys. Rev. **A29**, 425 (1984).
12. D. Cho, D. Sangster, and E. A. Hinds, Phys. Rev. Lett. **63**, 2559 (1989).
13. S. A. Murthy, D. Krause, Jr., Z. L. Li, and L. R. Hunter, Phys. Rev. Lett. **63**, 965 (1989).
14. K. Abdullah, C. Carlberg, E. D. Commins, H. Gould, and S. B. Ross, Phys. Rev. Lett. **65**, 2347 (1990).
15. E. D. Commins, S. B. Ross, D. DeMille, and B. C. Regan, Phys. Rev. **A50**, 2960 (1994).
16. S. K. Lamoreaux, J. P. Jacobs, B. R. Heckel, F. J. Raab, and N. Fortson, Phys. Rev. Lett. **59**, 2275 (1987).
17. J. P. Jacobs, W. M. Klipstein, S. K. Lamoreaux, B. R. Heckel, and E. N. Fortson, Phys. Rev. **A52**, 3521 (1995).
18. B. C. Regan, these proceedings; B. C. Regan, E. D. Commins, C. J. Schmidt, and D. P. DeMille, Phys. Rev. A, to be published.
19. M. V. Romalis, W. C. Griffith, J.P. Jacobs, and E. N. Fortson, Phys. Rev. Lett. **86**, 2505 (2001).
20. G. L. Kane, *Perspectives on Supersymmetry*, World Scientific, Singapore, 1998 p. xv; J. H. Schwartz and N. Seiberg, Rev. Mod. Phys. **71**, S112 (1999).
21. R. Barbieri, A. Romanino, and A. Strumia, Phys. Lett. **B369**, 283 (1996); M. Brhlik, G. J. Good, and G. L. Kane, Phys. Rev. **D65**, 115004 (1999); S. Dimopoulos and L. Hall, Phys. Lett. **B344**, 185 (1995).
22. T. Falk, K. A. Olive, M. Pospelov, and R. Roiban, Nucl. Phys. **B560**, 3 (1999); M. Pospelov and A. Ritz, Phys. Rev. Lett. **83**, 2526 (1999).
23. A. Riotto and M. Trodden, Annu. Rev. Nucl. Part. Sci. **49**, 35 (1999).
24. C. Bouchiat, Phys. Lett. **57B**, 284 (1975); E. A. Hinds, C. E. Loving, and P. G. H. Sandars, Phys. Lett. **62B**, 97 (1976).
25. I. B. Khriplovich, in *Atomic Phys. 11*, edited by S. Haroche, J. C. Gay, and G. Grynberg, World Scientific, Singapore (1989).
26. N. Davidson, H. J. Lee, C. S. Adams, M. Kasevich and S. Chu, Phys. Rev. Lett. **74**, 1311 (1995); M. J Bijlsma, B. J. Verhaar, and D. J. Heinzen, Phys. Rev. **A49**, R4285 (1994); D. Weiss, (private communication); H. Gould (private communication). .
27. Y. Takehashi and B. P. Das, (private communication); V. Sevak, N. Auerbach, and V. V. Flambaum, Phys. Rev. **C56**, 1357 (1997).
28. M. V. Romalis and E. N. Fortson, Phys. Rev. **A56**, 4547 (1999).
29. R. Golub and S. Lamoreaux, Phys. Rep. **237**, 1 (1991).
30. M. A. Rosenberry and T. E. Chupp, Phys. Rev. Lett. **86**, 22 (2001); R. E. Stoner, M. A. Rosenberry, J. T. Wright, T. E. Chupp, E. R. Oteiza, and R. L. Walsworth, Phys. Rev. Lett. **77**, 3971 (1996).
31. M. V. Romalis and M. P. Ledbetter, Phys. Rev. Lett. **87**, 067601 (2001).
32. P. G. H. Sandars, Phys. Rev. Lett. **19**, 1396 (1967).
33. E. A. Hinds, Physica Scripta **T70**, 34 (1997); B. E. Sauer, Jun Wang, and E. A. Hinds, J. Chem. Phys. **105**, 7412 (1996).
34. D. P. DeMille, these proceedings.

New Search for a Permanent Electric Dipole Moment of ^{199}Hg.

M. V. Romalis, W. C. Griffith, J. P. Jacobs[1], and E. N. Fortson

Department of Physics, University of Washington, Seattle, Washington 98195

Abstract. We give a detailed description of a recently completed search for a permanent electric dipole moment of ^{199}Hg. The measurement is done in an entirely new apparatus that improves the statistical sensitivity per unit time by an order of magnitude compared with the previous version of the experiment. Our measurements give $d(^{199}Hg) = -(1.06 \pm 0.49 \pm 0.40) \times 10^{-28} e\,\mathrm{cm}$, which we interpret as an upper limit $|d(^{199}Hg)| < 2.1 \times 10^{-28} e\,\mathrm{cm}$ (95% C.L.). This improves the previous limit on the ^{199}Hg EDM by a factor of 4. The result sets new limits on $\bar{\theta}_{QCD}$, chromo-EDMs of the quarks, and CP violation in Supersymmetric models. We also briefly describe subsequent upgrades of the experiment which improved our sensitivity by another factor of 2.

INTRODUCTION

A search for a permanent electric dipole moments (EDM) is ideally suited to look for new physics that could explain why the Universe is made almost entirely of matter. A finite EDM can exist only if T and CP symmetries are violated. All cases of CP violation observed so far are well explained by a single CP-violating phase in the CKM matrix of the Standard Model. However, this CP-violating phase only induces EDMs in high-order diagrams involving all three generations of quarks, and they are far too small to be detected in the foreseeable future. CP violation in the CKM matrix is also insufficient to account for the matter-antimatter asymmetry of the Universe. Looking at physics beyond the Standard Model, one almost invariably finds that new particles induce a large EDM if additional CP-violating phases are present. Thus, a finite EDM would give a background-free signal of CP violation beyond the Standard Model that could explain the matter - antimatter asymmetry of the Universe.

If there are new sources of CP violation, they are likely to generate an EDM in more than one type of particles. In fact, three quite different kinds of experiments have comparable sensitivities: search for the neutron EDM [1], search for the electron EDM utilizing paramagnetic atoms or molecules, the most sensitive of which is done with Tl atoms [2], and search for an EDM of diamagnetic atoms or molecules, the most sensitive of which is done with ^{199}Hg atoms. The exact size of the signal in each of the experiments would depend on the details of the CP violating effect, and comparison between different measurements would be very important if an EDM signal were ever

---

[1] Permanent Address: Department of Physics and Astronomy, University of Montana, Missoula, MT 59812

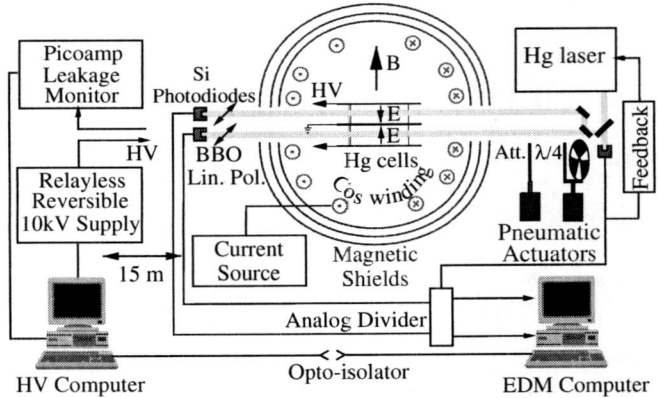

FIGURE 1. Schematic of the apparatus used to search for a permanent EDM of ^{199}Hg atoms.

observed.

Here we describe our recently completed search for an EDM of ^{199}Hg [3] and provide more experimental details. Our present experimental setup is quite different from the previous generation of the experiment [4]. We now use a UV laser instead of discharge lamps for optical pumping and detection. We also changed the basic measurement technique, the preparation procedure of the ^{199}Hg cells, and nearly all hardware components. We discuss the implications of the new ^{199}Hg EDM bound for CP violation in Supersymmetry and other models. We also briefly describe the present status of the experiment, which has been further upgraded to include four ^{199}Hg cells, and discuss future plans.

The ^{199}Hg atoms have no electron spin and a nuclear spin $I = 1/2$. Hence the atomic EDM of ^{199}Hg is primarily sensitive to nuclear T and CP-violating effects [5, 6, 7]. If there is a non-zero electric dipole moment d, the interaction with electric and magnetic fields is given by

$$H = -(dE + \mu B)\frac{\mathbf{I}}{I}. \qquad (1)$$

To detect the EDM we measured the Zeeman precession frequency of ^{199}Hg atoms placed in parallel electric and magnetic fields. To reduce the frequency noise due to magnetic field fluctuations the measurements were simultaneously performed in two cells with oppositely directed electric fields, as shown in Figure 1. The difference between the Zeeman spin precession frequencies in the two cells is given by

$$\hbar(\omega_1 - \omega_2) = 4dE + 2\mu\Delta B, \qquad (2)$$

where ΔB is the difference between the average magnetic fields in the two cells due to the magnetic field gradients. To measure the electric dipole moment d we frequently reversed the direction of the electric field and looked for a correlated change of the frequency difference between the two cells. Part of the challenge of the experiment was to keep ΔB constant as the electric field was reversed.

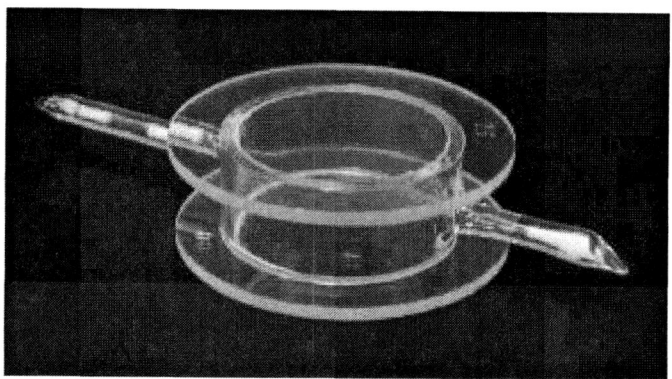

FIGURE 2. Photograph of the ^{199}Hg vapor cell. ^{199}Hg vapor is initially condensed into the thinner stem. After the wax in the cell is remelted to make a uniform coating, excess wax is moved into the stems. Conductive SnO_2 coating covers the entire inner surfaces of the disks.

EXPERIMENTAL APPARATUS

^{199}Hg Cells

Perhaps the most crucial component of the experiment are the cells containing ^{199}Hg vapor. A photograph of one of the cells used in the experiment is shown in Figure 2. The cells were constructed from Suprasil quartz tubing and disks. The materials were thoroughly cleaned using several rinses in chromic acid and organic solvents in order to reduce leakage currents. The inner diameter of the cells was 25 mm and the distance between the electric field plates was 11 mm. The cells had two stems, one for holding a small excess of ^{199}Hg to maintain the density of atoms close to the room temperature vapor pressure ($[Hg] = 5 \times 10^{13} cm^{-3}$), the other to hold paraffin ($C_{32}H_{66}$) for the surface coating. A conductive SnO_2 coating was chemically deposited on the inside surfaces of the disks and they were glued to the cell body using a low-outgassing adhesive [8]. The cells were pumped out to 10^{-7} torr and baked for several days at 100°C. A small amount of paraffin was distilled into the cells using a torch. An excess of isotopically enriched ^{199}Hg (92% ^{199}Hg) was condensed into one of the stems by cooling it in liquid nitrogen. Then the cells were filled with a previously purified mixture of 450 torr of N_2 gas and 50 torr of CO gas. After the cells were sealed the paraffin was remelted to obtain a thin transparent coating on all surfaces.

The coating dramatically increased the nuclear spin relaxation time of the ^{199}Hg atoms. After remelting of the paraffin it was typically about 300 sec. To determine the dominant relaxation mechanism, we made extensive measurements of the longitudinal spin relaxation time T_1 as a function of the magnetic field, temperature, buffer gas pressure and composition. These measurements indicate that the spin relaxation was dominated by surface effects, most likely an interaction of ^{199}Hg nuclear spins with paramagnetic sites on the surface. In the operation of the experiment we have observed slow changes of the spin lifetime when the cells were exposed to resonant UV light. The lifetime would sometimes increase at first, but after a week of continuous UV exposure the lifetime would drop to below 100 sec. It could be restored by remelting the paraffin coating. We believe this behavior was due to damage of the coating surface by collisions with Hg atoms in the metastable 6^3P_0 state to which they are quickly quenched by N_2 after optical excitation. N_2 is very ineffective in quenching ^{199}Hg atoms from the 6^3P_0 state to the ground state, while CO has a much higher cross-section for this process [9]. We have recently made cells containing only CO gas and found them to be much more stable under UV exposure. The paraffin coating also improves the uniformity of the electric field in the cells [10, 11].

The cells were placed in a sealed vessel that was filled with SF_6 gas to reduce leakage currents and electrical breakdown. Initially, the vessel was constructed from teflon and coated on all surfaces with a thin aluminum film. This eliminated the magnetic field noise due to Johnson thermal currents [12]. Later we found that teflon, as well as many other materials, contain significant amounts of ferromagnetic impurities. These impurities can change their magnetization and cause a permanent shift in the magnetic field if subjected to a large magnetic field spike. We have seen several such jumps in our signal when large sparks occurred inside the vessel. This behavior was further investigated by placing small coils near the ^{199}Hg cells and discharging a capacitor through them. Although we found no evidence that such jumps could occur during the normal operation of the experiment, we rebuilt the vessel using a conductive carbon-filled polyethylene "Tyvar". This material has a resistivity of about 10^5 Ωcm, making the Johnson noise negligible, and is also relatively free from ferromagnetic impurities. The electric fields inside the vessel were modeled using a commercial field simulation program [13]. We calculated mutual capacitances between different conductors in the vessel and found them to be in excellent agreement with the values determined from the charging currents. The vessel was located inside a three layer cylindrical magnetic shield with a shielding factor of 5×10^4 in the vertical direction. A uniform vertical magnetic field of 17 mG was applied inside the shields, giving a Larmor spin precession frequency of 13 Hz. The field was created using a coil with the density of turns proportional to the cosine function. The magnetic field gradient ΔB between the cells was on the order of 1 μG. The current for the coil was provided by an ultra-low noise current source based on the design described in [14]. It used a Hg battery as a voltage reference and a low noise JFET transistor for active control. On a time scale of 100 sec the field was stable to 25 ppb.

Signal Detection

Optical pumping and detection were done using a laser operating at the 253.7 nm $6^1S_0 \rightarrow 6^3P_1$ transition of Hg. To obtain this wavelength with low amplitude noise and long term stability, we quadrupled the output of a semiconductor Master Oscillator - Power Amplifier (MOPA) laser operating at 1015 nm [11, 15]. The laser used two bow-tie doubling cavities, the first with a $KNbO_3$ crystal and the second with a BBO crystal. At full power of the MOPA laser (500 mW) we obtained 6 mW of UV light, however only about 1 mW was needed for operation of the experiment. The laser could be continuously tuned over tens of GHz by changing the frequency of the Master Oscillator. The laser intensity noise was $10^{-4}/\sqrt{\text{Hz}}$ at 10 Hz. An additional feedback system adjusted the current of the Power Amplifier to provide long term stability of the intensity.

The laser beam was split into two equal parts and sent through the cells perpendicular to the direction of the magnetic and electric fields. The same laser beam was used for both optical pumping and detection. For optical pumping we used a synchronous pumping technique to build-up transverse spin polarization in the rotating frame. The laser was circularly polarized and tuned to the center of the $F = 1/2$ hyperfine line of the $6^1S_0 \rightarrow 6^3P_1$ transition. The light was chopped with a mechanical chopper at the Larmor frequency of ^{199}Hg spins with a duty cycle of 30%. The rate of build-up of the transverse polarization in the cells was about $(15 \text{ sec})^{-1}$ with 70 μW of UV light incident on each cell.

Optical rotation was used to measure the frequency of Larmor precession. The polarization of the laser was switched to linear, the frequency detuned from resonance by 20 GHz and the intensity attenuated to about 7 μW with a neutral density filter. The mechanical chopper, quarter-wave plate, and the neutral density filter were mounted on pneumatic actuators that moved them in and out of the laser beams to switch between pumping and probing modes. For maximum signal-to-noise ratio the optical excitation rate during the probe phase should be equal to the intrinsic spin relaxation rate. However, we used a somewhat slower excitation rate of $(500 \text{ sec})^{-1}$ to reduce the degradation of the spin lifetime in the UV light. Spin polarization of ^{199}Hg atoms rotated the plane of polarization of the probe light by an angle $\phi \propto \mathbf{I} \cdot \mathbf{k}$, where $\mathbf{k}$ is the direction of the laser beam propagation [16]. As the spins precessed in the horizontal plane, the rotation angle ϕ oscillated at the Larmor frequency with an amplitude of 3-4°. The oscillations were detected by passing the laser beams through BBO Glan-laser linear polarizers rotated by an angle $\alpha \sim 12°$ from complete extinction. The intensity of the light transmitted through the polarizers,

$$I(t) = I_0 \sin^2(\alpha + \phi(t)), \qquad (3)$$

had a modulation depth of about 50% and was measured using silicon photodiodes with quantum efficiency of about 55% at 254 nm. The signals from the two cells were divided by the incident laser intensity using analog dividers and then digitized with 16-bit resolution at 200 Hz.

Figure 3 shows the data recorded during a typical measurement. It consisted of a 30 sec pumping phase and 100 sec probe phase. During the pumping phase we also reversed the direction of the electric field. The high voltage (HV) applied to each cell was typically switched between 10 kV and -10 kV in 5 to 20 sec. Several measures were taken to prevent a spurious correlation between the direction of the electric field and other parameters. We used a solid-state reversible relayless HV power supply and placed it in a magnetic shield to reduce magnetic fields correlated with HV. All HV-related equipment was located 15 m away from the main apparatus and was controlled by a separate computer linked to the main computer through an optical isolator. We also occasionally skipped a HV reversal to guard against correlations with periodic fluctuations. The currents flowing from the HV electrodes through each cell were collected separately at virtual ground and measured by current-to-voltage converters using CMOS

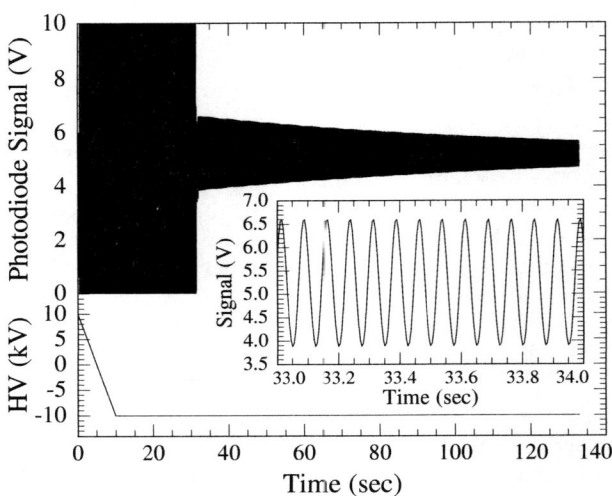

FIGURE 3. Photodiode signal recorded during a single measurement (top panel). During the pumping phase the intensity of the light was increased and it was chopped at the spin precession frequency. At the same time the HV was linearly swept between +10 kV and -10 kV (bottom panel). The inset shows a 1 sec segment of the data in the probe phase.

operational amplifiers with a bias current less than 50 fA. The steady-state leakage currents where on the order of 1pA. The electrical paths through the vessel were designed to be as symmetrical as possible to reduce the projection of magnetic fields due to the charging and leakage currents onto the vertical magnetic field. Even the charging currents, which were on the order of 1 nA, did not produce an observable EDM signal when the electric field was reversed during the probe phase.

In addition to the precession signals from the two cells, 12 other signals were continuously monitored to look for possible correlations with the electric field. All three components of the magnetic field outside of the shields were measured using a flux-gate magnetometer. The position of the laser beam transmitted through the cell was measured using a four-quadrant detector. We also monitored the power, current, and wavelength of the laser, the temperature of the shields and other miscellaneous parameters. A typical run lasted about 24 hrs. and consisted of hundreds of individual measurements with the electric field reversed between each measurement.

DATA ANALYSIS

The data analysis was performed in the following steps. Each of the spin precession signals was linearized by inverting Eq. (3) to extract the rotation angle ϕ, filtered using a bandpass FFT filter, and fit to an exponentially-decaying sine wave. The fit contained 4 adjustable parameters: the Larmor precession frequency, the initial phase

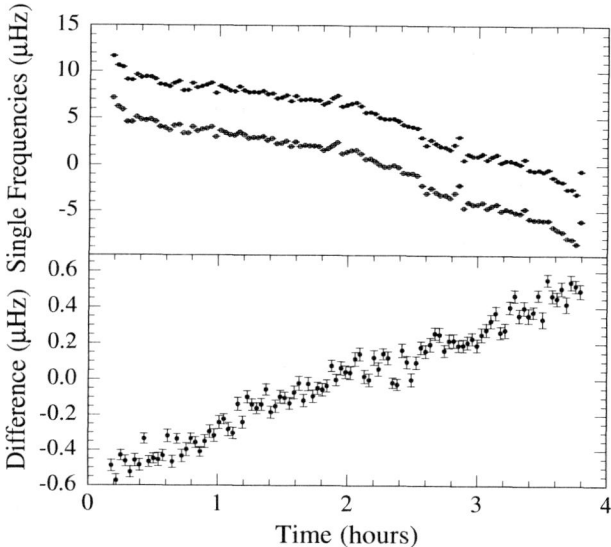

FIGURE 4. Larmor precession frequencies measured in the two cells (top panel) and their difference (bottom panel). The error bars reflect frequency uncertainty due to the phase noise. The data are offset by a constant frequency to show them on the same plot.

of the precession, the amplitude of the rotation modulation, and the spin relaxation time. To analyze the imperfections in the signal we also split it into 100 separate slices and fit each slice individually allowing only the phase and the amplitude to vary. This gave us a measurement of the phase noise and of the deviations of the signal amplitude from a perfect exponential decay. The uncertainty in the fit value of the frequency was calculated from the phase noise. We found that it was within 50% of the shot-noise level estimated from the currents in the photodiodes. The deviations of the amplitude from an exponential decay, on the order of 10^{-3}, were mainly due to laser frequency fluctuations and had a negligible correlation with the Larmor frequency measurements.

Figure 4 shows a typical set of precession frequencies measured in the two cells while reversing the electric field between each point. Magnetic field drifts were canceled to a large extend by taking a difference between the frequencies in the two cells, reducing the scatter by a factor of 5. The correlation between the Zeeman frequency difference and the direction of the electric field, which is proportional to the EDM, was calculated by analyzing groups of several consecutive measurements. We used groups of 3 to 6 points with relative weighting set to eliminate frequency drifts up to the highest possible order. For example, using a 3-point correlation one can eliminate linear drifts in the frequency difference,

$$d = \frac{\hbar}{4E} \frac{(\omega_1[n] - \omega_2[n]) - 2(\omega_1[n+1] - \omega_2[n+1]) + (\omega_1[n+3] - \omega_2[n+3])}{4}. \quad (4)$$

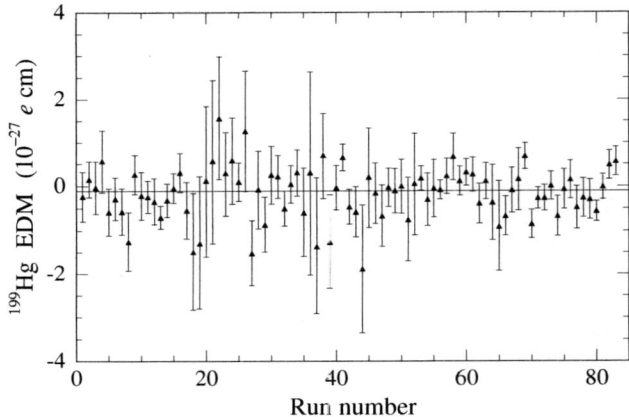

FIGURE 5. ^{199}Hg EDM signal as a function of run number. The solid line shows the average of the data. Runs with larger errors were done in non-optimal configurations.

We mostly used overlapping correlations in the analysis, increasing the measurement number n by one between successive correlations. The errors were then rescaled to take into account the fact that the correlations were not statistically independent. We also tested non-overlapping correlations. All methods gave consistent results and the final numbers were calculated using 3-point overlapping correlations.

The scatter between successive frequency measurements, which determines the uncertainty in the EDM correlation, is due to both phase noise and frequency noise [17]. The frequency uncertainty due to the phase noise can be calculated directly from the fit and was used for weighting of individual measurements and for an initial estimate of the error in the EDM correlation. A typical run included several hundred independent measurements of the EDM correlation and had a $\chi^2/n.d.f.$ on the order of 2. The excess scatter was due to the frequency noise caused by magnetic field gradient fluctuations, and it could not be estimated directly from individual frequency measurements. Instead, we increased the error on the average EDM correlation for each run by $\sqrt{\chi^2/n.d.f.}$ to reflect the actual data scatter in that run. We studied the frequency spectrum of the frequency fluctuations and performed a number of simulations to verify that this data analysis procedure gave an accurate estimate of the error.

The data were collected over a period of 6 months in 2000. A total of 70000 frequency measurements were made (about 50% running efficiency), of which 40000 were used for EDM measurements and the rest for diagnostics and systematic studies. Frequent reversals and changes were done during the experiment to monitor for systematic effects. We periodically reversed the data acquisition channels for the two cells and the direction of the magnetic field, which should change the sign of the EDM signal. We also frequently changed the EDM cells and their orientation in the vessel. In addition, the paraffin in the cells was remelted and the outside surfaces cleaned each time the cells were changed, which would likely change the path of the leakage currents. Over

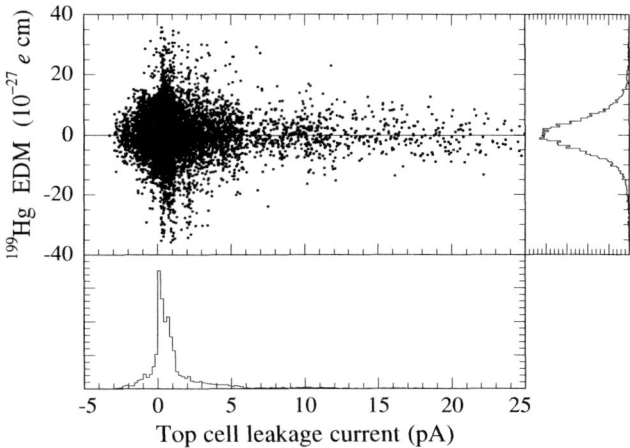

FIGURE 6. Correlation between the leakage current and the EDM signal. Histograms of the leakage current and the EDM data are also shown. The solid line is a linear fit giving a correlation of $(-0.4 \pm 2.0) \times 10^{-29} e\,\text{cm/pA}$.

the course of the experiment we used two different vessels and changed other components of the setup. Figure 5 shows the results of all EDM runs. The weighted average of all data gives $d(^{199}Hg) = -(1.06 \pm 0.49) \times 10^{-28} e\,\text{cm}$. We do not observe any excess data scatter between runs due to changes in the experiment and $\chi^2/n.d.f.$ is equal to 0.95. The statistical error corresponds to a frequency difference between the two cells of 0.4 nHz, a factor of 5 smaller than in the previous experiment [4]. The statistical sensitivity per unit time has been improved by a factor of 10.

SYSTEMATIC EFFECTS

We looked for systematic effects by changing the operating parameters of the experiment, looking for correlations among different parameters, and exaggerating certain imperfections. The leakage currents are a potentially serious source of systematic errors because they can produce magnetic fields that are correlated with the electric field and mimic an EDM signal. It should be noted that only leakage currents flowing in a helical path around the cell will contribute to first order. For a fixed helical path the EDM-like signal cannot be reversed by changing the orientation of the cell. Figure 6 shows a scatter plot of the EDM signal vs. the leakage current in one of the cells. No statistically significant correlation was observed. The average cell leakage currents were about 0.6 pA. From the error on the correlation slope we can set a limit on the contribution of the leakage currents to the EDM signal of $0.14 \times 10^{-28} e\,\text{cm}$. However, the correlation slope fit is dominated by a few runs with large leakage currents. We estimate the error more conservatively by calculating the magnetic field created by a leakage current mak-

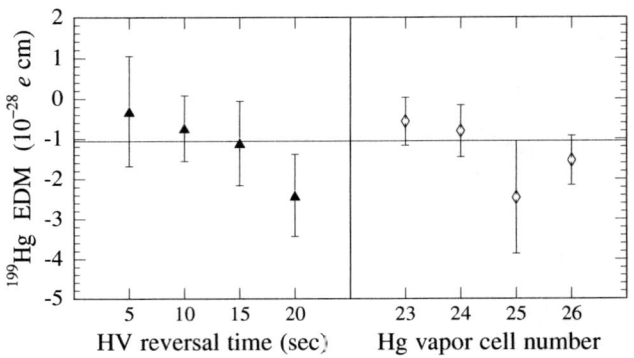

FIGURE 7. The left panel shows the dependence of the EDM signal on the HV reversal time. The right panel shows the EDM signal obtained with each of the EDM cells. The solid line is an average of all data.

ing one complete loop around the cell. This rather unlikely path would give an average EDM signal of 0.25×10^{-28} e cm. As can be seen in Figure 6, the leakage currents were sometimes negative. We believe this effect was due to changes in the mutual capacitance caused by redistribution of charges on HV insulators. If the HV was not reversed for a long time, the leakage currents became positive and approached a steady state value of about 0.1 pA. A total of 4 vapor cells were used in the experiment in various pairs. The right panel of Figure 7 shows that the EDM data taken with each cell are consistent with the average.

We looked for correlations with the electric field of 30 other variables, such as the magnetic field outside the shields, the amplitude and relaxation time of the precession signals, the position of the laser beam, etc. We found no statistically significant correlations. We also determined the cross-correlation between each of these variables and the EDM signal using random fluctuations of the variables. This allowed us to set limits on possible EDM signals coming from correlations with other variables that are 10 to 100 times smaller than the statistical error.

For positive direction of the magnetic field the average EDM signal was $d(B+) = -(1.78 \pm 0.70) \times 10^{-28}$ e cm and for negative direction $d(B-) = -(0.36 \pm 0.69) \times 10^{-28}$ e cm. The two results are within 1.4 σ of each other. A systematic effect that does not reverse with the magnetic field as a true EDM signal does would show up in the difference but cancel in the average of the two results. To study possible frequency shifts due to magnetization of the magnetic shields caused by the charging currents, we varied the high voltage reversal time from 5 to 20 sec. The dependence of the EDM signal on the HV reversal time is shown in the left panel of Figure 7. The statistical significance of a non-zero slope in this data is 1.5σ. We did not resolve any common-mode correlations of the individual Larmor frequencies ω_1 and ω_2 with the electric field, although their error bars are a factor of 6 larger than the statistical error on $\omega_1 - \omega_2$.

We looked for effects proportional to E^2 in separate runs by applying the HV to only one of the two cells and alternating it between 0 and ± 10 kV. The frequency shift between the two cells was less than 2 nHz. We checked that the electric field in the cells was uniform and was reversible with an accuracy of 1.5% [11]. This sets a limit on a possible EDM signal due to reversal imperfections to less than $7 \times 10^{-30} e\,\text{cm}$. Although the average velocity of the atoms in the cells is equal to zero, residual $v \times E$ effects [2] can exist if the surface relaxation on the walls is not uniform. In this case the ^{199}Hg atoms would continue to diffuse through the cell and contribute to the optical rotation signal until they hit an area of enhanced relaxation, and that would result in a non-zero average velocity. We looked for these effects by taking data with the magnetic field intentionally misaligned by 5° from the electric field using a small magnetic field in the horizontal plane. Both orthogonal directions of the magnetic field were tested and no EDM signal was seen at the level of $1.5 \times 10^{-28} e\,\text{cm}$. During normal operation the electric and magnetic fields were parallel within 1° and the $v \times E$ effect was constrained to be less than $0.3 \times 10^{-28} e\,\text{cm}$. By using the optical rotation technique to probe the precession of the spins with an off-resonant linearly polarized light we significantly reduced possible light shift effects. Among various frequency shifts caused by the probe light the most significant was due to the magnetic dipole and electric quadrupole transitions in an electric field. This effect is odd in the E field and can mimic an EDM signal. Based on previous calculations [18] we estimate that it is suppressed to about $3 \times 10^{-30}\,e\,\text{cm}$ because the laser beam is directed perpendicular to the magnetic field. Our tests for the $v \times E$ effect would also detect an E-field odd light shift of this type.

In summary, no statistically significant systematic effects that mimic an EDM signal were observed, although in several cases our systematic studies were limited by statistics. We estimate the total systematic uncertainty to be $0.40 \times 10^{-28}\,e\,\text{cm}$ by adding in quadrature the limits on systematic effects due to the leakage currents, the $v \times E$ effect, the light shifts, the E^2 shift, and the cross-correlations. Thus we obtain $d(^{199}\text{Hg}) = -(1.06 \pm 0.49 \pm 0.40) \times 10^{-28}\,e\,\text{cm}$ and interpret the result as an upper limit on the ^{199}Hg EDM,

$$|d(^{199}\text{Hg})| < 2.10 \times 10^{-28}\,e\,\text{cm}\ (95\%\ \text{C.L.}). \qquad (5)$$

IMPLICATIONS OF THE EXPERIMENT

This limit can be used to place new constraints on a number of hadronic and semi-leptonic CP-violating effects. The EDM of the ^{199}Hg atom is proportional to the Schiff moment of the ^{199}Hg nucleus S, which is a measure of the difference between the distributions of the electric charge and electric dipole moment in the nucleus. Using a Hartree-Fock calculation for Hg atomic wavefunctions [19] and a simple nuclear shell model [5, 7], the Schiff moment was calculated with an uncertainty of about 30-50%: $d(^{199}\text{Hg}) = -3.1 \times 10^{21} S\,\text{cm}^{-2}$[5]. The Schiff moment can be due to EDMs of the nucleons, but a larger contribution comes from a CP-violating nucleon-nucleon interaction of the type $\xi G_F(\bar{p}p)(\bar{n}i\gamma_5 n)/\sqrt{2}$. It was calculated in [7] using Woods-Saxon potentials and neglecting many-particle correlations. The result is $S = -1.8 \times 10^{-7} \xi e\,\text{fm}^3$ with an uncertainty of about 50%. Possible enhancements of the Schiff moment due to collective octupole nuclear excitations have been considered recently in [20], although no definite estimates exist. As shown in [21, 22], this CP-odd nucleon-

TABLE 1. Summary of limits (95%C.L.) set by the ^{199}Hg EDM and other experiments on model-independent and "naturalness" parameters.

Parameter	Limit from ^{199}Hg	Best other limit		Theory Ref.
$\bar{\theta}_{QCD}$	1.5×10^{-10}	6×10^{-10}	n [1]	[22, 24]
$\tilde{d}_d$ (cm)	7×10^{-27}	1.1×10^{-25}	n [1]	[22, 25]
d_e (e cm)	1.5×10^{-26}	4×10^{-27}	Tl [2]	[29]
C_T	1×10^{-8}	5×10^{-7}	TlF [28]	[5]
C_S	3×10^{-7}	4×10^{-7}	Tl [2]	[5]
ε_q^{SUSY}	2×10^{-3}	1×10^{-2}	n [1]	[26]
ε^{Higgs}	$0.4/\tan\beta$	$0.7/\tan\beta$	Tl [2]	[26]
x^{LR}	1×10^{-3}	1×10^{-2}	n [1]	[26]

nucleon interaction is dominated by π^0 exchange and is proportional to the pion-nucleon CP-odd coupling constant $\bar{g}_{\pi NN}$.

The constraints on various model-independent and model-specific CP-violating parameters are summarized in Table 1. A limit on $\bar{g}_{\pi NN}$ can be used to directly constrain the CP-violating QCD vacuum angle $\bar{\theta}_{QCD}$ [23]. We obtain $|\bar{\theta}_{QCD}| < 1.5 \times 10^{-10}$, which is smaller than the limit set by the neutron EDM [1, 24] by a factor of 4. We also set a limit on a linear combination of quark chromo-EDMs [22],

$$e|\tilde{d}_d - \tilde{d}_u - 0.012\tilde{d}_s| < 7 \times 10^{-27} e\,\text{cm}. \quad (6)$$

This limit can be compared with a constraint on a different combination of EDMs and chromo-EDMs set by the neutron EDM experiment [1, 25],

$$|e(\tilde{d}_d + 0.5\tilde{d}_u) + 1.3d_d - 0.3d_u| < 1.1 \times 10^{-25} e\,\text{cm}. \quad (7)$$

In most extensions of the Standard Model, including Supersymmetry, EDMs and chromo-EDMs of the quarks have comparable size [26]. We also place new constraints on semileptonic CP-violating parameters C_S and C_T, which are significant for certain multi-Higgs models [27]. These parameters are not affected by the hadronic uncertainties. Even though ^{199}Hg does not have an unpaired electron, it has some sensitivity to the electron EDM through hyperfine interactions [29]. We include here the limit on the electron EDM for completeness, even though its not competitive with the limit from the Tl experiment.

In addition to the model-independent constraints discussed above, one can set limits on specific CP-violating parameters in various extensions of the SM. For example, in the Minimal Supersymmetric Standard Model (MSSM) the limit on the ^{199}Hg EDM can be used to set tight constraints on a linear combination of CP-violating phases [22]. Figure 8 shows a comparison of the constraints on the two CP-violating phases from neutron, electron, and ^{199}Hg EDM limits in a simplified MSSM model with all mass scales set equal to $M = 750$ GeV. Even for such heavy masses of supersymmetric particles the combination of EDM experiments constraints both CP violating phases to a small region near zero. Its important to emphasize that in general different EDM experiments constrain different linear combinations of CP-violating phases and the combination of all experiments excludes the possibility of accidental cancellations. In another recent

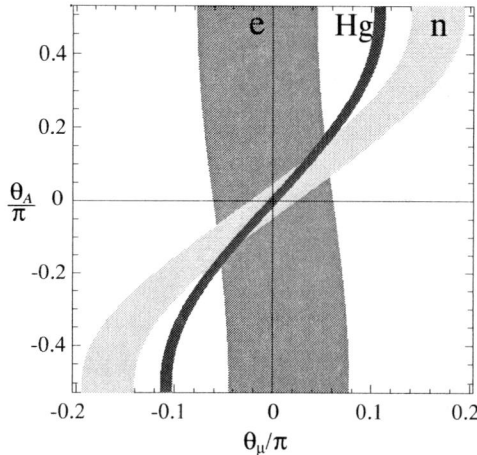

FIGURE 8. Allowed regions of CP violating phases θ_A and θ_μ that are consistent with the limits on the electron, ^{199}Hg and neutron EDMs. All MSSM mass scales are set to $M = 750$ GeV and $\tan\beta = 2$. Adapted from [22] and [25].

analysis of an mSUGRA model with a realistic range of parameters [30] it was found that EDM experiments exclude 99% of the parameter space for possible CP violation. In Table 1 we give general limits for simple "naturalness" parameters, as defined in [26], for Supersymmetric, multi-Higgs, and Left-Right symmetric models. They represent the degree of fine-tuning required to bring predictions of these theories in line with the experimental limits. For example, in Supersymmetry ε_q^{SUSY} is roughly equal to the CP-violating phase if the masses of the sypersymmetric particles are on the order of 100 GeV.

RECENT DEVELOPMENTS

Since finishing this measurement we have completed another major upgrade of the experiment. The most significant change is the addition of two more ^{199}Hg cells to the apparatus. An overall schematic of the current EDM setup is shown in Figure 9. The two additional cells are placed inside the HV electrodes and do not have any electric field applied to them, serving only as sensitive magnetometers. Two more laser beams, directed along the axis of the shields, are used to measure the precession frequencies in these cells. Using the outer cells we can independently measure a linear magnetic field gradient and subtract it from the EDM correlation. Another linear combination of the four Larmor frequencies allows us to look for evidence of magnetic fields produced by the leakage currents. As was already mentioned, we also changed the buffer gas composition in the cells, improving the stability of the spin lifetime under UV illumination. With longer spin lifetimes we were able to increase the probe time of the measurement

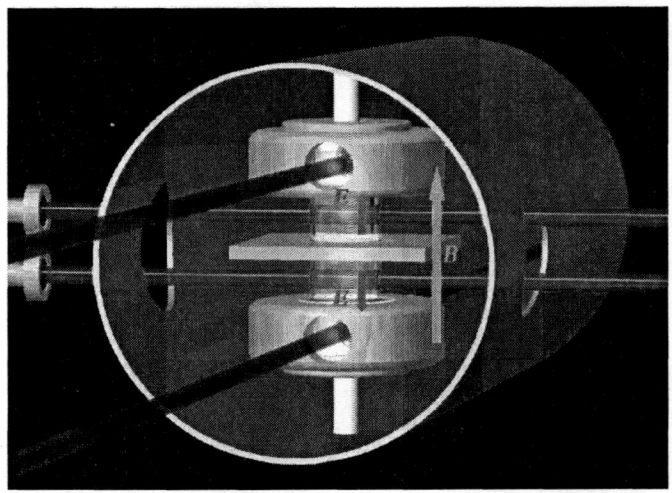

FIGURE 9. Four-cell EDM setup. The two outer cells are contained inside high voltage electrodes. The magnetic shield is not drawn to scale.

to 200 sec. These upgrades have already improved the statistical sensitivity of the experiment per unit time by a factor of 2. We are beginning to collect data and plan to further reduce the error on Hg EDM by a factor of 2 to 4.

ACKNOWLEDGEMENTS

We would like to thank Warren Nagourney for help with the initial design of laser quadrupling system and Blayne Heckel for helpful discussions. This work was supported by NSF Grant No. PHY-9732513.

REFERENCES

1. P.G. Harris *et al.*, Phys. Rev. Lett. **82**, 904 (1999).
2. E.D. Commins, S.B. Ross, D. DeMille, and B.C. Regan, Phys. Rev. A **50**, 2960 (1994). See also C. Regan contribution in these proceedings.
3. M.V. Romalis, W.C. Griffith, J.P. Jacobs, and E.N. Fortson, Phys. Rev. Lett. **86**, 2505 (2001).
4. J.P. Jacobs, W.M. Klipstein, S.K. Lamoreaux, B.R. Heckel, and E.N. Fortson, Phys. Rev. A **52**, 3521 (1995).
5. I.B. Khriplovich and S.K. Lamoreaux, *CP violation without Strangeness*, Springer, Berlin (1997).
6. V.A. Dzuba, V.V. Flambaum, and P.G. Silvestrov, Phys. Lett. **154**B, 93 (1985).
7. V.V. Flambaum, I.B. Khriplovich, and O.P. Sushkov, Phys. Lett. **162**B, 213 (1985).
8. NOA-88 from Norland, www.norlandprod.com
9. J.S. Deech, J. Pitre, L. Krause, Can. J. Phys. **49**, 1976 (1971).
10. L.R. Hunter, D. Krause, Jr., S. Murthy, and T.W. Sung, Phys. Rev. A **37**, 3283, (1988).
11. D.M. Harber and M.V. Romalis, Phys. Rev. A **63**, 013402 (2001).

12. T. Varpula and T. Poutanen, J. Appl. Phys. **55**, 4015 (1984).
13. Maxwell 2D Field Simulator from Ansoft Corp., www.ansoft.com.
14. C. Ciofi, R. Giannetti, V. Dattilo, and B. Neri, Proc. IEEE Instr. Measur. Tech. Conf., Ottawa, 1486 (1997).
15. C. Zimmermann, V. Vuletic, A. Hemmerich, T.W. Hansch, Appl. Phys. Lett. **66**, 2318, (1995).
16. W. Happer and B. Mathur, Phys. Rev. **163**, 12 (1967).
17. J.A. Barnes *et al.*, IEEE Trans. Instrum. Meas., **IM-20**, 105 (1971).
18. M.V. Romalis and E.N. Fortson, Phys. Rev. A **59**, 4547 (1999), S.K. Lamoreaux and E.N. Fortson, Phys. Rev. A. **46**,7053 (1992).
19. A.M. Martensson-Pendrill, Phys. Rev. Lett. **54**, 1153 (1985).
20. J. Engel, J.L. Friar, A.C. Hayes, Phys. Rev. C **61**, 035502 (2000).
21. V.M. Khatsymovsky, I.B. Khriplovich, and A.S. Yelkhovsky, Ann. Phys. **186**, 1 (1988).
22. T. Falk, K.A. Olive, M. Pospelov, and R. Roiban, Nucl. Phys. B**560,** 3 (1999).
23. R.J. Crewther, P. Di Vecchia, G. Veneziano, and E. Witten, Phys. Lett. **88**B**,** 123 (1979); **91**B, 487(E) (1980).
24. M. Pospelov and A. Ritz, Phys. Rev. Lett. **83**, 2526 (1999).
25. M. Pospelov and A. Ritz, Phys. Rev. D **63**, 073015(2001).
26. S. M. Barr, Int. J. Mod. Phys. A **8**, 209 (1993).
27. S. M. Barr, Phys. Rev. Lett. **68**, 1822 (1992).
28. D. Cho, K. Sangster, and E. A. Hinds, Phys. Rev. Lett. **63**, 2559 (1989).
29. V.V. Flambaum and I.B. Khriplovich, Sov. Phys. JETP **62**, 872, (1985).
30. V.Barger, T.Falk, T.Han, J.Jiang, T.Li, T.Plehn, Phys. Rev. D, (in press), (2001).

Measuring the electron electric dipole moment in YbF

B. E. Sauer, J. J. Hudson, M. R. Tarbutt, E. A. Hinds

Sussex Centre for Optical and Atomic Physics, University of Sussex, Falmer, Brighton BN1 9QH, UK

Abstract. We describe the current status of the experiment at Sussex to measure the electron's permanent electric dipole moment d_e using the paramagnetic molecule YbF. We consider a number of systematic effects which could mimic d_e and conclude that they are far below the current experimental sensitivity. As of June 2001 we have accumulated over 20 hours of data, which combined give a measurement $d_e = 0.6 \pm 4 \times 10^{-26}$ e.cm.

INTRODUCTION

For over half a century, it has been recognised that an elementary particle cannot have a permanent electric dipole moment (edm) unless the discrete symmetries parity (P) and time reversal (T) are broken [1]. Since the early 1950's measurements have been made on a variety of elementary particles, atoms and molecules, all with null results suggesting that T violation is not exhibited in ordinary matter [2,3]. T violation does occur within the standard model of elementary particle physics through complex Yukawa couplings between the quarks (the KM mechanism [4]), indeed, this seems to explain the well-known CP-violating behaviour of kaons [5], but the edms predicted for ordinary matter are extremely small. This makes edm measurements very powerful in the search for new physics since any nonzero result immediately implies physics beyond the standard model [6,7].

PRINCIPLE OF THE EXPERIMENT

The electron edm's interaction with a total electric field $\mathbf{E}$ is described by the effective Hamiltonian $H_{E1} = -d_e(1-\beta)\mathbf{\Sigma} \cdot \mathbf{E}$, where β and $\mathbf{\Sigma}$ are standard Dirac matrices and the weak magnetic interaction is ignored (see §4 of [8] or §III of [9]). The explicit form of the interaction makes it clear that the effect is relativistic,

$$\langle H_{E1} \rangle = \langle \psi_0 | \begin{matrix} 0 & 0 \\ 0 & 2d_e \mathbf{\sigma} \cdot \mathbf{E} \end{matrix} | \psi_0 \rangle,$$

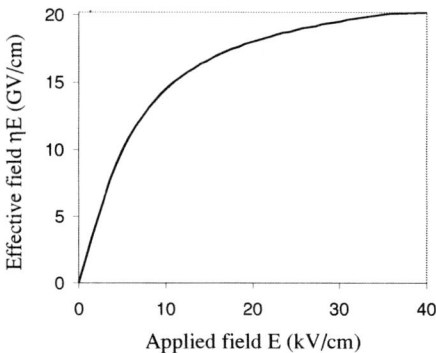

FIGURE 1. The effective field seen by the electron spin in YbF as a function of the applied lab field. Note the different scales, the enhancement η is about 10^6. The asymptotic value of the effective field is calculated to be 26 GV/cm [11]. The current experiment at Sussex applies a field of 8kV/cm, which gives about half of the maximum effective field. The form of the curve, $\langle \hat{\sigma} \cdot \hat{\lambda} \rangle$, is that of the polarization of a rigid rotor.

as only the small components of the Dirac wavefunction ψ_0 are important. The effect is therefore large in heavy atoms or molecules- an external electric field $E_{ext}\hat{\lambda}$ can polarize the wavefunction near the heavy nucleus, yielding an energy $\langle H_{E1} \rangle = d_e \eta E_{ext} \hat{\sigma} \cdot \hat{\lambda}$ ($\hat{\sigma}$ is a vector along the electron's spin). As $\hat{\lambda}$ is a polar vector this interaction clearly violates both P and T symmetry. Sandars [10] was the first to note that the effective field ηE_{ext} is much larger than the applied field for heavy atoms. Recent calculations of the enhancement factor for a large number of atoms and molecules have been summarized by Commins [9]. A few years after Sandars discovered the atomic enhancement he also pointed out that heavy polar molecules, being far more polarizable than atoms, have a great advantage over atoms in that ηE_{ext} can actually saturate in quite modest laboratory fields [11]. We have chosen YbF for our edm experiment as it is experimentally the most tractable of the simple heavy polar molecules. The polarization of YbF is shown in figure 1, where the asymptotic value of ηE_{ext} has been calculated using several different methods to be about 26 GV/cm [12].

YbF has a simple hyperfine structure in the ground electronic, rotational and vibrational state used by the edm experiment. The electron spin and the $I = \frac{1}{2}$ fluorine nuclear spin combine to produce an $F=1$ triplet and an $F=0$ singlet, separated by about 170 MHz. An applied electric field lifts the degeneracy between the $|F, m_F\rangle = |10\rangle$ and the $|1 \pm 1\rangle$ states, although T symmetry requires that the $|11\rangle$ and $|1-1\rangle$ states remain degenerate, as shown in figure 2.

A magnetic field $\vec{B}$ lifts this degeneracy by producing an energy shift $-\mu_B \hat{\sigma} \cdot \vec{B}$ dependent on the direction of the electron spin (μ_B is the electron's magnetic moment). In a P- and T-violating analogue of this, the edm interaction $d_e \hat{\sigma} \cdot \eta \vec{E}_{ext}$

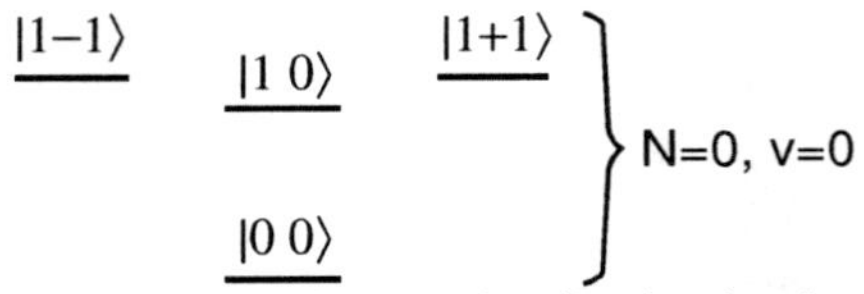

|1−1⟩ |1 0⟩ |1+1⟩
         |0 0⟩   } N=0, v=0

FIGURE 2. The hyperfine levels of the ground electronic, vibrational and rotational state of YbF (The rotational quantum number is N). The F=1 to F=0 splitting is about 170MHz; the tensor splitting in the F=1 manifold is shown in detail in figure 5.

also induces a level splitting which is what we seek to measure. This shift is of course very small and not accessible by normal spectroscopy. The Sussex YbF experiment therefore use an interferometric technique, illustrated in figure 3. We start by preparing a superposition state of two hyperfine levels $\frac{1}{\sqrt{2}}(|1 1\rangle+|1-1\rangle)$. When this state evolves for a time T in combined electric and magnetic fields, the two parts of the wavefunction develop a relative phase shift $2\varphi = 2(d_e\eta E_{ext}+\mu_B B)T/\hbar$. Probing the state in a way sensitive to this phase gives a signal proportional to $\cos^2\varphi$; by setting the magnetic phase to be $\pi/4$ we maximize the edm signal produced when E is reversed relative to B.

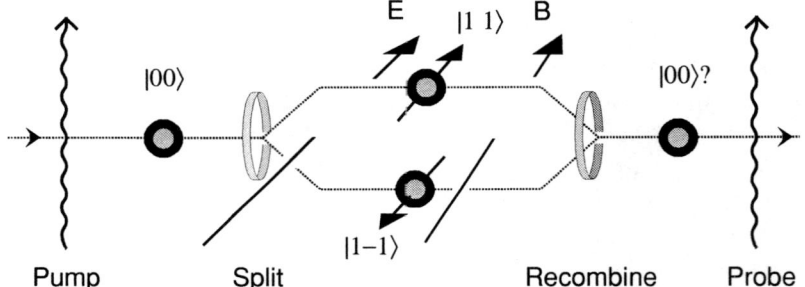

FIGURE 3. A schematic of the spin interferometer. The splitter and recombiner are driven by separate rf sources which have no mutual phase coherence.

EXPERIMENTAL DETAILS

The YbF apparatus

The YbF molecular beam is enclosed in an outer layer of magnetic shielding about 1.4 m long. YbF is produced by the reaction of Yb metal with AlF_3 in a Mo crucible heated to 1500K. The resulting thermal beam has a most probable velocity **v** of 440 ms^{-1}. The YbF molecules first interact with ~552 nm laser light tuned to the electronic transition $P_{12}(2)$ exciting the F=1 and F=2 states of the N=2 rotational manifold and transferring the population by spontaneous decay to the ground state N=0 shown in figure 2. A second laser beam tuned to the Q(0) F=1 transition partially overlaps the first, this transfers population out of the ground state F=1 manifold [13]. Two independent dye lasers are required to generate this light; their relative frequencies are stabilized via an external cavity lock. To the extent that the electronic transitions conserve the vibrational state this is a closed system, unfortunately the Franck-Condon factors for YbF have not been carefully measured but the vibrational leakage is expected to be small. The net effect of the laser transitions is to transfer molecular population to the $|00\rangle$ state and to leave the F=1 triplet unpopulated. The beam then enters an inner cylindrical magnetic shield which encloses the electric and magnetic interaction region.

The electric fields are produced by three pairs of plates, two short guard regions at either end and a ~60 cm high field region. The plates are formed from 1.5mm aluminum sheet which was cut to size, electropolished and glued to glass backing plates. The backing plates are separated by precision glass spacers- the electrode spacing is uniform to 25 μm. This arrangement has the advantage that it proved very easy to produce a uniform electric field region, unfortunately the presence of the glass dielectric limits the peak field in the center region to 8.3 kV/cm, which in turn limits the effective electric field to about half of its maximum value (figure 1). Referring to figure 3, the 'splitter' and 'recombiner' of the interferometer are rf loops which drive M1 π-flip transitions between $|00\rangle$ and the coupled state $\frac{1}{\sqrt{2}}(|11\rangle + |1-1\rangle)$ in the ~3.3 kV/cm guard fields. A magnetic field parallel to the electric field is produced by 4 wires glued to the inside of the inner magnetic shield at locations which produce a B_z field uniform to 1% across the beam [14].

After the YbF molecules leave the inner magnetic shield the population of the $|00\rangle$ state is probed by collecting laser induced fluorescence from the F=0 part of the Q(0) transition. The interference curve as a function of B_z is shown in figure 4. Note that the depth of the interference is only about 1% of the total count rate; the large background is caused by oven blackbody light, by scattered probe laser light and by overlapping molecular transitions.

As noted previously, the interferometer is most sensitive to small changes in phase when the magnetic field B_{Z0} is set to the steepest point on either side of the central interferometer fringe. During data acquisition the slope of the curve at this point is measured by applying a small step in magnetic field. We record the fluorescence at eight points $(\pm E_z, \pm(B_{Z0} \pm \delta B_z))$ and take appropriate combinations to extract the edm, which is odd in $\vec{E}\cdot\vec{B}$, the slope, and various other diagnostic signals. As the applied magnetic fields are well characterized the experiment essentially measures the ratio d_e/μ_B. A typical block of data consists of 1024 points where each point is measured for 50ms; the duty cycle per block is 65%, since additional time is added between points to allow switching transients to die away. The fields are switched using patterns which reject low frequency noise [15].

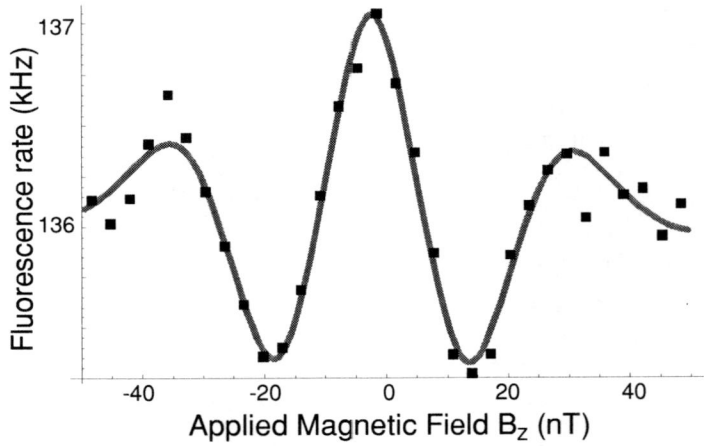

FIGURE 4. $F = 0$ probe fluorescence signal rate (kHz) vs. applied B_z in the interference region. Each point represents 12s of integration time. The curve is a velocity averaged calculation whose only free parameters are the overall normalization and a magnetic field offset.

SYSTEMATIC EFFECTS

Stray magnetic fields are a problem in all edm experiments because the electron's magnetic dipole moment μ_B is so much larger than the edm d_e one hopes to measure. In general the YbF experiment is less sensitive than atomic experiments to stray magnetic fields because the enhancement of T violating effects is larger. In other words, the ratio $\mu_B B_{stray} / d_e \eta E_{ext}$ is smaller because η is larger by roughly three orders of magnitude. Any magnetic field which reverses with the applied E field is dangerous; some well known effects are B fields due to leakage currents from the high voltage plates and the magnetic field generated by the motion of the molecules through the applied E field, which we discuss next.

The $E \times v/c^2$ motional magnetic field

A polar molecule like YbF is highly anisotropic. Because it is strongly aligned along the direction of the applied electric field, $\vec{E}_{ext}$, which defines the z axis, it responds very differently to external fields parallel or perpendicular to $\vec{E}_{ext}$. This has the consequence that the edm signal is remarkably insensitive to transverse magnetic fields, in particular the motional field $\vec{B}_E = \frac{v}{c^2} E_{ext} \hat{x}$ which causes so much trouble in the Tl experiment. We estimate the size of this motional field effect for our experiment as follows: (The same argument applies to transverse B fields generated by leakage currents.) Consider the effective Hamiltonian

$$\hat{H} = d_e \eta E_{ext} \hat{\sigma}_z + \mu_B B_\| \hat{\sigma}_z + \mu_B (B_\perp + B_E) \hat{\sigma}_x,$$

where the applied magnetic field makes an angle α with E_{ext}, i.e. $B_\| = B_{applied} \cos\alpha$, $B_\perp = B_{applied} \sin\alpha$. The edm signal is proportional to the energy difference U_{12} between the $|11\rangle$ and $|1-1\rangle$ states through the phase $\varphi = \frac{1}{2} U_{12} T / \hbar$. The lowest order term is diagonal in $|F, m_F\rangle$ and picks out the parallel parts of the Hamiltonian, $U_{12}^{(0)} = 2U_\| = 2(d_e \eta E_{ext} + \mu_B B_\|)$. The field $\vec{E}_{ext}$ stark shifts the $|10\rangle$ state by an energy Δ relative to the $|11\rangle$ state, as shown in figure 5. There is a second order mixing through this state with an energy given by

$$U^{(2)} = \frac{|\langle 1\pm 1|\mu_B(B_\perp + B_E)\hat{\sigma}_x|10\rangle|^2}{E_{|1\pm1\rangle} - E_{|10\rangle}} = \frac{|M|^2}{E_{|1\pm1\rangle} - E_{|10\rangle}}.$$

The energy denominator is approximately $\pm U_\| + \Delta$ which means that this second order shift is slightly different for the two states. The total energy difference is

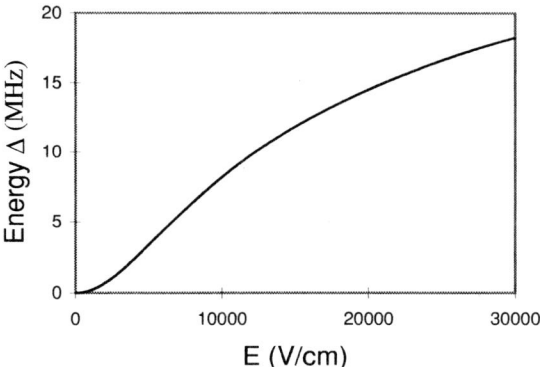

FIGURE 5. The energy splitting between the $|11\rangle$ and $|10\rangle$ hyperfine levels in an applied electric field.

$U_{12} = 2U_\|(1 - |M|^2/\Delta^2)$, to order $|M|^2/\Delta^2$. Writing out the matrix element explicitly, we find there is a cross term $\mu_B^2 B_\perp B_E$ which is potentially disastrous because it reverses with both E and B. However, it is strongly suppressed by the denominator Δ^2, which is large because the molecule is very anisotropic and polarisable. If the applied E and B fields are exactly parallel there is no effect regardless of B_E, but even for large angles the systematic effect is strongly suppressed. For example, with a large misalignment of $\alpha=10°$ in our experiment, where $B_{applied} \cong 200\mu G$, this effect produces only a small false edm of the order 10^{-36} e.cm. In contrast to atomic experiments, where the energy levels usually respond to the magnitude of the total magnetic field, our interferometer scheme is exceptionally good at canceling this effect because it is sensitive to the energy *difference* between the $|1 1\rangle$ and $|1 -1\rangle$ states, which are affected almost identically by perpendicular fields. This order of magnitude estimate is confirmed by numerically integrating the Schödinger equation for the full wavefunction and the actual field distribution in the experiment. Also note that the systematics resulting from both the motional field and the leakage current fields will have functional dependencies on the applied E which are almost certain to be different to the true edm signal, which follows the form of the YbF polarization curve shown in figure 1.

Leakage currents

Leakage currents from the high voltage plates which polarize the beam are a worry as they could produce a reversing magnetic field near the molecular beam. The z component couples to the electron's magnetic moment to mimic the edm whereas the transverse component is suppressed by the same mechanism that reduces the $\mathbf{E} \times \mathbf{v}$ systematic. As a worse case estimate, a current which flows 1 cm from the beam produces an edm given by $d_e[\text{e.cm}] < 10^{-19} I_{leakage}[\text{A}]$, e.g. the leakage for our aluminium/glass plates $I_{leakage} < 10\text{nA}$ gives a false edm of $<10^{-27}$ e.cm. In reality it seems unlikely that leakage currents are this effective at generating a false edm as there is not an obvious leakage path so close to the molecular beam which produces the correct orientation of magnetic field.

Intensity shifts

So far, only one systematic effect has been significant. It results from change in beam intensity when the E field is reversed, together with an imperfect reversal of B due to slow drifts of the ambient magnetic field. The beam intensity change is due to a dependence of the RF "splitter" and "recombiner" transition efficiencies on the direction of the E field. We believe that the field does not reverse exactly, perhaps due to surface charges on the field plates or on nearby insulators. This inexact reversal leads to RF transition efficiencies that depend on the direction of E, through the Stark shift of the hyperfine levels. The inexactitude of the reversal, calculated from the size of the systematic edm, is around 5 V/cm. This is one of the reasons we are currently replacing the field generating electrodes.

On its own, this difference in RF efficiency is not enough to generate a false edm. However, in combination with a non-zero residual magnetic field inside the machine, a phoney edm results. The effect is best explained with reference to figure 6. The interferometer lineshape is rather crudely drawn, the solid and dashed lines represent the different RF efficiencies for the two directions of E. When we measure the edm, the magnetic phase is set to $\pi/4$ and the E field is reversed (points A and B). This is repeated with the magnetic phase set to $-\pi/4$ (points C and D). The edm is proportional to $(A+D-B-C)$. Figure 6(a) shows the lineshape in the absence of any residual magnetic fields - the lineshape is centered on $B_z = 0$. Clearly in this case the false edm is zero, regardless of the RF efficiencies. Figure 6(b) shows the lineshape with the centre biased from zero. The edm, $(A+D-B-C)$, is no longer zero.

It is straightforward to extract both the difference in RF efficiencies for the two E directions and the residual B_z field from the data. The expected false edm can be calculated easily from these two quantities. This is then subtracted from the measured edm for that dataset to yield a corrected edm.

After the effect was discovered, we implemented a system to actively track the residual magnetic field, since if this field can be reduced to zero the effect disappears, provided that only the size and not the shape of the curve changes when E is reversed. During data acquisition the residual field is periodically calculated from the quantity $(A+B-C-D)$ with recently acquired data. The strength of a canceling magnetic field is adjusted accordingly by a computer controlled current supply. With the magnetic field tracking active, the correction due to this intensity dependence is negligible.

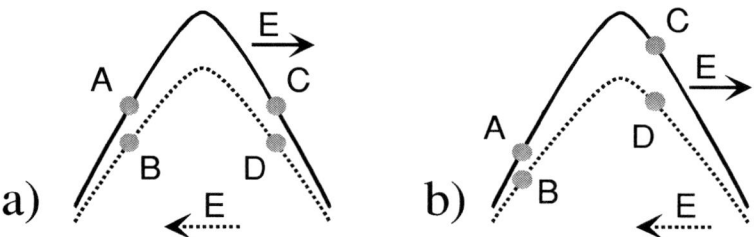

FIGURE 6. Simplified interferometer curves, showing points for the four combinations of $(\pm E, \pm B_z)$ where data is acquired. A change in interferometer efficiency when E is reversed (solid/dashed lines), combined with a non-symmetric B reversal (b), gives a false edm. See text for details.

RESULTS AND FUTURE PROSPECTS

We have accumulated about 23 hours of data with the current apparatus. The results are shown in figure 7; these data average to give a value of the edm $d_e = 0.6 \pm 4 \times 10^{-26}$ e.cm, where the uncertainty is the 1σ statistical error.

The limit to the precision of our measurement of d_e is entirely due to counting statistics. We are currently pursuing several possibilities which might increase the edm signal size and reduce the unwanted background. The electrodes for the electric field region will be replaced by a set with a new design. These new field plates are aluminium coated with gold to reduce surface charging effects, which should reduce the problems associated with intensity shifts. The support structure for the plates has also been improved so that higher field strengths will be possible- we expect to reach ~80% polarization rather than 50%.

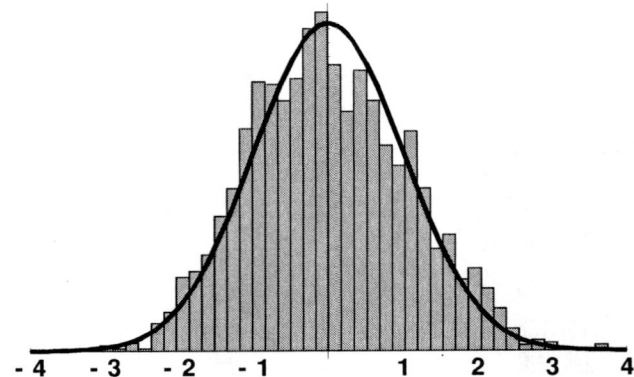

FIGURE 7. A histogram of our YbF edm data, where each 50s measurement has been normalized to its standard deviation. The curve is a Gaussian of unit width, centered on zero, whose overall height has been normalized to the data.

We have also begun to explore the possibility of producing a rotationally cold YbF beam. Our thermal source populates hundreds of rotational levels of the ground electronic state, so that only about 10^{-4} of the YbF molecules which leave the oven are in the proper N=0 state to participate in the edm experiment. It is a well established technique to produce rotationally cold refractory molecules by laser ablation of a suitable precursor into a gas whose expansion provides cooling. We are currently trying to implement this scheme for YbF. We have had preliminary success in producing YbF by ablating a solid target of AlF_3 and Yb and are working towards a robust design for a laser ablation/supersonic expansion source. Success should increase the YbF experiment's sensitivity by two or three orders of magnitude.

ACKNOWLEDGEMENTS

This work is support by the UK research councils EPSRC and PPARC.

REFERENCES

1. E. M. Purcell and N. F. Ramsey, Phys. Rev. **78,** 807 (1950).
2. I. B. Khriplovich and S. K. Lamoreaux, *CP violation without Strangeness*, (Springer Verlag, Berlin, 1996).
3. For a review, see E. A. Hinds in *Atomic Physics 11*, (World Scientific Singapore 1989).
4. M. Kobayashi and T. Maskawa, Prog. of Theor. Phys. **49** 652 (1972).
5. *CP violation* edited by C. Jarlskog (World Scientific Singapore 1989).
6. W. Bernreuther and M. Suzuki, Rev. Mod. Phys. **63** 313 (1991).
7. S. M. Barr, Int. J. Mod Phys. A **8** 209 (1993).
8. E. A. Hinds, Physica Scripta, **T70** 34 (1997).
9. Eugene D. Commins, "Electric dipole moments of leptons", Ad. At. Mol. Opt. Phys. **40**, 1 (1999).
10. P. G. H. Sandars, Phys. Lett. **14** 194 (1965).
11. P. G. H. Sandars, Phys. Rev. Lett. **19**, 1396 (1967); The use of a polar molecule to search for d_e is implicit in this paper. The first explicit statement of the idea appears to be in O. P. Sushkov and V. V. Flambaum, Zh. Eksp. Theo. Fiz **75**, 1208 (1978) [Sov. Phys. JETP **48**, 608 (1978)].
12. M. G. Kozlov, V.F. Ezhov, Phys. Rev. **A49** 4502 (1994); M. G. Kozlov, J. Phys. B **30** L607 (1997); A. Titov, M. Mosyagin, V. Ezhov, Phys. Rev. Lett. **77** 5346 (1996); H. M Quiney, H. Skaane, I. P. Grant, J. Phys. B **31** L85 (1998) (after correcting for the trivial factor of 2 between s and σ their result becomes 26 GV/cm); F. A. Parpia, J. Phys. B **31** 1409 (1998); N. Mosyagin, M. Kozlov, A. Titov, J. Phys. B **31** L763 (1998).
13. B. E. Sauer, Jun Wang, E. A. Hinds, J. Chem. Phys. **105** 7412 (1996).
14. G. D. Redgrave, D. Phil. Thesis, University of Sussex (1998), unpublished.
15. M. A. Player and P. G. H. Sandars, J. Phys. B **3** 1620 (1970).

Search for the Electric Dipole Moment of the Electron Using Metastable PbO

D. DeMille,[1] F. Bay,[1] S. Bickman,[1] D. Kawall,[1]
L. Hunter,[2] D. Krause, Jr.,[2] S. Maxwell,[2] and K. Ulmer[2]

[1]*Physics Department, Yale University, New Haven, CT 06520*
[2]*Physics Department, Amherst College, Amherst, MA 01002*

Abstract. The metastable excited state $a(1)[^3\Sigma^+]$ of PbO is proposed as a candidate system in which to search for an electron electric dipole moment (EDM). It is shown that the sensitivity to an electron EDM (d_e) could reach 10^{-31} e·cm—an improvement of $>10^4$ over the current limit. Observation of an electron EDM would provide definitive evidence for physics beyond the standard model, and many currently favored theories predict $|d_e| > 10^{-31}$. The sensitivity of the proposed technique arises from a combination of several unique properties of PbO, which simultaneously provide a large EDM enhancement factor, narrow magnetic resonance lines, and high counting rates. The structure of the nearly degenerate Ω-doublet system in PbO permits the use of unique and powerful techniques for the rejection of systematic effects.

WHY SEARCH FOR EDMS?

The EDMs of elementary particles such as the electron have been the subject of nearly continuous searches for over four decades. These searches are motivated by the fact that an observable value of an EDM of the electron (or the neutron or a diatomic atom such as Hg) would be unambiguous evidence for physics beyond the Standard Model (SM): while EDMs are non-zero in the SM, they are too small to be observed by any conceivable technique.[1] However, in almost all viable extensions to the SM, EDMs are much larger than in the SM. In fact, there are serious theoretical grounds on which to expect a non-zero signal within the next few orders of magnitude of sensitivity. Thus, any such improvements in sensitivity could have a profound impact on our understanding of elementary particle physics.

The sensitivity of EDMs to physics beyond the SM can be understood qualitatively. The existence of an EDM of any elementary particle requires non-invariance under both spatial inversion (parity, P) and time-reversal (T). While P-violation is an integral part of the structure of the SM, T-violation is not. Of course, T-violation can be incorporated into the SM, via a complex phase in the CKM matrix. From the modern theoretical point of view, however, the SM is notable in that this is the only source of T-violation in the theory. Indeed, this uniqueness leads to the very small predicted values of the EDMs in the SM: a 4-loop level of perturbation theory is required to "access" the CKM phase and generate (for example) an electron EDM.[2]

By contrast, virtually every conceived extension of the SM includes additional scalar fields, which allow new complex phases—and thus new sources of T-violation. Because of these new complex fields, the electron EDM generally appears at the one- or two-loop level of perturbation theory, leading to a dramatically enhanced effect. It is difficult to justify any significant suppression of the new complex phases in such theories, for it is already known from observations in neutral kaons that T is not a good symmetry of nature; moreover, it is generally accepted that the observed dominance of matter over antimatter in the universe actually requires additional sources of T-violation, beyond the CKM phase.[3] The implications of EDMs for a variety of models have been reviewed recently.[4,5] Each review stresses that EDM limits impose severe constraints—often much stronger than from direct accelerator searches—on models including supersymmetry, left-right symmetry, multiple Higgs bosons, leptoquarks, composite fermions, etc. In the simple words of Steven Weinberg: "electric dipole moments… offer one of the most exciting possibilities for progress in particle physics."[6]

DETECTING THE ELECTRON EDM WITH PBO

Here we describe the principle of our proposed experiment using PbO, and show that the first generation of this experiment can plausibly achieve a sensitivity 100 times beyond the current limit on $|d_e|$ (from the Berkeley experiment of Commins *et al.*,[7] using atomic Tl). We also will mention future developments which could allow yet another 100-fold improvement using PbO.

Overview of EDM experiments

We begin by defining a figure of merit to compare different electron EDM experiments. Any EDM experiment searches for a linear Stark shift arising from a term in the Hamiltonian of the form $H' = -d\vec{J} \cdot \vec{E}$, where $\vec{J}$ is the angular momentum of the system, $\vec{E}$ is an applied electric field, and d is the permanent electric dipole moment of the system. The ultimate sensitivity of an EDM experiment depends both on the size of the energy shift ΔE, and the ability to accurately measure this small quantity. The energy resolution can be parameterized easily: for shot-noise limited detection of a signal with analyzing power of unity, coherence time τ, counting rate dN/dt and unit observation time T, the energy resolution δ(ΔE) is given by

$$\delta(\Delta E) = \frac{\hbar}{\tau \sqrt{T \cdot dN/dt}}. \quad (1)$$

Now consider the size of the energy shift to be measured. For an electron EDM experiment, it is optimal to use a system with an unpaired valence electron, so that the angular momentum $\vec{J}$ has a contribution from the electron spin. In this case, the Stark energy shift ΔE can be written as

$$\Delta E = d_e \cdot E_{eff}, \tag{2}$$

where E_{eff} is the effective internal electric field experienced by a valence electron. This effective field is nonzero because of relativistic effects, and thus increases rapidly with atomic number Z. It has long been known that there is a qualitative difference between the effective fields in atoms and molecules.[8,9] The effective field E_{eff} can be expressed as $E_{eff} = QP$, where Q is a factor which includes both the relativistic effects and details of atomic/molecular structure, and P is the degree of polarization of the system by the external field. For both atoms and molecules Q is typically[10,11] ~ (6-60)×10^9 V/cm × (Z/80)^3. The difference in intrinsic sensitivity for molecules and atoms arises primarily in the degree of polarization P. For atoms, P does not approach unity even with the largest possible laboratory electric fields: $P_{atom} \sim 10^{-3} \times [E/(100\,kV/cm)]$. By contrast, with polar molecules, $P \approx 1$ can be achieved with laboratory-scale fields (see more on this below). Hence, the intrinsic sensitivity of heavy, polar, paramagnetic molecules to an electron EDM can be 100-1000 times greater than for atoms.

Using molecules to search for EDMs

Why do experiments using molecules not *already* give the best limits on d_e? The answer is that—up to now—it has seemed that the gain in intrinsic sensitivity using molecules would be offset by enormous losses in counting rate. There are two reasons for this. First, the paramagnetic molecules needed for optimal sensitivity to d_e are chemical radicals, which usually require extreme thermal and chemical conditions for production. Second, the Boltzmann distribution spreads the molecular population over many rotational sublevels (~10^4 under typical conditions T ≳ 1000 K), while only one of the lowest levels is useful for an EDM experiment (which requires states of well-defined energy and angular momentum). The difficulty in even producing molecular radicals deterred many in the field. Recently the group of E. Hinds at Sussex (U.K.) has begun the first experiment that makes use of such molecules; their experiment uses a beam of YbF (see a contribution by B. E. Sauer et al in these Proceedings).

The promise of PbO

We will use the paramagnetic metastable excited state a(1) [$^3\Sigma^+$, $|\Omega|=1$] of PbO to search for an electron EDM.[12] Crude models of PbO have led to the conclusion that this state has an effective field value that is on the low end of the typical range for heavy molecules:[13]

$$E_{eff}\,[PbO, a(1)] = (6\times 10^9 \text{ V/cm}) \cdot P \tag{3}$$

(but see further comment on this below). The particular attractiveness of PbO arises from the possibility to perform the experiment in the high-density environment of a vapor cell rather than a beam, which leads to dramatically increased counting rates. The ability to work in a vapor cell is entirely novel for molecular EDM experiments, and arises only because several conditions can be simultaneously met with PbO. PbO (in its diamagnetic X $^1\Sigma$ ground state) is thermodynamically and chemically stable and is easily vaporized. The a(1) state can be selectively populated by laser excitation, which replaces the step of non-equilibrium chemistry previously considered necessary for using paramagnetic molecules. Finally, the a(1) state requires only remarkably small electric fields $E \geq 15$ V/cm to achieve $P \approx 1$; the modest required fields have been achieved routinely in heated metal-vapor cells.[14] The extremely high polarizability of the a(1) state is a consequence of the very small energy splitting (~12 MHz) between the Ω-doublet opposite parity levels in this state.[15]

Consider the counting rates which can be achieved using a vapor cell of PbO. The coherence time for measurements in the a(1) state will be limited by the lifetime[16,17]

$$\tau_a \sim 80 \ \mu s. \tag{4}$$

The useful vapor density will be limited by the constraint that the collisional relaxation time of coherences between a(1)-state Zeeman sublevels be longer than the natural lifetime. With a conservative estimate of the relaxation cross-section ($\sigma \sim 10^{-14}$ cm^2), the total density can be n ~ 3×10^{13} cm^{-3}, corresponding to the saturated vapor pressure 3×10^{-3} Torr at a temperature[18] of ~690°C. The fraction of molecules in the lowest rotational level is approximately f ~ B/kT ~ 3×10^{-4}, where B = 4×10^{-5} eV is the rotational energy constant.[19] With cell dimensions chosen to limit wall-collision quenching at the same level (L ~ $2\tau v_{rms}$ ~ 5 cm), the number of molecules in the lowest rotational level is N_0 ~ fnL^3 ~ 10^{12}. If these molecules are excited to and detected from the a(1) state with total efficiencies ε_e and ε_d respectively, the counting rate will be

$$dN/dt = \varepsilon_e \varepsilon_d N_0 / \tau_a \sim \varepsilon_e \varepsilon_d \cdot 10^{16}/s. \tag{5}$$

For a given set of excitation and detection efficiencies, we are now in a position to estimate the sensitivity of the PbO experiment to an electron EDM. From Eqns. (1)-(5), one finds that the uncertainty in the determination of the electron EDM is given by

$$\delta(d_e) = \frac{1.4 \times 10^{-32}}{\sqrt{\varepsilon_e \varepsilon_d}} e \cdot cm, \tag{6}$$

where we have assumed complete molecular polarization $P \approx 1$ and a total integration time T = 100 hours. It should be noted that this estimate assumes shot-noise limited signals and perfect analyzing power of the signal; however, we expect both of these to introduce additional degradation factors on the order of unity.

We will proceed with the EDM experiment in two phases, corresponding to two levels of excitation/detection efficiency. Phase I will use the simplest possible methods for exciting and detecting the a(1) state. Excitation will proceed via direct pumping of the X-a transition, and detection will be accomplished by viewing a-X decay fluorescence. As shown below, this scheme leads to $\varepsilon_e \varepsilon_d$ ~ 10^{-5}. From Eqn. (6)

(and taking into account likely imperfections), we can thus expect an electron EDM sensitivity of $\delta(d_e) \lesssim 10^{-29}$ e·cm, an improvement by a factor of ~100 over that achieved by the Berkeley experiment.[7] We emphasize that this estimate depends only on known data for PbO, along with a few conservative assumptions (e.g. for the relaxation cross-section). A few parameters must be determined to proceed with Phase I, but none of these is critical to the anticipated EDM sensitivity, and these measurements will arise naturally from the development of techniques for the EDM search. These preliminary measurements are described later.

Phase II of the EDM search will use more complex excitation and detection schemes to increase the efficiency of each step. As discussed below, dramatic gains can likely be made by taking advantage of strong electronic transitions between the a(1) state and other excited states of PbO. We believe $\varepsilon_e \varepsilon_d \sim 10^{-1}$ is ultimately possible, leading to a potential statistical sensitivity at the level $\delta(d_e) \lesssim 10^{-31}$ e·cm.

A novel and powerful means for rejecting systematic errors

The large value of E_{eff} for PbO is, by itself, of advantage for rejection of systematic errors: unlike the EDM effect, most systematic effects are not enhanced in molecules. However, we stress here that the PbO EDM experiment has an additional, unique feature, which should allow unprecedented control over systematic errors. This feature arises from the fact that it is possible to perform the EDM measurement independently on each energy level of the Ω-doublet.

The importance of this statement can be understood from the following argument. The observable molecular EDM arises from d_e via mixing of the Ω-doublet levels in first-order perturbation theory.[8] Since the energy denominators are equal and opposite for the two members of the pair, these two levels have equal but opposite values of E_{eff}. This same conclusion can be reached from an alternative argument. In the external electric field, the Ω-doublet levels mix and repel each other. The states which move lower in energy must correspond to molecular polarization along the external field ($\vec{P} \| \vec{E}$), while those which increase in energy have opposite polarization ($-\vec{P} \| \vec{E}$). Since the internal effective field $\vec{E}_{\text{eff}}$ is parallel to $\vec{P}$, these two states have opposite values of E_{eff}.

Measurement of the EDM in two states with opposite values of E_{eff} is conceptually similar to the use of so-called "co-magnetometers" in other EDM experiments. These have typically employed two different species in the same apparatus, which are chosen to have different sensitivity to T-odd effects (e.g., by choosing one heavy atom and one light atom). By contrast, in PbO the unique near-degeneracy of the Ω-doublet levels allows both levels to be addressed independently with only a minor change in the experimental parameters (see below). This eliminates the difficulties of producing, populating, and probing two chemically different species in the same apparatus.

Measurement of the difference between the EDMs of the two members of the Ω-doublet is an extraordinarily powerful means to isolate systematic effects. The Ω-doublet levels are nearly perfect mirror images of each other, and switching between them takes full advantage of the P-odd nature of the EDM interaction. For instance, the e and f members of the Ω-doublet should have magnetic g-factors that are identical (up to corrections $\sim \mu_{rot}/\mu_{el} \sim 10^{-3}$). Thus, magnetic fields such as those due to leakage currents will induce the same energy shifts in both levels; this effect will cancel under "doublet reversal." In fact, the doublet states will have nearly identical values for a wide variety of parameters that can give rise to systematic errors (e.g. lifetimes, electric quadrupole moments, quadratic Stark shifts, etc).

TECHNICAL DETAILS OF THE PBO EXPERIMENT

Basic experimental design for Phase I

The Phase I EDM experiment will measure the linear Stark shift between the J = 1, $m_J = \pm 1$ levels of the a(1) state, using the method of quantum beats in fluorescence. The entire measurement will take place in parallel static electric and magnetic fields $\vec{E} = E\hat{z}$ ($E \sim 100$ V/cm) and $\vec{B} = B\hat{z}$ (B ~0.1 G). In such an electric field, the m = ±1 components of the J=1 Ω-doublet in the a(1) state will be completely mixed; the m=0 components are not mixed because the Clebsch-Gordan coefficient <1,0;1,0|1,0> vanishes. The level diagram for this system is shown in Fig. 1.

A pulse of z-polarized light will pump (for example) the X(J=0)→a(J=1⁻) transition, resulting in population only of the |J = 1⁻, m = 0> component of the a(1) state. Selective excitation of either parity component e(-) or f(+) of the lowest-lying (J=1) Ω-doublet of the a(1) state is possible (see Fig. 2): the individual rovibrational

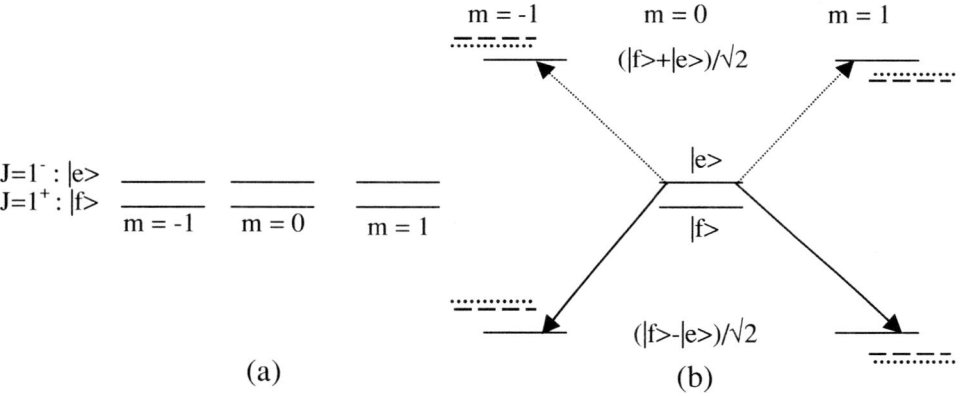

Figure 1: Level structure of the lowest-lying (J=1) Ω-doublet of the a(1) state. Levels in the absence of applied fields are shown in (a); the splitting is ~12 MHz. The structure in the presence of a large electric field (≫10 V/cm) is shown in (b). Dashed-line levels show Zeeman shifts. Dotted-line levels show the additional linear Stark shift which would arise from a non-zero EDM. (Both linear effects are greatly exaggerated.) Note the opposite effect of the EDM in the upper and lower levels. Arrows correspond to alternate choices for Δm = ±1 magnetic resonance transitions from the initial |e, m=0> state. The "doublet reversal" described in the text corresponds to switching between solid arrows and dotted arrows.

components of the X→a band can be resolved in the presence of the cell Doppler width and laser bandwidth ($\Delta\nu \sim 0.8$ GHz).[15]

Following the laser pulse, a pulse of oscillating magnetic field $\vec{\beta}_1 = \beta_1 \cos(\omega t)\hat{x}$, with ω tuned to the resonance with the $\Delta m = \pm 1$ transitions, will be applied for a time $\tau \sim 1$ μs. The strength of β_1 will be adjusted to provide a $\pi/2$ pulse; subsequently, the system will be in a state aligned along the y-axis (i.e., in a superposition of $m = \pm 1$). This state emits in a dipole radiation pattern, with preferential emission along one axis in the x-y plane. Interaction with the applied fields will cause the $m = \pm 1$ components to acquire different phases, due to their different Zeeman and linear Stark shifts. This corresponds to precession of the alignment about the z-axis, at angular frequency $\omega' = 2(\mu B \pm d_e E_{\text{eff}})/\hbar$ (where $\mu \sim \mu_B$ is the magnetic dipole moment of the a(1) level). The sign between the terms is determined by the relative directions of $\vec{E}$ and $\vec{B}$. Fluorescence detectors viewing a fixed direction in the x-y plane will see an exponentially decaying signal modulated at frequency ω' (i.e., quantum beats).[20] This technique is insensitive to the inevitable pulse-to-pulse fluctuations in laser power and spectral distribution. The EDM effect will thus be distinguished by the dependence of the modulation frequency on the explicitly T- and P- odd quantity $\Delta\omega' \propto \vec{E}\cdot\vec{B}$. The measurement can be performed in the other level of the Ω-doublet (which, as discussed above, also reverses the sign of E_{eff}) simply by changing the rf frequency by ~ 12 MHz, in order to excite the other degenerate pair of $m = \pm 1$ levels (see Fig. 1).

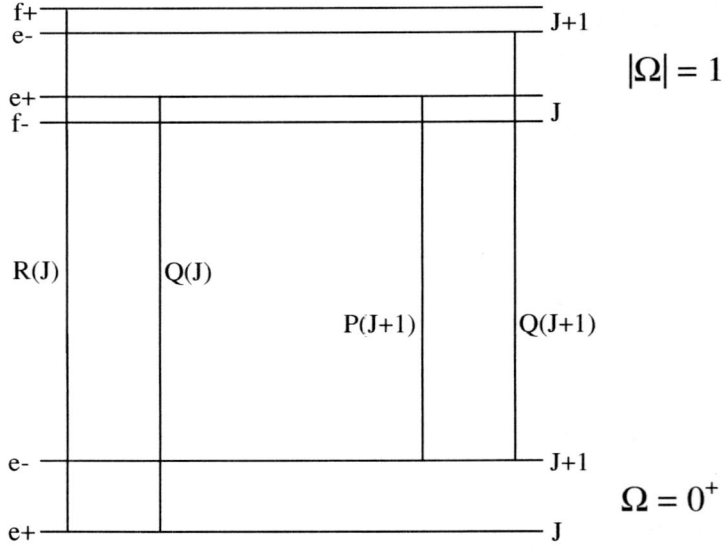

Figure 2: Level diagram and selection rules for a $0^+ \to 1$ transition such as the X→a transition in PbO. E1 selection rules require $\Delta J = 0, \pm 1$ and a change in parity. The level parity is given by $(-1)^J$ for "e" levels, and $(-1)^{J+1}$ for "f" levels. For the Q branch ($\Delta J = 0$), only e→f and f→e are allowed; for the P and R branches ($\Delta J = -1, +1$ respectively), only e→e and f→f. For clarity, both rotational splittings and Ω-doublet splittings are exaggerated. Adapted from: J. Brom and W. Beattie, J. Mol. Spect. **81**, 445 (1980).

Estimates for excitation and detection efficiencies

In our earlier discussion, we stated that Phase I of the EDM experiment will have overall efficiency $\varepsilon_e \varepsilon_d \sim 10^{-5}$. Here we derive this estimate. In Phase I, we will populate the a(1) state by direct excitation of the X-a transition. The excitation cross-section at $\lambda = 562$ nm is[21] $\sigma = (\lambda^2/8\pi) \times (\Gamma_{partial}/\Gamma_{Doppler})$, where $\Gamma_{Doppler} \sim 2\pi \times 800$ MHz and $\Gamma_{partial} = (1/\tau_a) \times$ B.R.[a(v'=4, J=1⁻)→X(v''=0, J = 0)/a(v'=4, J=1⁻)→all]. This branching ratio is not known accurately, because a-X Franck-Condon (FC) factors and Hönl-London (HL) factors (which express B.R.'s among rotational levels) have not been measured (and are difficult to calculate for forbidden transitions such as this). However, our qualitative observations of relative transition strengths among vibrational levels, along with some general arguments concerning FC and HL factors, make it unlikely that this branching ratio is less than ~10%. We then find $\sigma \gtrsim 3 \times 10^{-17}$ cm². Our current laser system is a pulsed dye laser delivering ~4 W average power at a repetition rate R = 100 Hz, with a linewidth closely matching that of the transition Doppler width. The transition probability per laser pulse is $P = \sigma F t$, where F is the photon flux and t is the pulse duration. In each pulse with energy E ~ 40 mJ, the quantity $Ft = (E/h\nu)(1/L^2)$, where $h\nu$ is the energy per photon, and $L^2 \sim 25$ cm² is the beam area. By retro-reflecting the beam through the cell once, P ~ 30% of the ground state molecules will be excited to the a(1) state in each pulse. The total overall efficiency of excitation is then $\varepsilon_e = PR\tau \sim 0.3\%$.

For detection, fluorescence from the a-X decay will be collected into two solid fused silica light pipes, each with radius r = 2.5 cm, located R = 5 cm from the center of the cell. The solid angle collected will thus be $\Omega \sim 0.1$. In order to suppress backgrounds from blackbody radiation and scattered laser light, we must use bandpass filters that transmit fluorescence accompanying decay to only a single vibrational level of X. We expect the branching ratio in a favorable channel to be $\gtrsim 20\%$, and filter transmission to be ~30%. The transmitted light will be detected by a large-area photodiode with quantum efficiency ~80%. Thus we expect total detection efficiency $\varepsilon_d \sim 0.5\%$ in Phase I.

The efficiency of both population and detection could be greatly enhanced in a second phase of the experiment. In Phase I, the excitation efficiency for the weak X-a transition is limited simply by available laser power. More efficient population can be achieved by using a two-step stimulated Raman transition (with each step a strong transition), rather than the weak direct X-a transition. We have completed a preliminary search for strong E1 couplings of the a(1) state, to all known states of PbO which also couple to the ground state X. Based on this search, we have identified the X→C'→a process as the most promising candidate.[17] The attainable efficiency depends on unknown Franck-Condon factors in each step (measurements of which are planned). With reasonable estimates for the FC factors, we find that the cross-section for the X→C' transition (τ~3 μs; λ~367 nm, depending on choice of vibrational level) is ~10× larger than for the direct X→a excitation; and the cross-section for the C'→a step (λ~1.11 μm) is another factor of 5 larger than this. Thus, with higher repetition-rate lasers of similar power, a 5-10× improvement in ε_e appears possible using a two-laser scheme.

The prospects for enhanced detection efficiency are also excellent. We know already that exciting the a→ C' transition makes more efficient detection possible. This infrared excitation induces fluorescence at λ~370-480 nm, thus eliminating the need for narrowband fluorescence filters and leading to a 10-fold increase in ε_d. However, even greater detection efficiency is likely to be possible. A sufficiently strong transition from the a(1) state would enable detection via laser absorption. If the vapor cell has column depth ~1 for the probe transition, ε_d ~1 is possible. Simple estimates show that this requires oscillator strength ~1; from the usual sum rule arguments, such transitions should exist. As for the pumping transition, this requirement can be relaxed if multiple passes of the probe beam can be arranged. A search for such a probe transition (to either a previously unobserved state of PbO, or to a recently observed Rydberg state[22]) is now underway.

In summary, the Phase I efficiency estimate relies only on general and conservative assumptions. Phase II requires some additional measurements for full optimization, but an increase in counting rate by 50-100× compared to Phase I is possible already, and an additional 100× seems likely with additional development.

Systematic effects

A preliminary analysis of systematic effects has considered both the effects important to previous EDM experiments, and some potential systematics novel to this system.[23] For example, we have analyzed effects due to motional magnetic fields $\vec{B} = \vec{v} \times \vec{E}/c$. These effects are dramatically suppressed in highly polarizable systems with J≥1,[24] and should be negligible for the PbO experiment.

Leakage currents (and the accompanying magnetic fields) associated with application of an electric field are a notorious source of systematic effects in all EDM experiments, and are of particular concern in the environment of the high-temperature cell required for this work. However, we have verified that BeO has sufficient resistivity at our operating temperature to maintain leakage currents ≪1 nA, and also that it does not interact chemically with PbO at our operating temperature. In addition, suitable high work function electrode materials should keep thermionic and photoelectric emission currents at a negligible level.[25] Current amplification due to discharge effects should be negligible under our conditions.

We have also considered possible effects due to multi-photon ionization of PbO by the population laser. Although our estimates of this effect are crude, we believe the ionization probability will be extremely small (≪10^{-7}). We have made provisions to temporally stretch our population laser pulses by a factor of ~30, to diminish the ionization probability if necessary. Any free charges produced by the population lasers will be swept out of the cell by the applied electric field in <10 μs, and the early-time data can be cut to avoid associated magnetic (current) and electric (space charge) effects.

We have also analyzed effects due to E-field gradients, in combination with electric quadrupole moments; effects due to stray $\vec{E}$-fields; geometrical phases associated with inhomogeneities in $\vec{E}$; effects due to fields induced by the PbO molecules themselves; and effects due to ordinary P-odd weak interactions, in combination with the (apparently T-odd) decay of the a(1) state. These all appear to be controllable at the level required to reach the final Phase II sensitivity $\delta(d_e) \sim 10^{-31}$ e·cm, given an understanding of stray and misaligned fields at a level comparable to that achieved in previous EDM experiments.

We note finally that the ability to fully polarize the molecules in the applied electric field provides an additional degree of control over systematic effects, which is absent in atomic EDM experiments. Since the EDM-induced energy shift is proportional to the molecular polarization (i.e., mixing of the Ω-doublet levels in the applied electric field), the desired signal will be linear at low electric fields but will saturate for "large" fields ($E \gtrsim 15$ V/cm). Several of the more dangerous systematic effects have very different dependences on E. For example, leakage currents will grow linearly (or faster) with E; conversely, an important effect arising from imperfect reversals of both $\vec{E}$ and $\vec{B}$ actually diminishes with increasing values of E. We believe that this extra non-trivial dependence on E will prove to be an important tool for eliminating systematics.

Current status, plans, and goals

We have an ongoing effort at Amherst College to determine relevant spectroscopic properties of PbO. Using a molecular beam apparatus, we have recently finished measurements of the hyperfine structure, isotope shifts, and Ω-doubling in the X-a transition, as well as of the polarizability of the a(1) state.[26] The hyperfine splitting due to the nuclear spin of ^{207}Pb is considerably larger than had been expected, based on the previous crude model of the a(1) state.[12,13] This in turn makes it likely that the relativistic enhancement of the applied electric field in PbO is larger than had been expected. All our sensitivity estimates above are based on the early, conservative estimate $E_{eff} = (6 \times 10^9$ V/cm); however, *our new data makes is likely that the effective field is in fact several times larger than this.*[27] Refined semi-empirical calculations of E_{eff} will use as input these data, in combination with values for the a(1) state g-factors.[28] These g-factors will be determined in the first measurements at Yale.

Phase I is underway at Yale, where we have focused our first efforts on cell construction. Unfortunately, PbO at our operating temperature destroys quartz and most metals. Through extensive testing, we have identified appropriate materials for the cell, and are now attempting to construct cells with sapphire windows, alumina and/or sapphire structural parts, Pt or ZrO_2 (conductive ceramic) electrodes, and BeO insulators. We have had good success with small cells sealed simply by contact between flat, optically polished surfaces. We soon expect to have a working vapor cell (without internal electrodes) based on this. Ultimately this method of cell construction will be unsatisfactory for the actual EDM measurements. The final design will likely require the ability to mechanically bond at least some of the cell components. We have had some limited success with diffusion bonding and/or active-metal brazing of ceramic parts, and plan to continue our investigation of these techniques. We remain optimistic that a PbO cell with all required characteristics can be constructed.

In the meantime, the laser, vacuum system, cell furnace, detection system, and magnetic shields (thanks to Eugene Commins' generous loan) are in place. Upon observation of our first signals, we will use the quantum-beat technique to make precise measurements of g-factors, Stark shifts, and Ω-doublet splittings in the J=1 doublet of the a(1) state. The dependence of the analyzing power on the vapor density will allow measurement of coherence-quenching cross-sections. These measurements are important for optimization of experimental conditions for the EDM experiment. The techniques developed in the course of these measurements will evolve naturally into the EDM search.

ACKNOWLEDGMENTS

We would like to thank Dmitry Budker for his original suggestion to us, many years ago, to consider PbO as a system with which to search for the electron EDM (as well as for many useful discussions in the intervening years). This work has been supported at Yale by National Science Foundation Grant PHY 9987846, a NIST Precision Measurement Grant, Yale University, a Research Corporation Cottrell Scholars Award, and the David and Lucile Packard Foundation; and at Amherst by NSF RUI grants PHY 9722611 and PHY 9987863, and Amherst College. D.D. is an Alfred P. Sloan Research Fellow.

Finally, we (D.D. and L.H.) would like to personally acknowledge the central role played by Eugene Commins in our development as a scientists. We cannot imagine a better mentor or colleague.

REFERENCES

1. See e.g. I.B. Khriplovich and S.K. Lamoreaux, "CP violation without strangeness: electric dipole moments of particles, atoms, and molecules," (Springer: New York, 1997).
2. M.E. Pospelov and I.B. Khriplovich, Sov. J. Nucl. Phys. **53**, 638 (1991).
3. P.Huet and E. Sather, Phys. Rev. D **51**, 379 (1995).
4. M. Suzuki, Rev. Mod. Phys. **63**, 313 (1991)
5. S. Barr, Int. J. Mod. Phys. A **8**, 209 (1993)
6. S. Weinberg, *Summary Talk at the XXVIth ICHEP* (Dallas, Texas, Aug. 12, 1992).
7. B. Regan *et al.*, in this volume (2001); E. Commins, S. Ross, D. DeMille, and B. Regan, Phys. Rev. A **50**, 2960 (1994); K. Abdullah, C. Carlberg, E. Commins, H. Gould, and S. Ross, Phys. Rev. Lett. **65**, 2347 (1990).
8. P. Sandars, Phys. Rev. Lett. **19**, 1396 (1967).
9. O. Sushkov and V. Flambaum, Sov. Phys. JETP **48**, 608 (1978).
10. Calculated atomic enhancement factors ($R \equiv E_{eff}/E$) and appropriate references are tabulated in Ref. [7].
11. See M. Kozlov and L. Labzowsky, J. Phys. B **28**, 1933 (1995).
12. This was first suggested in: V. Flambaum, unpublished doctoral dissertation, Inst. of Nucl. Phys., Novosibirsk, Russia, 1987; see also: L. Barkov, M. Zolotorev, and D. Melik-Pashaev, Sov. J. Quant. Electr. **18**, 710 (1988).
13. M. Kozlov and T. Titov, private communication; see also Refs. [12] and [11].

14. See e.g. P. Bucksbaum, E. Commins, and L. Hunter, Phys. Rev. D **24**, 1134 (1981); I. Davies, P. Baird, and J. Nicol, J. Phys. B **21**, 3857 (1988). In both of these, fields >200 V/cm were achieved in hot metal-vapor cells.
15. F. Martin et. al., Spectrochim. Acta **44A**, 889 (1988).
16. W. Beattie, M. Revelli, and J. Brom, J. Mol. Spectrosc. **70**, 163 (1978).
17. D. DeMille, F. Bay, S. Bickman, D. Kawall, D. Krause, S. Maxwell, and L. Hunter, Phys. Rev. A **61**, 052507 (2000).
18. R. Weast (ed.): *CRC Handbook of Chemistry and Physics* (Chemical Rubber Company, Cleveland, 1967).
19. K. Huber and G. Herzberg: *Molecular Spectra and Molecular Structure, Vol. IV: Constants of Diatomic Molecules* (Van Nostrand Reinhold, New York, 1979).
20. For reviews of quantum beats, see S. Haroche, "Quantum Beats and Time-Resolved Fluorescence Spectroscopy," in *Topics in Applied Physics Vol. 13: High-Resolution Laser Spectroscopy*, ed. K. Shimoda (Springer, Berlin, 1976); J. Dodd and G. Series, "Time-Resolved Fluorescence Spectroscopy," in *Progress in Atomic Spectroscopy, Part A*, ed. W. Hanle and H. Kleinpoppen (Plenum, New York, 1978). For a recent demonstration of precision measurement using quantum beats, see e.g. S. Rochester, C. J. Bowers, D. Budker, D. DeMille, and M. Zolotorev, Phys. Rev. A **59**, 3480 (1999).
21. I. Sobelman: *Atomic Spectra and Radiative Transitions* (Springer-Verlag, New York, 1979).
22. H. Bredohl et al., J. Mol. Spectr. **199**, 1 (2000).
23. D. Budker and D. DeMille, unpublished manuscript.
24. M. Player and P. Sandars, J. Phys. B **3**, 1620 (1970).
25. A.M. Howatson: *An Introduction to Gas Discharges (2nd Edition)* (Pergamon, Oxford, 1976).
26. S.E. Maxwell, Undergraduate Thesis, (Amherst College, 2000); K. Ulmer, Undergraduate Thesis, (Amherst College, 2001), L. R. Hunter et al., to be published.
27. M. G. Kozlov, private communication. Kozlov cautions, however, that the current models require some unexpected ratios of coefficients in the MO–LCAO expansion, and also that the uncertainties in these models are extremely large. He expects that in order to build a truly reliable semiempirical model one will need a better understanding of the structure of the two-electron wavefunction. To this end, even relatively simple ab initio calculations may help. The quantum chemistry group of A. Titov is planning to undertake such calculations.
28. M. G. Kozlov, Sov. Phys. JETP **62**, 1114 (1985).

Progress towards fundamental symmetry tests with nonlinear optical rotation

D. F. Kimball[1], D. Budker[1,2,†], D. S. English[1], C.-H. Li[1], A.-T. Nguyen[1],
S. M. Rochester[1], A. Sushkov[1], V. V. Yashchuk[1], and M. Zolotorev[3]

[1] *Department of Physics, University of California at Berkeley, Berkeley, California 94720-7300*
[2] *Nuclear Science Division, Lawrence Berkeley National Laboratory, Berkeley, California 94720*
[3] *Center for Beam Physics, Lawrence Berkeley National Laboratory, Berkeley, California 94720*
† e-mail = budker@socrates.berkeley.edu

Abstract. Magneto-optical (Faraday) rotation is a process in which the plane of light polarization rotates as light propagates through a medium along the direction of a magnetic field. In atomic vapors where ground state atomic polarization relaxes very slowly (relaxation rates $\lesssim 1$ Hz), there arise ultranarrow, light-power-dependent (nonlinear) features in the magnetic field dependence of Faraday rotation. The shot-noise-limited sensitivity of a magnetometer based on nonlinear Faraday rotation can exceed 10^{-11} G/$\sqrt{\text{Hz}}$, corresponding to a sensitivity of $\sim 10^{-6}$ Hz/$\sqrt{\text{Hz}}$ to Zeeman sublevel shifts. Here we discuss recent progress in magnetometry based on nonlinear optical rotation and consider the application of these methods to searches for fundamental-symmetry-violating interactions.

INTRODUCTION

Professor Commins's heroic efforts in testing fundamental symmetries with atoms have inspired many of his students and "grandstudents" to follow in his footsteps. For example, a number of Professor Commins's former students are pursuing searches for the permanent electric dipole moment (EDM) of the electron in their own laboratories – Larry Hunter, who completed his first search for the electron EDM using optically-pumped cesium in a vapor cell filled with buffer gas [1] a full year before Professor Commins's initial EDM result [2]; Steven Chu, who has been working to develop an EDM experiment using optically-trapped cesium atoms [3]; David DeMille, who is pursuing an EDM measurement in metastable states of lead oxide [4]; and our own group, whose work is described in this paper. However, for the past decade no one has been able to match the sensitivity of Professor Commins's first electron EDM search using thallium [2] – let alone his two subsequent improvements [5, 6]!

In this contribution, we describe recent progress in our efforts to use nonlinear optical rotation for precision magnetometry. In addition, we discuss how the techniques developed for magnetometry can be applied to fundamental symmetry tests in atomic systems.

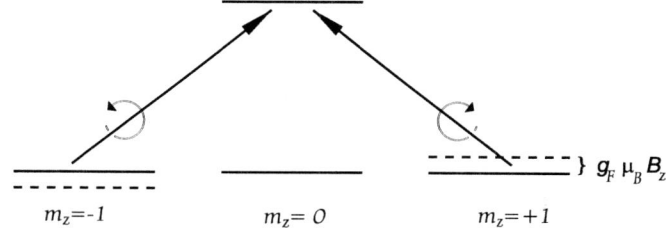

FIGURE 1. Energy level diagram for an $F = 1 \to F' = 0$ atomic transition.

LINEAR FARADAY ROTATION: THE MACALUSO-CORBINO EFFECT

When linearly-polarized light propagates through a medium immersed in a magnetic field, the plane of light polarization at the output is rotated; this effect was observed by Michael Faraday almost one hundred and fifty years ago [7]. In 1898, Italian physicists Macaluso and Corbino discovered that Faraday rotation was resonantly enhanced near atomic absorption lines [8].

The origin of this linear (light-power-independent) Faraday rotation (the Macaluso-Corbino effect) can be understood by considering, for example, an $F = 1 \to F' = 0$ atomic transition (Fig. 1); F and F' are the total angular momenta of the lower and upper state, respectively. Linearly polarized light incident on an atomic sample can be decomposed into left- (σ_+) and right- (σ_-) circularly polarized components. When a magnetic field B_z is applied to the sample along the direction of light propagation (the longitudinal direction, $\hat{z}$), the Zeeman shifts between adjacent magnetic sublevels ($= g_F \mu_B B_z$, where g_F is the Landé factor and μ_B is the Bohr magneton) cause the refractive indices for σ_+ and σ_- light to differ (circular birefringence). This, in turn, causes the circular components of the linearly-polarized light to change their relative phase as they propagate through the medium – leading to optical rotation (Fig. 2).

For the Doppler-free case with narrow-band light, the complex indices of refraction $n_\pm(\omega)$ for $\sigma_\pm$ light can be described by Lorentzian lineshape functions:

$$n_\pm(\omega) \approx 1 + 2\pi\chi_0 \cdot \left(\frac{\gamma_0}{2(\omega - \omega_0 \mp g_F \mu_B B_z) + i\gamma_0} \right), \quad (1)$$

where χ_0 is the amplitude of the linear optical susceptibility, ω is the light frequency, ω_0 is the resonant frequency of the atomic transition, and γ_0 is the natural linewidth. The difference between the refractive indices for σ_+ and σ_- light is given by:

$$n_+(\omega) - n_-(\omega) \approx -2\pi\chi_0 \cdot \frac{4 g_F \mu_B B_z / \gamma_0}{(2 g_F \mu_B B_z / \gamma_0)^2 + \left(1 - 2i(\frac{\omega - \omega_0}{\gamma_0})\right)^2}. \quad (2)$$

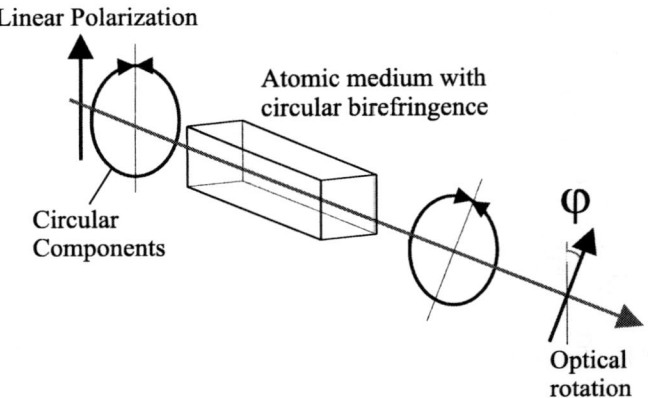

FIGURE 2. A medium possessing circular birefringence causes rotation of the plane of light polarization.

On resonance ($\omega = \omega_0$), this leads to optical rotation φ with a dispersively-shaped magnetic field dependence given by

$$\begin{aligned}\varphi &= \frac{\omega_0 \ell}{2c} \cdot \text{Re}[n_-(\omega_0) - n_-(\omega_0)] \\ &\approx \frac{\ell}{2\ell_0} \cdot \frac{2g_F\mu_B B_z/\gamma_0}{1 + (2g_F\mu_B B_z/\gamma_0)^2},\end{aligned} \quad (3)$$

where ℓ is the path length through the sample, c is the speed of light in vacuum, and $\ell_0 = (4\pi\chi_0\omega_0/c)^{-1}$ is the unsaturated absorption length on resonance. For media where the linewidth of the atomic transition is primarily determined by Doppler broadening, the indices of refraction are described by Gaussian functions, and thus optical rotation on resonance is given by

$$\varphi \approx \frac{\ell}{2\ell_0} \cdot \frac{2g_F\mu_B B_z}{\Gamma_D} \cdot e^{-(g_F\mu_B B_z/\Gamma_D)^2}, \quad (4)$$

where Γ_D is the Doppler width.

Any physical mechanism that causes a relative shift of the energies of the $\pm m_z$ sublevels will lead to optical rotation. For example, in Refs. [9, 10] it was proposed that the analog of linear Faraday rotation in the presence of a longitudinal electric field could be used to search for an EDM.

The sensitivity $\delta\Delta$ of an optical rotation measurement to a shift of the Zeeman sublevels Δ (where in the case of magneto-optical rotation, Δ is the Zeeman shift) is given by:

$$\delta\Delta = \left(\frac{\partial\varphi}{\partial\Delta}\right)^{-1}\delta\varphi, \qquad (5)$$

where $\delta\varphi$ is the sensitivity to light polarization rotation (measured in, e.g., rad/$\sqrt{\text{Hz}}$) and $\partial\varphi/\partial\Delta$ is the slope of the optical rotation with respect to the energy shift Δ. The shot-noise limit of $\delta\varphi$ is inversely proportional to the square root of the light power transmitted through the atomic vapor [11]. For an atomic transition whose linewidth is dominated by Doppler broadening, we find from Eqs. (4) and (5) that for small Δ ($\Delta \ll \Gamma_D$)

$$\delta\Delta \approx e^{\ell/2\ell_0} \cdot \sqrt{\frac{\pi}{P_0}} \cdot \frac{\ell_0}{\ell} \cdot \Gamma_D, \qquad (6)$$

where P_0 is the incident light flux in photons per second (we ignore factors of order unity related to the polarimetry method). Note that the same result is obtained for Doppler-free media from Eq. (3), except that Γ_D is replaced by γ_0.

One can also see from Eq. (6) that the optimal sensitivity is achieved when

$$\frac{\partial}{\partial\ell}\left(\frac{e^{\ell/2\ell_0}}{\ell}\right) = \frac{e^{\ell/2\ell_0}}{\ell}\left(\frac{1}{2\ell_0} - \frac{1}{\ell}\right) = 0, \qquad (7)$$

or when $\ell = 2\ell_0$.

It is also important to note that at high light powers, saturation effects become important, and therefore the sensitivity cannot be arbitrarily improved by increasing P_0.

COHERENCE EFFECTS IN NONLINEAR FARADAY ROTATION

At sufficiently high light powers, there arise light-power-dependent (nonlinear) features in the magnetic field dependence of Faraday rotation which are related to optical pumping (for reviews, see, e.g., Refs. [12, 13]). Such nonlinear magneto-optical rotation (NMOR) can arise due to the formation of Bennett structures in the atomic velocity distribution [14] and (for transitions involving states with $F > 1/2$) due to the evolution of optically-pumped atomic polarization (see, e.g., Ref. [15]). Figure 3 shows a density matrix calculation [16, 17] of the magnetic field dependence of Faraday rotation for the case of the 571 nm line $(1 \rightarrow 0)$ in samarium (studied experimentally in Ref. [15]).

In Doppler-broadened media, the linear effect has a width determined by Γ_D, the Bennett-structure-related effect has a width determined by γ_0 [17], and NMOR due to the evolution of optically-pumped, ground state polarization has a width determined by the relaxation rate of coherences between ground state Zeeman sublevels γ_{rel}. γ_{rel} can be made very small by employing buffer gases or using vapor cells with anti-relaxation wall coatings. As can be seen from Fig. 3, nonlinear effects greatly enhance the small-field optical rotation.

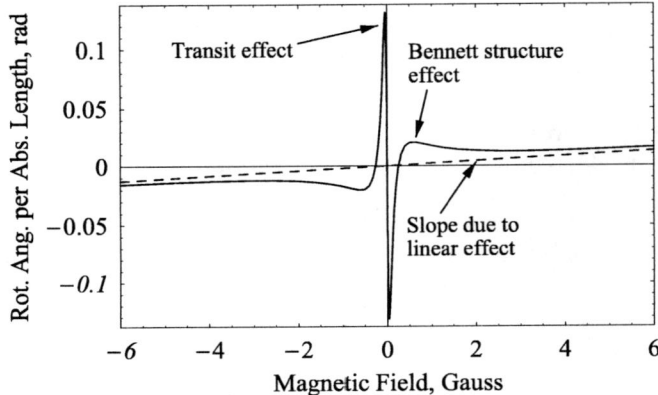

FIGURE 3. Density matrix calculation of light polarization rotation as a function of longitudinal magnetic field for the case of the samarium 571 nm line, under the conditions of the experiment described in Ref. [15]. Under these conditions, γ_{rel} is determined by the transit rate of atoms through the laser beam (diameter ~ 1 mm), hence the coherence effect in this case is termed the "transit effect."

Low light power: linear dichroism

First we consider relatively low light powers for which the optical pumping saturation parameter $\kappa \ll 1$, where

$$\kappa = \frac{d^2 E_0^2}{\gamma_0 \gamma_{rel}}, \qquad (8)$$

d is the transition dipole moment, and E_0 is the amplitude of the optical electric field. Under these conditions, NMOR related to the coherence effects is due to precession of optically-pumped atomic alignment (see, e.g., Refs. [18, 19, 20] and references therein).

Consider a closed $F = 1 \rightarrow F' = 0$ transition (Fig. 4). Resonant, linearly polarized (along x) light optically pumps the atomic sample into dark states (the $m_x = \pm 1$ states) – the vapor thus acquires an axis of linear dichroism (it is transparent to x-polarized light and strongly absorbs y-polarized light). The $\hat{z}$-directed magnetic field causes the atomic alignment to precess (Fig. 5). Since the axis of linear dichroism is no longer along the light polarization, the polarization plane of light rotates as the light propagates through the medium.

In a magnetic field, the optically-pumped atomic polarization precesses at the Larmor frequency $g_F \mu_B B_z$ and relaxes at a rate γ_{rel}. The aligned atoms can be treated as linear polarizers which rotate the light polarization by an angle $d\varphi \propto \sin(2g_F \mu_B B_z)$. On resonance for an optically thin medium ($\ell/\ell_0 \ll 1$), the overall rotation φ produced by the entire collection of atoms is given by

$$\begin{aligned}
\varphi &= \eta \cdot \frac{\ell}{2\ell_0} \frac{\int_{t=0}^{\infty} dt \, e^{-\gamma_{rel} t} \sin(2 g_F \mu_B B_z t)}{\int_{t=0}^{\infty} dt \, e^{-\gamma_{rel} t}} \\
&= \eta \cdot \frac{\ell}{2\ell_0} \frac{2 g_F \mu_B B_z / \gamma_{rel}}{1 + (2 g_F \mu_B B_z / \gamma_{rel})^2},
\end{aligned} \qquad (9)$$

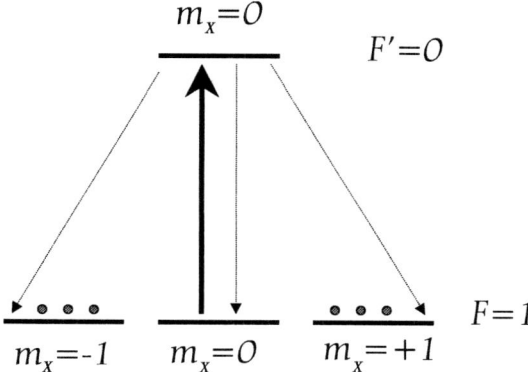

FIGURE 4. For an $F = 1 \to F' = 0$ transition, optical pumping by linearly polarized light along x creates an atomic ensemble consisting of an incoherent mixture of atoms in the $m_x = \pm 1$ states. Dots represent the population of the different ground state sublevels.

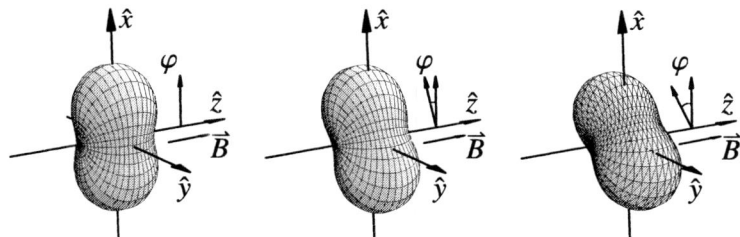

FIGURE 5. Sequence showing the evolution of optically-pumped, ground state atomic alignment in a longitudinal magnetic field for an $F = 1 \to F' = 0$ transition at low light powers, time proceeds from left to right. The distance from the surface to the origin represents the probability of finding the projection $m = F$ along the radial direction (see Ref. [25] for a detailed description of this method of representing atomic polarization). In the first plot, the atoms have been optically pumped into an aligned state by x-polarized light (they have been pumped out of the "bright" state, which interacts with the light field, into "dark" states which do not). Note that, just as in Fig. 4, the highest probability of finding $m = F$ is obtained for $\pm\hat{x}$. The magnetic field along $\hat{z}$ creates a torque on the polarized atoms, causing the alignment to precess (second and third plots). This rotates the medium's axis of linear dichroism, which is observed as a rotation of the polarization of transmitted light by an angle φ with respect to the initial light polarization.

where t is the time between pumping and probing for a given atom, and $\eta < 1$ is a factor that accounts for the efficiency of optical pumping and probing in the system.

The sensitivity of a measurement of optical rotation produced by the low-light-power coherence effect to small shifts of the Zeeman sublevels ($\Delta \ll \gamma_{\text{rel}}$) is given, in analogy with Eq. (6), by

$$\delta\Delta \approx \frac{1}{\eta} \cdot e^{\ell/2\ell_0} \sqrt{\frac{\pi}{P_0} \cdot \frac{\ell_0}{\ell}} \cdot \gamma_{\text{rel}}. \qquad (10)$$

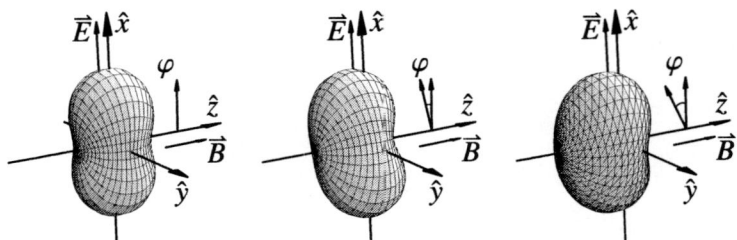

FIGURE 6. Sequence showing the evolution of optically-pumped, ground state atomic alignment in a longitudinal magnetic field for an $F = 1 \to F' = 0$ transition for high light powers, time proceeds from left to right. In the first plot, the atoms have been polarized along the axis of light polarization as in Fig. 5. If the atomic alignment is parallel to the optical electric field, the ac Stark shifts have no effect on the atomic polarization – they merely shift the energies of the bright and dark states relative to each other. However, when the magnetic field along $\hat{z}$ causes the alignment to precess, the atoms evolve into a superposition of the bright and dark states (which have different energies due to the ac Stark shifts), so optical-electric-field-induced quantum beats occur. These quantum beats produce atomic orientation along $\hat{z}$, which appears in the second plot and grows in the third plot, causing optical rotation due to circular birefringence.

Once again, as in the case of linear optical rotation, the maximum sensitivity is obtained for $\ell = 2\ell_0$. It is also important to note that for $\kappa \gtrsim 1$ saturation effects become important and a different physical mechanism can play an important role in optical rotation (see below). The maximum P_0 that can be realized without loss of sensitivity corresponds to $\kappa \sim 1$, so from Eqs. (8) and (10), we find that $\delta\Delta \propto \sqrt{\gamma_{\rm rel}}$.

As will be discussed below, in paraffin-coated alkali vapor cells, $\gamma_{\rm rel}$ can reach $\sim 2\pi \times 1$ Hz [21, 22, 23, 20]. Thus the coherence effects enhance the sensitivity of a measurement of optical rotation to level shifts by several orders of magnitude (compared to linear optical rotation)! In fact, our measurements of NMOR in paraffin-coated alkali vapor cells [20, 24, 26] (described below) indicate that the sensitivity of an NMOR-based magnetometer can compete with current state-of-the-art optical pumping magnetometers [27] and superconducting quantum interference device (SQUID) magnetometers [28].

High light power: alignment-to-orientation conversion

At relatively high light powers, where $\kappa \gtrsim 1$ (Eq. (8)), a different physical mechanism can become important for NMOR [29]. Under such conditions, the evolution of the optically-pumped ground state atomic alignment involves both Larmor precession due to the magnetic field and Stark beats due to the presence of the strong optical electric field. This combined action of the magnetic and optical electric fields causes atoms to acquire orientation along the direction of the magnetic field (alignment-to-orientation conversion (AOC) [30, 31, 32, 33, 34]). A sample oriented along the direction of light propagation has a difference in the populations of the $\pm m_z$ sublevels. An oriented sample causes optical rotation due to circular birefringence, since the refractive indices for σ_+ and σ_- light are different. The evolution of atomic polarization for the NMOR coherence

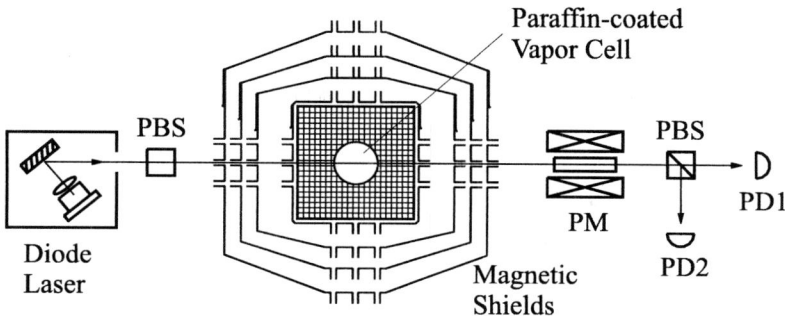

FIGURE 7. Schematic diagram of the apparatus used for nonlinear Faraday rotation measurements. PBS = polarizing beamsplitter; PM = polarization modulator; PD1,2 = photodiodes.

effect for an $F = 1 \rightarrow F' = 0$ transition at high light power is illustrated in Fig. 6.

The atomic orientation produced by AOC is proportional to $\vec{d}_{ind} \times \vec{E}_0$, where $\vec{d}_{ind}$ is the induced electric dipole moment caused by the interaction of atoms (whose optically pumped alignment has precessed in the magnetic field) with the optical electric field $\vec{E}_0$. The quantity $\vec{d}_{ind} \times \vec{E}_0$ is proportional to the ac Stark shift, which has an antisymmetric dependence on detuning of the light from the atomic resonance. Thus, net orientation in the cell can only be produced when light is detuned from resonance.

EXPERIMENTS WITH ANTI-RELAXATION COATED CELLS

As discussed above, the sensitivity of nonlinear optical rotation to interactions that shift the energies of the Zeeman sublevels can be considerably enhanced by reducing the relaxation rate of atomic polarization. In our experiments [20, 24, 26], we obtain ultranarrow NMOR resonances ($\gamma_{rel} \approx 2\pi \times 1$ Hz) by using evacuated alkali vapor cells with high quality paraffin coating [27]. In such cells, atoms can undergo many thousand wall collisions without depolarizing.

Experimental Apparatus

A schematic diagram of the experimental apparatus used in the present measurements is shown in Fig. 7, and is essentially the same as that described in Refs. [20, 24, 26]. The alkali atoms are contained in buffer-gas-free, paraffin-coated cells. The cells are made by our collaborators in St. Petersburg, Russia (E. B. Alexandrov and M. V. Balabas), and the cell preparation procedure is described in Refs. [35, 36].

A cell is placed inside a four-layer magnetic shield with a shielding factor of 10^6 in all directions [24]. Three mutually perpendicular magnetic coils allow for compensation of residual magnetic fields inside the shields to a level of 0.1 μG (averaged over the cell volume) and application of arbitrarily-directed, well-controlled magnetic fields to the vapor cell. The gradients inside the shields have been measured by NMOR using a spherical (10 cm diameter) paraffin-coated cell containing ^{87}Rb on a movable mount, and the gradients were found to be ≈ 5 μG/cm and essentially independent of temperature (they changed by $\lesssim 1$ μG/cm over a 10°C variation in cell/shield temperature). Since, under typical conditions, atoms undergo $\sim 10^3$ wall collisions between pumping and probing, the contribution of the gradients to γ_{rel} is significantly suppressed due to motional narrowing. Estimates and measurements of "relaxation in the dark" [37] as a function of leading magnetic field indicate that gradients contribute no more than $\sim 2\pi \times 0.5$ Hz to γ_{rel}.

We use tunable extended cavity diode lasers to produce light at 795 nm for the Rb D1 line ($^2S_{1/2} \rightarrow {^2P_{1/2}}$), 780 nm for the Rb D2 line ($^2S_{1/2} \rightarrow {^2P_{3/2}}$), and 852 nm for the Cs D2 line ($^2S_{1/2} \rightarrow {^2P_{3/2}}$). NMOR signals are detected using the technique of polarization-modulation polarimetry (see, e.g., Ref. [38]). After passing through the coated Rb vapor cell, light goes through a Faraday rotator which modulates the direction of linear polarization at a frequency of $\Omega_m = 2\pi \times 1$ kHz with a 5 mrad amplitude. The polarization of the light is subsequently analyzed with a polarizing beamsplitter with one of the transmission axes aligned with the initial light polarization. The signal from the photodiode PD1 (which detects light from the dark port of the polarizing beamsplitter) at the first harmonic of Ω_m is measured with a lock-in amplifier. Transmitted light intensity is the sum of the light detected in the bright channel (PD2) and the dark channel (PD1) of the analyzer. The ratio of the first harmonic signal from PD1 to the transmitted light intensity is a measure of the optical rotation angle (see, e.g., [38]).

Sensitivity of nonlinear optical rotation measurements to level shifts

Figure 8 shows optical rotation as a function of magnetic field for light resonant with the ^{85}Rb D2 $F = 3 \rightarrow F'$ transition (this data was taken using a 10 cm diameter spherical cell). As discussed above, the width of this dispersively-shaped feature is determined by the relaxation rate of atomic polarization in the cell. The data shown in Fig. 8 are fit by Eq. (9), and for this particular data set we find that $\gamma_{\text{rel}} \approx 2\pi \times 0.9$ Hz. The slope of the the optical rotation with respect to B_z in the linear region near $B_z = 0$ determines the shot-noise-limited sensitivity of NMOR to level shifts (Eq. (5)).

We have studied the sensitivity of an NMOR-based ^{85}Rb magnetometer as a function of light power and light frequency [26], and found that the optimum sensitivity occurred at a light intensity of ≈ 5 mW/cm^2 tuned ≈ 400 MHz to the high-frequency side of the $F = 3$ hyperfine component of the ^{85}Rb D2 line. Similar sensitivity is obtained for the ^{85}Rb D1 line at a light intensity of ≈ 1 mW/cm^2 tuned ≈ 600 MHz to the low-frequency side of the $F = 3$ hyperfine component. The optimum sensitivity corresponds to a shot-noise-limited sensitivity to Zeeman sublevel shifts of $\sim 10^{-6}$ Hz/$\sqrt{\text{Hz}}$. It is interesting to note that this sensitivity is close to the fundamental shot-noise limit for an ideal setup with the given number of atoms in the vapor cell ($\sim 10^{12}$ at room temperature ≈ 20 °C) and rate of ground state relaxation ($\sim 2\pi \times 1$ Hz). Thus under optimum conditions, the shot noise due to photons is about the same as the shot noise

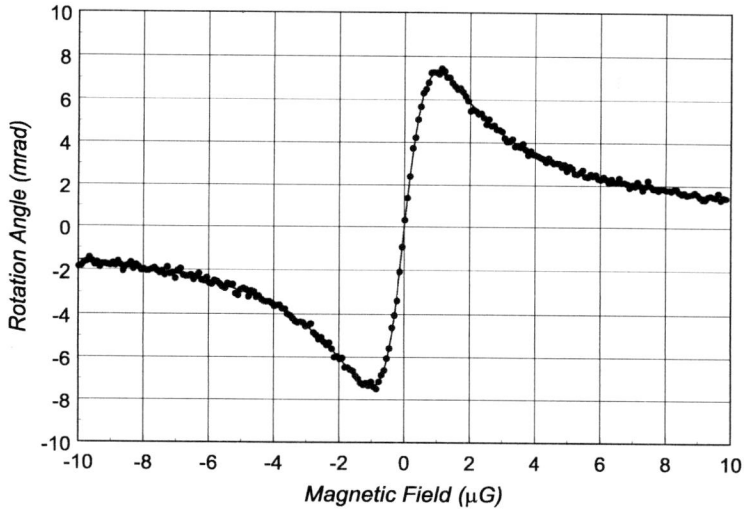

FIGURE 8. Magnetic field dependence of nonlinear Faraday rotation in a paraffin-coated cell (diameter ≈ 10 cm). Laser light is tuned to the high frequency slope of the ^{85}Rb D2 $F = 3 \rightarrow F'$ transition where maximum rotation occurs, light intensity ≈ 50 μW/cm^2, beam diameter ≈ 2 mm. The temperature of the cell was ≈ 19°C and the vapor density of ^{85}Rb was ≈ 4×10^9 cm^{-3}. Dots are experimental data and the solid line is a fit to Eq. (9).

due to atoms. The ability of nonlinear optical rotation to reach the fundamental shot-noise limit for an ideal experiment demonstrates the great potential of this technique for precision measurements.

Light-induced atomic desorption of Rb and Cs from paraffin coating

An interesting feature of anti-relaxation coated cells is that upon illumination by nonresonant light, the alkali vapor density inside the cells increases. For example, when a cylindrical cell (diameter ≈ 5 cm, height ≈ 2 cm) containing both Rb and Cs was exposed to fluorescent room light, the vapor densities of both Rb and Cs increased by a factor of ∼ 2. Since exposure to the light caused no measurable change in the cell temperature ($< 0.1°C$), the increase in atomic density did not result from heating of the vapor cell. When the lights were turned off, the atomic density in the cell returned to its original value after a few minutes.

This effect is caused by light-induced desorption of atoms from the paraffin coating, which is discussed in more detail in Ref. [36]. Such nonthermal, light-induced desorption of atoms has previously been observed from a wide variety of materials: sapphire surfaces [39], silane-coated glass [40, 41, 42, 43, 44], and even from a superfluid ^{4}He film [45]. The effect depends on the wavelength and power of the desorbing light, and in most cases appears to involve both light-induced desorption of atoms from the surface of the material and light-assisted diffusion of atoms inside the material.

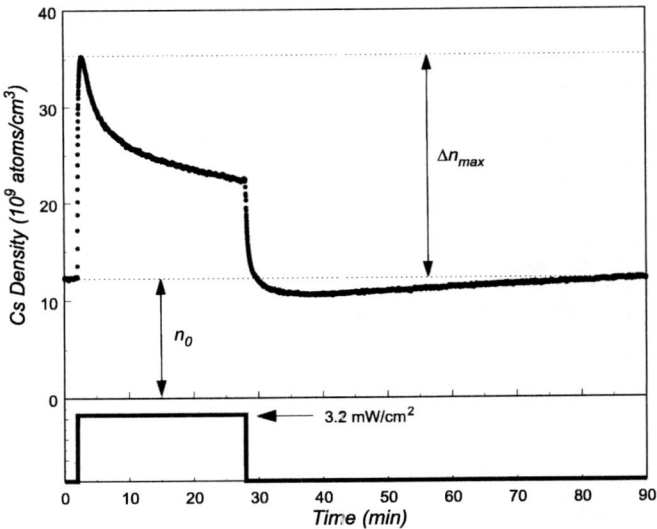

FIGURE 9. Upper plot shows Cs density as a function of time. The Cs density was measured by monitoring the transmission of a weak probe beam (intensity ≈ 10 μW/cm^2, diameter ≈ 2 mm) through the alkali vapor cell. n_0 is the initial density before exposure to light, Δn_{max} is the maximum increase in density. Lower plot shows intensity of desorbing light as a function of time, beam diameter ≈ 6.4 cm^2.

Figure 9 shows the time dependence of the Cs density before, during, and after the cylindrical cell was exposed to 546-nm light from an Ar$^+$-pumped dye laser (Coherent CR-699 using Rhodamine 560). Measurements of the time and light power dependence of atomic desorption were used to characterize properties of the paraffin-coated cells, such as the adsorption probability, adsorption energy, and the surface density of adsorbed atoms.

SEARCH FOR THE PERMANENT ELECTRIC DIPOLE MOMENT OF THE ELECTRON

As was pointed out in Refs. [9, 10], if an atom possesses a permanent electric dipole moment (EDM, for reviews, see e.g., [46, 47]), a longitudinal electric field will cause an atomic vapor to become optically active, leading to rotation of the plane of light polarization – the analog of Faraday rotation. In Ref. [48], it was demonstrated that nonlinear optical rotation is significantly more sensitive than linear optical rotation to the EDM of an atom or molecule.

We are presently investigating the possibility of performing a search for the electron EDM d_e using nonlinear optical rotation. For this experiment, a longitudinal electric field (along $\hat{z}$) would be applied to a paraffin-coated vapor cell containing both Rb and Cs. Rb, whose enhancement factor for the electron EDM is much smaller than that of Cs, would be used as a "co-magnetometer." If Cs possessed a permanent dipole moment

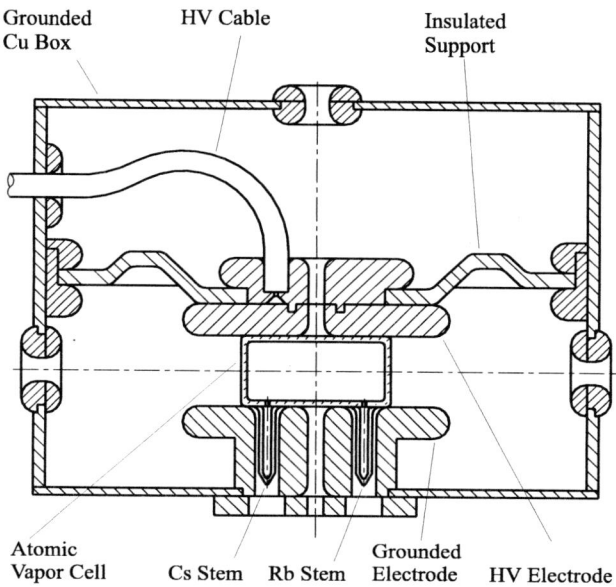

FIGURE 10. High-voltage electrode assembly for application of electric fields to paraffin-coated vapor cell. Outer shell and bottom plate (made of copper) are grounded, high voltage is applied to the upper plate. Windows allow laser beam access to the antirelaxation-coated vapor cell in three orthogonal directions. Metal corners are rounded to prevent discharges.

$\vec{d}$, the atomic polarization would precess in the longitudinal electric field $\vec{E}$ because of the $\vec{d} \times \vec{E}$ torque. The change in atomic polarization would in turn modify the light polarization, just as in the case of NMOR in a magnetic field.

There were, in fact, a number of experimental searches for d_e [49, 50, 51] carried out at the University of Washington using alkali atoms contained in paraffin-coated cells in the 1960's and early 1970's that set a limit of $d_e < 1.6 \times 10^{-23}$ e·cm [51]. In these experiments, the atoms were optically pumped into oriented states by circularly-polarized light propagating at 45° to the direction of applied electric ($\approx \pm 300$ V/cm) and magnetic fields (≈ 27 mG). An rf-field was applied perpendicularly to the electric and magnetic fields, which, in conjunction with the pump light, produced an oriented atomic sample that precessed about the fields (this is the configuration of a Dehmelt oscillator [52]). The precession frequency was measured by monitoring the modulation of light absorption. To look for the effect of an EDM, the modulation frequency of the transmitted light power was measured for $\pm$ electric field values. The statistical sensitivity of these experiments was limited by photon shot noise, while the primary systematic effect limiting the experiments was a change in the cell properties when the electric fields were applied.

For the proposed EDM experiment using nonlinear optical rotation, we expect to achieve significantly better statistical sensitivity [24, 26] than the previous experiments in paraffin-coated cells [49, 50, 51]. Our estimated shot-noise-limited sensitivity to d_e

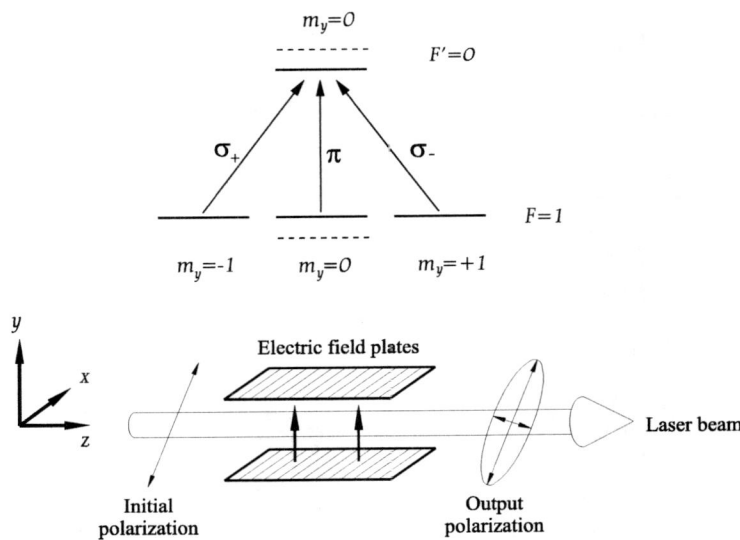

FIGURE 11. Schematic diagram of NEOE measurement for an $F = 1 \to F' = 0$ transition. In the presence of an electric field along the quantization axis ($\hat{y}$), the $m_y = 0$ Zeeman sublevels are shifted by the Stark effect (dashed lines). As a result there arises a difference in the resonance frequencies for y-polarized (π) and x-polarized light (which can be represented as a superposition of $\sigma_{\pm}$ light). As the light propagates through the atomic medium in the presence of the electric field, the initially linearly-polarized light develops ellipticity.

is $\approx 10^{-26}$ e·cm/$\sqrt{\text{Hz}}$ (for a 10 kV/cm electric field and taking into account the EDM enhancement factor of ≈ 120 for Cs [53, 54, 55, 56]). This statistical sensitivity should enable a nonlinear-optical-rotation-based EDM search to compete with the best present limits on d_e from measurements in Cs [1] and even in Tl [6].

However, in order to reach this projected sensitivity, there are several problems that must be overcome. The first is a change in the atomic density when electric fields are applied to the cell. The second is a coupling of the atomic polarizations of Cs and Rb via spin-exchange collisions, which would prevent Rb from functioning as an independent co-magnetometer. These effects are discussed in more detail in the following sections.

Application of electric fields to paraffin-coated cells and measurements of nonlinear electro-optical effects

The high-voltage electrode assembly used in our present measurements is shown in Fig. 10. The electrode assembly, containing a cylindrical, paraffin-coated Rb-Cs cell (diameter ≈ 5 cm, height ≈ 2 cm), is placed inside the 3D coils and magnetic shields shown in Fig. 7.

We have performed measurements of the electric field inside the cell using a nonlinear electro-optical effect induced by Stark shifts [57]. This effect can be understood by considering, e.g., an isolated $F = 1 \to F' = 0$ transition, where we choose the quantization axis along the $\hat{y}$ direction (Fig. 11). Light that is initially linearly polarized at 45° to the x-axis can be decomposed into x-polarized (a superposition of $\sigma_{\pm}$) and y-polarized (π) light. If a $\hat{y}$-directed electric field is applied to the sample, the quadratic Stark effect causes a relative shift in the resonance frequencies for x- and y-polarized light, resulting in a difference in the corresponding refractive indices. A linear electro-optical effect arises in this case since as near-resonant light propagates through the atomic medium, y-polarized light changes its phase relative to the x-polarized light. Thus the transmitted light acquires elliptical polarization due to linear birefringence.[1] Like in the case of NMOR, there arise nonlinear electro-optic effects due to the formation of Bennett structures in the atomic velocity distribution for atoms in particular ground state sublevels [17] and optically-pumped ground state coherences. These effects can considerably enhance the induced ellipticity with respect to the linear case.

To measure the nonlinear electro-optical effects, light resonant with the Rb D2 transition is directed through the cell perpendicular to the direction of the applied electric field. Before entering the cell, the light passes through a linear polarizer oriented at 45° to the direction of the applied electric field (Fig. 11). After the light passes through the cell, the degree of ellipticity ε is determined by a circular analyzer consisting of a quarter-wave plate with fast axis along y and a polarizing beamsplitter.

Figure 12 shows the spectrum of induced ellipticity for the ^{85}Rb D2 line when an electric field of ~ 5 kV/cm is applied to the cell. The light power is ≈ 3.5 μW and the laser beam diameter is ≈ 1 mm. The density of ^{85}Rb is $\approx 10^9$ atoms/cm^3, somewhat lower than the saturated density at room temperature (20°C) [59]. Note that the quadratic Stark shifts are larger in the upper state of the D2 transition, since the total electronic angular momentum J is $1/2$ for the lower state (so ground state Stark shifts arise only due to the hyperfine interaction). Density matrix calculations [16, 17] indicate that the measured ellipticity in Fig. 12 is primarily due to Bennett structures, and not coherence effects or the linear electro-optical effect.[2] These data show that atoms inside the vapor cell are experiencing close to the full 5 kV/cm electric field that is applied.

During the electric field measurements, we observed a dramatic change in the alkali vapor density when the electric field polarity was switched. Figure 13 shows the Cs vapor density and applied electric field as a function of time. If the electric field amplitude is changed without changing the polarity, there is no significant change in alkali vapor density. This effect is probably similar to the electric-field-related degradation of paraffin coated cells observed in the Seattle EDM experiments [49, 50, 51].[3] One characteristic

---

[1] The major axis of the ellipse is also rotated with respect to the initial light polarization due to linear dichroism. However, due to the different spectral lineshapes of rotation and ellipticity, Doppler broadening suppresses optical rotation relative to ellipticity by a factor of Γ_D/γ_0.

[2] The present density matrix calculation ignores the fact that the atoms can undergo collisions with the cell walls without depolarizing, which can affect the magnitude of the induced ellipticity by a factor estimated to be of order unity.

[3] It is interesting to note that similar behaviour has been observed for ^{199}Hg atoms cohabiting an ultracold neutron storage container (with quartz walls and teflon-coated electrodes) in the neutron EDM experiment at the Institut Laue-Langevin (ILL) [58]. In the ILL experiment, both the Hg density and atomic polarization relaxation rate changed when the polarity of the electric field was reversed.

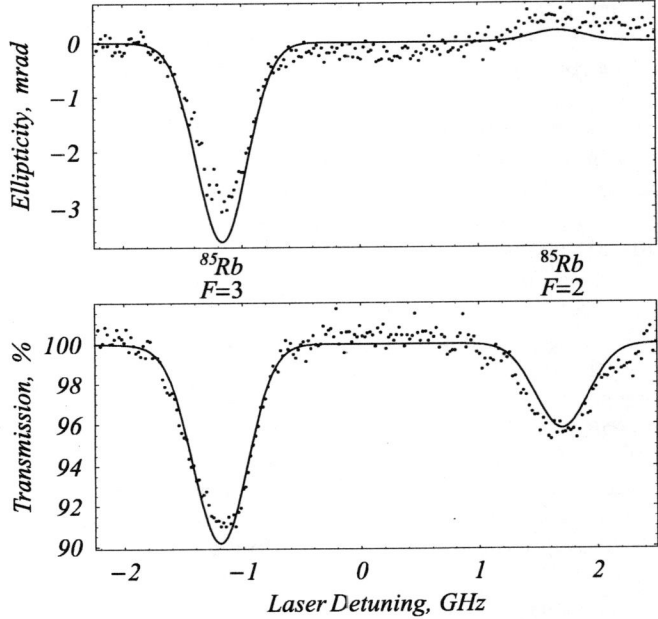

FIGURE 12. Upper plot compares a measurement of electric-field-induced ellipticity for the ^{85}Rb D2 line to a density matrix calculation. Lower plot shows the experimental and theoretical transmission spectra.

of this effect is the very slow recovery rate of the alkali vapor density after switching the electric field (~ 20 min). This time scale is similar to the recovery times observed in light-induced desorption measurements (Fig. 9) when the paraffin surface has been depleted of adsorbed alkali atoms. Also, we have found that the vapor density recovers more quickly when the stems (containing solid samples of Cs and Rb) are heated. These characteristics suggest that the electric field may be somehow depleting the paraffin of adsorbed atoms – thus changing the relative fluxes of atoms being adsorbed and desorbed from the paraffin coating. This effect is presently under investigation, and the results will be reported elsewhere.

NMOR with frequency-modulated light

An essential feature of the proposed EDM search using nonlinear optical rotation is the use of Rb as a "co-magnetometer," so in effect the experiment will measure a difference in the linear Stark shifts for Rb and Cs. In order for the Cs and Rb atoms to act independently, "locking" of the Cs polarization to the Rb polarization via spin-exchange collisions must be avoided.

The transfer of longitudinal (along the magnetic field direction) spin polarizations

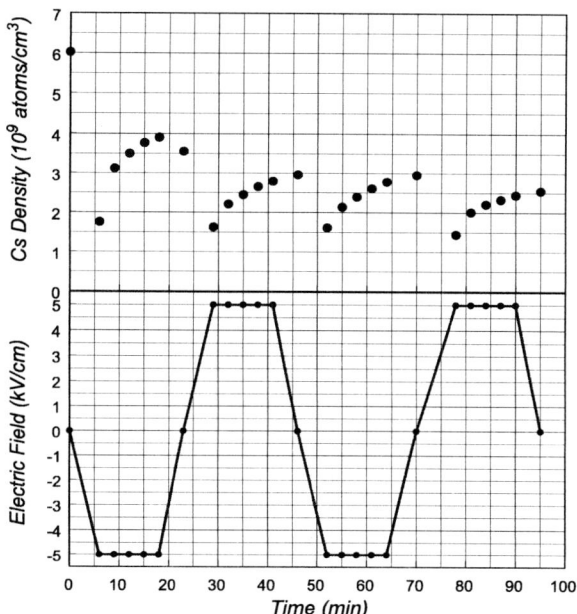

FIGURE 13. Upper plot shows Cs vapor density in a paraffin-coated cell as a function of time, lower plot shows applied electric field.

between species via spin-exchange collisions can be quite efficient: the Rb-Cs spin-exchange cross section is $\approx 2 \times 10^{-14}$ cm^2 [60]. Transverse spin polarization is more difficult to transfer because the transverse spin components of Cs and Rb dephase due to the difference between their Larmor frequencies [61, 62]. Nonetheless, at small magnetic fields (where the difference between the Cs and Rb Larmor frequencies is less than the relaxation rate of atomic polarization), efficient polarization transfer can occur. In the case where only transverse spin polarization is present, "locking" of the Cs polarization to the Rb polarization can be avoided by applying a bias magnetic field. However, using the NMOR techniques previously discussed, measurements can only be performed in small magnetic fields ($B_z < \gamma_{\text{rel}}/g_F \mu_B$).

Recently, we have developed a technique that can extend the dynamic range of an NMOR-based magnetometer to the Earth field range [63]. In this setup (Fig. 14), light polarization modulation (see Fig. 7) is replaced by frequency modulation of the laser, and the time-dependent optical rotation is measured at the first harmonic of the light modulation frequency Ω_m (FM NMOR). The frequency modulation affects both optical pumping and probing of atomic polarization, causing resonances to arise in the magnetic field dependence when Ω_m coincides with twice the Larmor frequency $2\Omega_L$. Additional resonances can be observed at higher harmonics, for example a resonance arises at the second harmonic of Ω_m when $\Omega_m = \Omega_L$.

The FM NMOR signals at the first harmonic of Ω_m as a function of longitudinal magnetic field are shown in Fig. 15. For these measurements, a spherical paraffin-coated

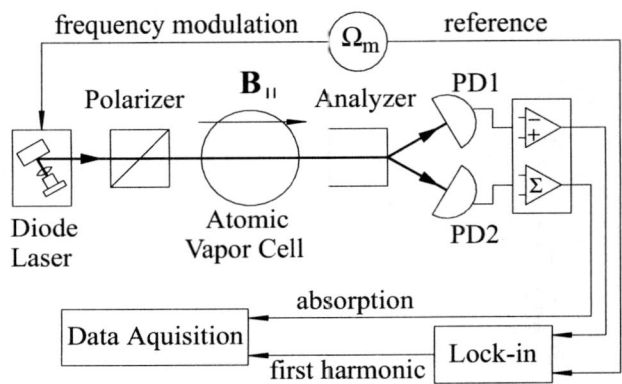

FIGURE 14. Schematic diagram for measurement of NMOR with frequency-modulated light. Paraffin-coated cell containing ^{87}Rb is placed between a polarizer and an analyzer oriented at $\approx 45°$ with respect to each other. For the present measurements, the laser frequency is modulated with a piezo actuator in the diode laser.

cell (diameter $\approx$ 10 cm) containing ^{87}Rb is employed. The laser frequency is modulated at $\Omega_m = 2\pi \times 1$ kHz with modulation amplitude $\Delta\omega = 2\pi \times 220$ MHz. The laser is tuned to the low-frequency slope of the $F = 2 \to F' = 1$ component of the D1 resonance where the signal at the first harmonic is maximal.

For low light powers, where optical pumping primarily produces atomic alignment (see Figs. 4 and 5), the origins of the FM NMOR resonances can be understood as follows. Resonances at the first harmonic of Ω_m occur at $\Omega_L = 0$ and when the modulation frequency of the pumping and probing light corresponds to $2\Omega_L$. As the laser frequency is modulated, the optical pumping rate changes depending on the instantaneous detuning of the laser from the atomic transition, given by $\omega(t) - \omega_0$ where $\omega(t) = \omega_l + \Delta\omega \cdot \sin(\Omega_m t)$ (ω_l is the central frequency of the laser). For a spectrally-isolated, Doppler-broadened transition, the time-dependent pumping (and probing) rates are

$$\propto e^{-\left(\frac{\omega(t)-\omega_0}{\Gamma_D}\right)^2}.$$

When the pumping rate is synchronized with the precession of atomic polarization, a resonance occurs and the atomic medium is pumped into an aligned state whose axis rotates at Ω_L. The optical properties of the medium are modulated at $2\Omega_L$, due to the symmetry of atomic alignment. Since the probability for probing is modulated, optical rotation due to the average axis of atomic alignment being at an angle to the light polarization (which has a resonant character) results in the zero-field resonance.

If alignment-to-orientation conversion occurs (see Fig. 6 and Ref. [29]), a longitudinal spin polarization arises, and spin-exchange collisions can then couple the Cs and Rb polarizations (as was observed for different ground state hyperfine levels of ^{85}Rb in Ref. [64]). This is significant, since in our studies of NMOR employing the light-polarization modulation technique, we found that the optimum sensitivity occurred for

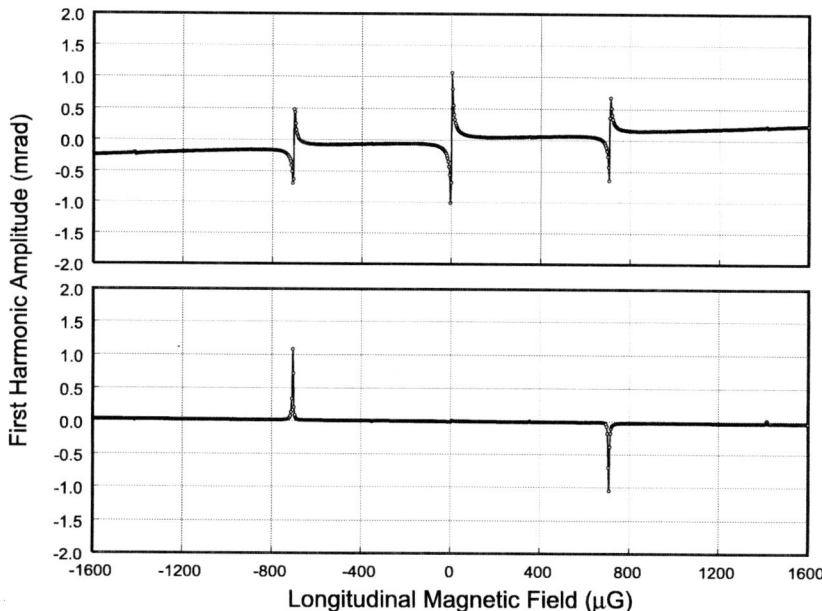

FIGURE 15. FM NMOR signals detected at the first harmonic of Ω_m as a function of longitudinal magnetic field. The laser power was 15 μW, beam diameter $\sim$ 2 mm, $\Omega_m = 2\pi \times 1$ kHz, $\Delta\omega = 2\pi \times 220$ MHz. Upper plot is the in-phase component (in-phase with the zero-field resonance) and the lower trace is quadrature component from the lock-in output. All resonances have widths corresponding to the slow rate of atomic polarization relaxation in the paraffin-coated cell.

conditions where AOC played an important role [26, 29]. We are presently investigating the role of AOC in FM NMOR.

ATOMS IN CELLS WITH COLD BUFFER GAS

As mentioned above, at the core of the considered experiments is the ability to obtain narrow (~ 1 Hz) atomic resonances due to the long lifetime of the ground state atomic polarization in an anti-relaxation coated cell [20, 24, 26]. An alternative method of linewidth reduction uses a buffer gas to decrease the rate of depolarizing collisions of atoms with cell walls (see, e.g. [65, 66, 67, 68] and references therein). Recently, with neon as the buffer gas, coherent dark resonances with linewidths as narrow as 42 Hz in cesium [69] and 30 Hz in rubidium [70] were observed in vapor cells at room temperature.

In order to obtain ultra-narrow resonances with sub-Hz width, we are exploring an extension of a buffer-gas method which relies on the properties of atomic scattering at low (cryogenic) temperatures, approaching the S-wave scattering regime, where spin relaxation should be suppressed. Theoretical estimates done in our group suggest that for the Cs-He case, a significant reduction of the relaxation cross-section, by a factor of $\sim 20-50$, may be expected already at liquid-Nitrogen temperatures.[4] These estimates were confirmed by Prof. Thad Walker of the University of Wisconsin, Madison [71]. Recently, it has been shown experimentally [45] that at temperatures below approximately 2 K, the spin relaxation cross-section of rubidium atoms in collisions with helium buffer gas atoms decrease by orders of magnitude in comparison with their room temperature values. It can be inferred from the data presented in Ref. [45] that relaxation times of minutes (corresponding to resonance widths on the order of 10 mHz) or even longer can be obtained.

The crucial experimental challenge is creating, in the cold buffer gas, atomic vapor densities comparable to those in our current room temperature experiments, i.e. of the order of $10^{10}-10^{12}$ atoms/cm^3. A group in Japan [45] relies on light-induced desorption of alkali atoms from the surface of the liquid He film inside their cell (presumably, this is similar to the effect observed in paraffin coated cells [36]). They can successfully inject Rb atoms into the He gas by irradiating the cell with about 200 mW of Ti:sapphire laser radiation (750 nm) for 10 s. But it turns out that the injection efficiency decreases with the repetitions of the injection cycle. It was found that the efficiency recovered by heating the cell to room temperature and then cooling again. They also report [72] that with their method they cannot inject Cs atoms, which are of particular interest among the alkali atoms for fundamental symmetry tests.

Our method for injecting atoms into the buffer gas differs significantly from that of [45, 72]. We plan to use laser evaporation of micron-sized droplets inside the cold He gas. The central elements of our experimental apparatus consist of: a small cryostat with optical access to the atomic sample (temperature range down to 1.4 K); a droplet generator - a system for injecting micron-size metal droplets into the cold He gas; and a system of lasers for evaporation of these droplets and for optical pumping and probing of the spin relaxation.

The design of the droplet generator is motivated by the "drop-on-demand" liquid micro-droplet generator concept [73] developed by the SLAC group of Prof. Martin Perl for their experiment searching for elementary particles with fractional electric charge [74, 75, 76, 77, 78]. However, the implementation of this method is restricted to elements whose melting point lies below the operational temperature of the dropper's piezo-ceramic transducer (the Curie temperature for some special piezo-ceramics can reach $\sim 360°C$). Fortunately, cesium (melting point of 29°C [79]) is one such element. We have recently proposed the extension of the method to the elements for which fine powders are available (such as Ag and Au). The idea is to load the dropper with powder and produce powder particles with the similar way as liquid drops. Preliminary experiments with a SLAC glass dropper (100 μm orifice diameter) and spherical silver powder with maximum size of ~ 20 μm (635 mesh) have shown generation of the powder clots with rather stable size (comparable with the dropper orifice diameter [80]).

Once the drop has reached the interaction region, it will be heated and vaporized by a sequence of two pulses from a 1-micron wavelength YAG laser. Estimates show that a

---

[4] Our estimates of low temperature dependence of Cs electron spin relaxation cross-section due to the spin-rotation interaction with He atoms are based on the approach developed in Ref. [67].

500 mJ, 10 ns (2 mm beam diameter) initial pulse will cause the droplet to explode into a collection of small clusters [81], and a subsequent pulse will evaporate these clusters, creating a dense atomic vapor in the He buffer gas. Large residual drops will quickly fall from the interaction region.

In the first stage of the experiment, we plan to investigate spin-relaxation processes of polarized silver atoms in He buffer gas using the relaxation in the dark method [37]. The injected atoms will be optically pumped with a silver hollow cathode lamp into a spin-polarized state, and the atomic vapor polarization will decay in the holding magnetic field created by coils outside the cryostat. After a period of evolution in the dark, the polarization will be optically probed, from which we will deduce the spin-relaxation lifetime and cross-section.

If the Ag-He spin relaxation cross section is as small as expected at low temperatures, then it will be possible to obtain spin-polarized atomic vapors with lifetimes of the order of minutes. The corresponding resonance widths will thus be on the order of mHz. Such narrow resonances may be applied to atomic tests of discrete symmetries, such as a search for the electron EDM. (Silver atoms have an EDM enhancement factor of ~ 50 [82], which is only a factor of two smaller than that for Cs ~ 120 [53, 54, 55, 56]. For gold atoms, the enhancement factor is even larger: ~ 250 [83, 84]). Narrow resonances may also be used for very sensitive measurements of magnetic fields. This method is being considered for detection of the electric-field-induced P,T-odd magnetization of a bulk ferro-magnetic material at low, ~ 50 mK, temperatures [85]. (An experiment of this type was first proposed by F. L. Shapiro in 1968 [86] and carried out by B. V. Vasiliev and E. V. Kolycheva in the end of the 1970's [87].)

TESTING PARITY CONSERVATION AND TIME-REVERSAL-INVARIANCE OF GRAVITY

Finally we would like to mention that nonlinear optical rotation can be used for other types of fundamental symmetry tests. For example, one idea that has been the subject of a few lunch-time conversations with Professor Commins is an experimental test of the symmetry properties of gravity. Gravity is by far the weakest and, it could be argued, least understood of the known fundamental interactions. Many authors have considered the possibility that gravity may violate parity and time-reversal invariance [88, 89, 90, 91, 92, 93]. In the nonrelativistic limit, the simplest Hamiltonian H_G describing such an interaction is given by

$$H_G = k\frac{\hbar}{c}\vec{g}\cdot\vec{S}, \tag{11}$$

where k is a dimensionless constant specifying the strength of the new interaction, $\vec{g}$ is the gravitational field due to, e.g., the earth, and $\vec{S}$ is the total spin of the system. The energy shift due to the interaction of an atom with the Earth is $\approx k \cdot (2 \times 10^{-23}$ eV), corresponding to a frequency shift of $\approx k \cdot (4 \times 10^{-9}$ Hz).

Presently, the most sensitive search for a long-range P- and T-violating gravitational interaction was performed by measuring the nuclear spin-precession frequencies of ^{199}Hg and ^{201}Hg as the direction of an applied magnetic field was reversed with respect to the earth's gravitational field [94] – this approach is similar to that employed in the Hg EDM experiments (see Ref. [95] and Refs. therein), except that the electric field is replaced by the earth's gravitational field. This experiment set a limit of $k < 70$.

An experiment employing nonlinear optical rotation appears capable of probing the particularly interesting region of $k \lesssim 1$. Consider, for example, an experiment measuring the precession frequencies of Cs atoms in the $F = 3$ and $F = 4$ ground state hyperfine levels,[5] for which $g_F \approx -1/4$ and $+1/4$, respectively. (Note that in such a measurement, corrections to the Landé factors due to the presence of a nonzero magnetic field must be taken into account.) The precession frequencies can be measured for different orientations of the magnetic field with respect to $\vec{g}$ using, e.g., the FM NMOR technique. The precession frequencies $\Omega_p(\pm)$ for the two ground state hyperfine levels in the presence of a magnetic field $\vec{B}$ along the direction of $\vec{g}$ are given by

$$\Omega_p(\pm) = \frac{k\hbar g}{c} \pm |g_F|\mu_B B. \tag{12}$$

We can then take the ratio of the two precession frequencies:

$$\frac{\Omega_p(+)}{\Omega_p(-)} \approx 2\frac{k\hbar g}{c|g_F|\mu_B B} - 1, \tag{13}$$

which constitutes a measurement of k.

The sensitivity of NMOR in paraffin coated cells ($\sim 10^{-6}$ Hz/$\sqrt{\text{Hz}}$), and, potentially, in cells filled with cryogenic buffer gas, offer a possibility of probing $k \lesssim 1$.

CONCLUSION

In conclusion, we have discussed nonlinear optical rotation of linearly polarized light related to the evolution of light-induced atomic polarization in magnetic and electric fields. In certain systems, such as atomic vapor cells coated with paraffin or filled with cold buffer gas, the relaxation rate of atomic polarization can be made very slow ($\lesssim 1$ Hz). Using these techniques, the sensitivity to Zeeman sublevel shifts is $\lesssim 10^{-6}$ Hz/$\sqrt{\text{Hz}}$, permitting a variety of atomic tests of fundamental symmetries. Hopefully, we can follow in Professor Commins's footsteps and bring these ideas to fruition!

ACKNOWLEDGMENTS

We would like to sincerely thank our collaborators in St. Petersburg, Russia, E. B. Alexandrov and M. V. Balabas, for providing us with the paraffin-coated vapor cells; L. Zimmerman and K. Kerner for assistance with some of the described experiments; D. DeMille, L. Hunter, and M. Romalis for fruitful discussions; A. Vaynberg, S. Bonilla, S. Butler, M. Solarz, and G. Weber for excellent machining of various

---

[5] For the ground states of alkali atoms $\vec{F} = \vec{S}$.

parts of the apparatus; and J. R. Davis for invaluable assistance with the electronics. We would like to thank Professor Commins for inspiration and many useful discussions.

REFERENCES

1. S. A. Murthy, D. Krause, Jr., Z. L. Li, and L. R. Hunter, Phys. Rev. Lett. **63**, 965 (1989).
2. K. Abdullah, C. Carlberg, E. D. Commins, H. Gould, and S. B. Ross, Phys. Rev. Lett. **65**, 2347 (1990).
3. C. Chin, V. Leiber, V. Vuletić, A. J. Kerman, and S. Chu, Phys. Rev. A **63**, 033401 (2001).
4. D. DeMille, F. Bay, S. Bickman, D. Kawall, D. Krause, Jr., S. E. Maxwell, and L. R. Hunter, Phys. Rev. A **61**, 052507 (2000); D. DeMille, in these Proceedings.
5. E. D. Commins, S. B. Ross, D. DeMille, and B. C. Regan, Phys. Rev. A **50**, 2960 (1994).
6. B. C. Regan, E. D. Commins, C. J. Schmidt, and D. DeMille, "New limit on the electron electric dipole moment," (to be published).
7. M. Faraday, *Experimental Research* **III** (London), 2164 (1855).
8. D. Macaluso and O. M. Corbino, Nuovo Cimento **8**, 257 (1898).
9. N.B. Baranova, Yu. V. Bogdanov, and B. Ya. Zel'dovich, Usp. Fiz. Nauk. **122-123**, 349 (1977) [Sov. Phys. Usp. **20**, 870 (1977)].
10. O.P. Sushkov and V.V. Flambaum, Zh. Eksp. Teor. Fiz. **75**, 1208 (1978) [Sov. JETP **48**, 608 (1978)].
11. Note, however, that with squeezed states of light the sensitivity to polarization rotation can in principle surpass the shot-noise-limit: see, e.g., P. Grangier, R.E. Slusher, B. Yurke and A. LaPorta, Phys. Rev. Lett. **59**, 2153 (1987).
12. W. Gawlik, in *Modern Nonlinear Optics* edited by M. Evans and S. Kielich, Advances in Chemical Physics Series, Vol. **LXXXV**, Pt. 3 (Wiley, New York, 1994).
13. D. Budker, D. J. Orlando, and V. Yashchuk, Am. J. Phys. **67**, (1999); D. Budker, V. Yashchuk, and M. Zolotorev, Sib. J. Phys. **1**, 27 (1999).
14. W.R. Bennett, Phys. Rev. **126**, 580 (1962).
15. L. M. Barkov, D. Melik-Pashayev, and M. Zolotorev, Opt. Commun. **70**, 467 (1989).
16. A detailed description of this density matrix calculation can found in S.M. Rochester, D.S. Hsiung, D. Budker, R.Y. Chiao, D.F. Kimball, and V.V. Yashchuk, Phys. Rev. A **63**, 043814 (2001).
17. D. Budker, D. F. Kimball, S. M. Rochester, and V. V. Yashchuk, "Nonlinear electro- and magneto-optic effects related to Bennett structures," (to be published).
18. S. I. Kanorsky, A. Weis, J. Wurster, and T. W. Hänsch, Phys. Rev. A **47**, 1220 (1993).
19. A. Weis, J. Wurster, and S. I. Kanorsky, J. Opt. Soc. Am. B **10**, 716 (1993).
20. D. Budker, V. Yashchuk, and M. Zolotorev, Phys. Rev. Lett. **81**, 5788 (1998).
21. H.G. Robinson, E.S. Ensberg, and H.G. Dehmelt, Bull. Am. Phys. Soc. **3**, 9 (1958).
22. M. A. Bouchiat and J. Brossel, Phys. Rev. **147**, 41 (1966).
23. E. B. Alexandrov and V. A. Bonch-Bruevich, Opt. Engin. **31**, 711 (1992).
24. V. Yashchuk, D. Budker and M. Zolotorev, in *Trapped Charged Particles and Fundamental Physics*, edited by D.H.E. Dubin and D. Schneider (American Institute of Physics, New York, 1999), pp. 177-181.
25. S. M. Rochester and D. Budker, Am. J. Phys. **69**, 450 (2001).
26. D. Budker, D. F. Kimball, S. M. Rochester, V. V. Yashchuk, and M. Zolotorev, Phys. Rev. A **62**, 043403 (2000).
27. E.B. Alexandrov, M.V. Balabas, A.S. Pasgalev, A.K. Vershovskii, and N.N. Yakobson, Laser Phys. **6**, 244 (1996).
28. J. Clarke in *The New Superconducting Electronics*, edited by H. Weinstock and R.W. Ralston, pp. 123-180 (Kluwer Academic, The Netherlands, 1993).
29. D. Budker, D. F. Kimball, S. M. Rochester, and V. V. Yashchuk, Phys. Rev. Lett. **85**, 2088 (2000).
30. M. Lombardi, J. Physique **30**, 631 (1969).
31. C. Cohen-Tannoudji and J. Dupont-Roc, Opt. Commun. **1**, 184 (1969).
32. M. Pinard and C.G. Aminoff, J. Physique **43**, 1327 (1982).
33. R.C. Hilborn, L.R. Hunter, K. Johnson, S.K. Peck, A. Spencer, and J. Watson, Phys. Rev. A **50**, 2467 (1994).

34. R.C. Hilborn, Am. J. Phys. **63**, 330 (1995).
35. M.V. Balabas, V.A. Bonch-Bruevich, and S.V. Provotorov, in *Proceedings of the First All-Union Seminar on Quantum Magnetometers* (S.I. Vavilov State Univ. Press, Leningrad, 1988), pp. 55-56.
36. E. B. Alexandrov, M. V. Balabas, D. Budker, D. S. English, D. F. Kimball, C.-H. Li, and V. V. Yashchuk, "Light-induced desorption of alkali atoms from paraffin coating," (to be published).
37. W. Franzen, Phys. Rev. **115**, 850 (1959).
38. G.N. Birich, Yu.V. Bogdanov, S.I. Kanorskii, I.I. Sobelman, V.N. Sorokin, I.I. Struk, and E.A. Yukov, J. of Russian Laser Research, **15**, 455 (1994).
39. I. N. Abramova, E. B. Aleksandrov, A. M. Bonch-Bruevich, and V. V. Khromov, Pis'ma Zh. Eskp. Teor. Fiz. **39**, 172 (1984) [JETP Lett. **39**, 203 (1984)].
40. A. Gozzini, F. Mango, J. H. Xu, G. Alzetta, F. Maccarrone, and R. A. Bernheim, Nuovo Cimento D **15**, 709 (1993).
41. M. Meucci, E. Mariotti, P. Bicchi, C. Marinelli, and L. Moi, Europhys. Lett. **25**, 639 (1994).
42. E. Mariotti, S. Atutov, M. Meucci, P. Bicchi, C. Marinelli, and L. Moi, Chem. Phys. **187**, 111 (1994).
43. J. H. Xu, A. Gozzini, F. Mango, G. Alzetta, and R. A. Bernheim, Phys. Rev. A **54**, 3146 (1996).
44. S. Atutov, V. Biancalana, P. Bicchi, C. Marinelli, E. Mariotti, M. Meucci, A. Nagel, K. A. Nasyrov, S. Rachini, and L. Moi, Phys. Rev. A **60**, 4693 (1999).
45. A. Hatakeyama, K. Oe, K. Ota, S. Hara, J. Arai, T. Yabuzaki, and A.R. Young, Phys. Rev. Lett. **84**, 1407 (2000).
46. I. B. Khriplovich and S. K. Lamoreaux, *CP Violation Without Strangeness: Electric Dipole Moments of Particles, Atoms, and Molecules* (Springer-Verlag, Berlin, 1997).
47. E. N. Fortson, in these Proccedings.
48. L. M. Barkov, M. S. Zolotorev and D. Melik-Pashayev, Sov. JETP Pis'ma **48**, 144 (1988).
49. E. S. Ensberg, Bull. Am. Phys. Soc. **7**, 534 (1962).
50. E. S. Ensberg, Phys. Rev. **153**, 36 (1967).
51. P. A. Ekstrom, Ph.D. Thesis (University of Washington, Seattle, 1971).
52. H. G. Dehmelt, Phys. Rev. **105**, 1924 (1957).
53. P.G.H. Sandars, Phys. Lett. **22**, 290 (1966).
54. W.R. Johnson, D.S. Guo, M. Idrees, and J. Sapirstein, Phys. Rev. A **34**, 1043 (1986).
55. B.P. Das, *Recent Advances in Many Body Theory* (Springer-Verlag, New York, 1988).
56. A. C. Hartley, E. Lindroth, and A.-M. Martensson-Pendrill, J. Phys. B **23**, 3417 (1990).
57. D. Budker, D. F. Kimball, S. M. Rochester, and V. V. Yashchuk, LBNL Preprint PUB5453 (Berkeley, CA, 1999).
58. K. Green, P. G. Harris, P. Iaydjiev, D. J. R. May, J. M. Pendlebury, K. F. Smith, M. van der Grinten, P. Geltenbort, and S. Ivanov, Nucl. Instr. and Meth. A **404**, 381 (1998).
59. I. S. Girgoriev and E. Z. Meilikhov, eds., *Handbook of Physical Quantities* (CRC Press, Boca Raton, FL, 1997).
60. H. G. Gibbs and R. J. Hull, Phys. Rev. **153**, 132 (1967).
61. S. Haroche and C. Cohen-Tannoudji, Phys. Rev. Lett. **24**, 974 (1970).
62. H. Ito, T. Ito, and T. Yabuzaki, J. Phys. Soc. Jpn. **63**, 1337 (1994); T. Ito, N. Shimomura, and T. Yabuzaki, J. Phys. Soc. Jpn. **64**, 2848 (1995).
63. D. Budker, D.F. Kimball, V.V. Yashchuk, and M. Zolotorev, "Nonlinear magneto-optical rotation with frequency-modulated light," (to be published).
64. V. V. Yashchuk, E. Mikhailov, I. Novikova, and D. Budker, "Nonlinear magneto-optical rotation with separated light fields in ^{85}Rb vapor contained in an antirelaxation-coated cell," Preprint LBNL-44762 (1999).
65. W. Happer, Rev. Mod. Phys. **44**, 169 (1972).
66. J. Vanier and C. Audoin, *The Quantum Physics of Atomic Frequency Standards, vol. 1* (Adam Hilger, Bristol and Filadelfia, 1989).
67. T. G. Walker, J. H. Thywissen, and W. Happer, Phys. Rev. A **56**, 2090 (1997).
68. D. K. Walter, W. Happer, and T. G. Walker, Phys. Rev. A **58**, 3642 (1998).
69. S. Brandt, A. Nagel, R. Wynands, and D. Meschede, Phys. Rev. A **56**, R1063 (1997).
70. M. Erhard, S. Numann, and H. Helm, Phys. Rev. A **62**, 061802(R) (2000).
71. T. G. Walker, private communication, March, 2001.
72. A. Hatakeyama, K. Enomoto, and T. Yabuzaki, in *ICAP 2000 (XVII International Conference on Atomic Physics, Firenze, Italy, June 4-9, 2000), Conference Abstracts*, (Universita di Firenze, 2000),

p.557.
73. D. Loomba, V. Halyo, E.R. Lee, I.T. Lee, P. C. Kim, and M. L. Perl, Rev. Scient. Instrum. **71**, 3409 (2000).
74. N. M. Mar, E. R. Lee, G. R. Fleming, B. C. K. Casey, M. L. Perl, E. L. Garwin, C. D. Hendricks, K. S. Lackner, and G. L. Shaw, Phys. Rev. D **53**, 6017 (1996).
75. M. L. Perl, E. R. Lee, Am. J. Phys. **65**, 698 (1997).
76. V. Halyo, P. C. Kim, E.R. Lee, I.T. Lee, D. Loomba, and M. L. Perl, Phys. Rev. Lett. **84**, 2576 (2000).
77. M. L. Perl, P. C. Kim, V. Halyo, E.R. Lee, I.T. Lee, D. Loomba, and K. S. Lackner, Int. J. of Modern Physics A **16**, 2137 (2001).
78. M. L. Perl, in these Proceedings.
79. C. C. Addison, *The chemistry of the liquid alkali metals* (John Wiley & Sons, New York, 1984).
80. V. V. Yashchuk, A. O. Sushkov, D. Budker, E.R. Lee, I.T. Lee, and M. L. Perl, to be published.
81. R. L. Armstrong, in *Optical effects associated with small particles. Edited by P.W.Barber, R.K.Chang.* (World Scientific, Singapore, 1988), p.p. 200-275.
82. Our estimate is based on the expression for the EDM enhancement factor given in I. B. Khriplovich, *Parity Nonconservation in Atomic Phenomena* (Gordon and Breach, Philadelphia, 1991).
83. W. R. Johnson, D. S. Guo, M. Idrees, J. Sapirstein, Phys. Rev. A **34**, 1043 (1986).
84. T. M. R. Byrnes, V. A. Dzuba, V. V. Flambaum, and D. W. Murray, Phys. Rev. A **59**, 3082 (1999).
85. S. K. Lamoreaux, private communication, July, 2001.
86. F. L. Shapiro, Usp. Fiz. Nauk **95**, 145 (1968) [Sov. Phys. Uspekhi **11**, 345 (1968)].
87. B. V. Vasiliev and E. V. Kolycheva, Zh. Eskp. Teor. Fiz. **74**, 466 (1978) [Sov. Phys. JETP **74**, 466 (1978)].
88. I. Y. Kobzarev and L. B. Okun, Zh. Eskp. Teor. Fiz. **43**, 1904 (1962) [Sov. Phys. JETP **16**, 1343 (1963)].
89. T. A. Morgan and A. Peres, Phys. Rev. Lett. **9**, 79 (1962).
90. J. Leitner and S. Okubo, Phys. Rev. **136**, B 1542 (1964).
91. N.D. Hari Dass, Phys. Rev. Lett. **36**, 393 (1976).
92. W. T. Ni, Phys. Rev. Lett. **38**, 301 (1977).
93. Y. N. Obukhov, Phys. Rev. Lett. **86**, 192 (2001).
94. B. J. Venema, P. K. Majumder, S. K. Lamoreaux, B. R. Heckel, and E. N. Fortson, Phys. Rev. Lett. **68**, 135 (1992).
95. M. V. Romalis, W. C. Griffith, J. P. Jacobs, and E. N. Fortson, Phys. Rev. Lett. **86**, 2505 (2001); M. V. Romalis, in these Proceedings.

Atomic Tests of Discrete Symmetries at Berkeley

D. S. English[1], D. F. Kimball[1], C.-H. Li[1], A.-T. Nguyen[1], S. M. Rochester[1], J. E. Stalnaker[a,b], V. V. Yashchuk[1], D. Budker[1,2], S. J. Freedman[a,b], and M. Zolotorev[3]

[1] *Department of Physics, University of California at Berkeley, Berkeley, California 94720-7300*
[2] *Nuclear Science Division, Lawrence Berkeley National Laboratory, Berkeley, California 94720*
[3] *Center for Beam Physics, Lawrence Berkeley National Laboratory, Berkeley, California 94720*

Abstract. Recent and ongoing experiments testing various fundamental discrete symmetries are discussed, including search for parity nonconservation in dysprosium and ytterbium, investigation of possibilities of searches for parity and time-reversal invariance violation in samarium, and a test of permutation properties of photons in a two-photon transition in barium.

EUGENE COMMINS AND ATOMIC TESTS OF DISCRETE SYMMETRIES

Professor Eugene Commins is a pioneer in a very tough business – testing fundamental symmetries of Nature with table-top atomic physics. A typical experiment of this kind takes anywhere from 8 to 15 years (and in some cases, even longer), which is a time scale not particularly well matched to that of funding agencies, or the time expected for a graduate student to complete his or her thesis. Unfortunately, not many experiments discover something unexpected, or even set limits for "new physics" at a desired level. Nevertheless, when such a result is eventually achieved, it is often of a scientific value that is hard to overestimate.

Professor Commins has succeeded in bringing several experiments of this kind to fruition, starting from the first measurement of atomic parity violation in a highly forbidden atomic transition [1, 2, 3], and culminating in the most stringent limit on the P,T-violating dipole moment of the electron (discussed in B. C. Regan's contribution elsewhere in these Proceedings).

In this paper, we review some of the recent and ongoing atomic tests of fundamental symmetries carried out at Berkeley by Eugene's colleagues, former students and "grand-students." Eugene has participated in much of this work either directly, or as an unlimited source of practical advice and theoretical expertise.

P- AND P,T-VIOLATION IN HEAVY ATOMS WITH CLOSELY SPACED OPPOSITE PARITY LEVELS

P- and P,T-violating interactions mix atomic levels of opposite parity. According to perturbation theory, the mixing is proportional to the matrix element of the symmetry-violating interaction, and inversely proportional to the difference in the energies of the two states (the energy denominator). This suggests an enhancement mechanism – the use of opposite-parity states that are nearly degenerate. Thus, many experimental attempts and proposals of atomic symmetry tests have concentrated on hydrogen (e.g. [4] and references therein) and hydrogenic ions (e.g. [5] and references therein).

Another enhancement, first pointed out by P. G. H. Sandars for P,T-odd interactions back in 1960's, and by M.-A. Bouchiat and C. Bouchiat for P-odd interactions in the 1970's, is present in heavy atoms. In fact, for neutral atoms, both the atomic electric-dipole moment (EDM) due to an EDM of the electron, and the dominant (nuclear-spin-independent) part of the parity-nonconserving (PNC) interaction scale approximately like Z^3, where Z is the atomic number.

Since mid-1980's [6, 7] there has been considerable interest in studying P- and P,T-odd effects in the rare-earth atoms, which seemed to have offered the best of both worlds – large $Z \sim 60 - 70$, and many cases of small spacing between opposite-parity states. Eugene Commins became interested in experimental work that was carried with samarium at Novosibirsk [8] during his visit to Siberia in the Summer of 1987, where he attended the Vavilov conference [3]. This is how this research eventually got transplanted from Novosibirsk to Berkeley.

PARITY NONCONSERVATION IN DYSPROSIUM

The idea of an enhancement due to closely spaced, opposite-parity levels is most compelling in the case of dysprosium (Dy; $Z = 66$). Here there are two levels whose energy separation is on the order of hyperfine-structure splittings and isotope shifts. These two states of even and odd parity (and designated as A and B, respectively) both have J= 10 and lie 19797.96 cm^{-1} above the J= 8 ground state (Fig. 1). Spectroscopic properties of Dy, and particularly of these states were studied in Refs. [9, 10, 11], including lifetimes which were found to be 7.9 μs for A and $> 200 \mu$s for B.

The smallness of the level separation grants one the opportunity (as well as dictates the necessity) of performing an entirely different kind of PNC measurement as compared to the traditional optical-rotation and PNC-Stark-interference-induced-dichroism PNC experiments. We apply a relatively weak magnetic field (~ 1.4 Gs), which brings energy separation between certain Zeeman sublevels of A and B close to zero, and observe quantum beats due to the presence of an external electric field of a few V/cm. Because the PNC effect is T-even, the electric field has to be time varying (for co-linear electric and magnetic fields) in order for there to be Stark-PNC interference [12]. For an oscillating electric field, the interference term in transition probability contributes at the same frequency as the the electric-field frequency (whereas the dominant signal due to the Stark-induced mixing oscillates at twice the frequency). Another PNC signature is

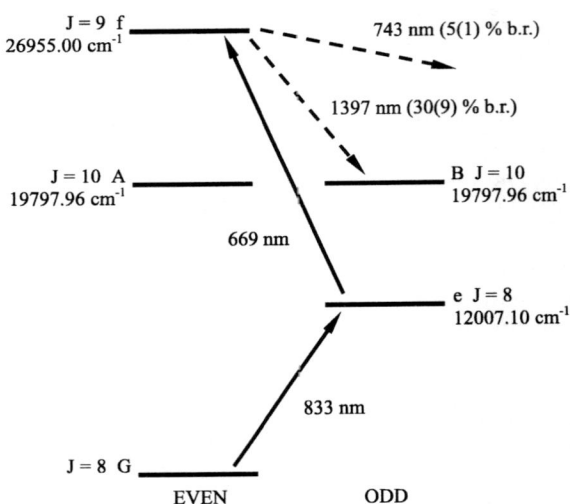

FIGURE 1. Partial level diagram of Dy showing the transitions in the current population scheme. Solid arrows indicate excitation; dashed arrows indicate spontaneous decay.

that it reverses with the sign of the residual energy separation (decrossing) as described by the P-odd, T-even rotational invariant:

$$\frac{d\mathbf{E}}{dt} \cdot (\mathbf{B} - \mathbf{B_c}), \qquad (1)$$

where $\mathbf{B_c}$ is the magnetic field required to cross the levels.

Our most recent PNC search [12] was originally motivated by theoretical estimates [13] predicting a substantial enhancement due to the small energy separation of the opposite-parity levels. These estimates had large uncertainties due to extreme complexity of the structure of the atomic states involved. Unfortunately, despite the fact that the experimental sensitivity to the PNC matrix element (H_w) considerably exceeds other PNC experiments performed so far (see, e.g., Ref. [14] for a review), no PNC effect was detected, and an upper limit of $|H_w| < 5$ Hz (68% C.L.) was established. This statistics-limited experiment used pulsed lasers with repetition rate of 10 Hz, which led to a low effective duty cycle ($\sim 10^{-4}$). Using cw lasers, we have developed an efficient population method of the nearly degenerate states [15] which will significantly improve the sensitivity of the PNC measurement (see below).

The basic setup for the current PNC measurement is depicted in Fig. 2. Atoms emerge from an effusive oven source operating at $T \sim 1500$ K and pass through several collimators. Then, they enter the electric- and magnetic-field interaction region where they are excited by laser beams and end up populating the odd-parity state B (the role of the cylindrical lenses will be described below). A sinusoidally varying electric field is applied with two grids consisting of $\sim 5 \times 10^{-3}$ cm diam. Be-Cu wire. Wire grids were chosen instead of plates in order to minimize stray surface charge. The magnetic field

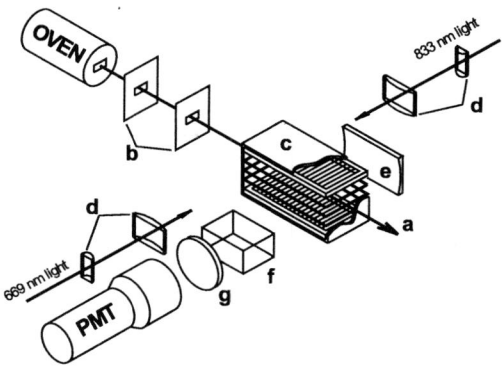

FIGURE 2. Current Dy PNC setup: a) atomic beam produced by effusive oven source at T= 1500 K; b) atomic beam collimators; c) interaction region of atoms with E-field ($\sim$ 4 V/cm) produced between wire grids and B-field ($\sim$ 1.4 Gs) produced by wire turns; mu-metal yoke provides high homogeneity; entire region is enclosed in a magnetic shield (not shown); d) cylindrical lens to diverge laser beams; e) mirror; f) light pipe; and g) interference filter.

is produced by wires forming a rectangular solenoid whose magnetic flux is "shorted" by a CO-NETIC yoke. The mirror currents due to the yoke add to the field produced by the coil, and the entire configuration leads to a rather homogeneous magnetic field in the interaction region similar to that of an infinitely long solenoid. The electric- and magnetic-field homogeneity is $\sim 10^{-3}$ within the volume of ~ 100 cm^3 where atoms interact with the laser beams, experience quantum beats and fluoresce. The interaction region is enclosed by a single layer of magnetic shielding (not shown). Fluorescence is directed by a light pipe onto a photomultiplier tube.

In the current population scheme, three transitions are required to reach the longer-lived, odd-parity state B (Fig. 1). The atoms are first excited by 833-nm and then by 669-nm light. The final step involves spontaneous decay at 1397 nm, with a measured branching ratio of 0.30(9) [16]. In order to populate a large fraction of the weakly collimated atomic beam, the laser frequency is effectively broadened by using a cylindrical lens to diverge the laser beam. For sufficiently large light intensities, this leads to an efficient and robust population inversion analogous to adiabatic passage used in magnetic resonance [17]: as the atoms pass through the beam, they experience a sweep in light frequency due to the Doppler effect that is slow compared to Rabi oscillations. In our case, the adiabatic criterion requires that the Rabi frequency be much larger than the transverse Doppler width [15].

With this scheme, we have achieved population transfer over a large fraction of the transverse-atomic-velocity distribution. For the 833-nm transition, a substantial portion of the transverse-velocity distribution underwent adiabatic passage. Although the second transition at 669 nm did not exhibit adiabatic passage, the transfer efficiency into state B is nevertheless large due the short lifetime of state f (Fig. 1) [15]. Further improvements are possible by increasing the power of the lasers used and by utilizing a laser at 1397 nm to stimulate the f $\rightarrow$ B transition. The achieved population efficiency translates into

$\sim 10^4$ times higher counting rate compared to the pulsed PNC experiment[12]. With a similar technique and a total integration time of 20 hours, this should allow us to reach a statistical sensitivity to the weak matrix element of ~ 10 mHz.

The future of dysprosium as a laboratory for PNC studies depends crucially on the results of the current phase of the experiment. If the effect is "around the corner," (i.e. $|H_w| \sim 1$ Hz) Dy can still contribute to the study of both nuclear-spin-independent effects (via isotopic comparisons of the PNC effect), and to the study of the nuclear-spin-dependent PNC (via comparison of the effect on different hyperfine transitions). If the effect is suppressed even more strongly, it appears that Dy would not have sufficient advantages over other systems for PNC studies. However, the unique situation in Dy could also be applied in other studies that benefit from near-degeneracy of long-lived opposite parity Zeeman sublevels, forming a well isolated two-level system with adjustable parameters (level spacing, projection of angular momenta of the crossing sublevels, their effective width, etc.).

PARITY NONCONSERVATION IN YTTERBIUM

Ytterbium (Yb; $Z = 70$) is another example of a unique system in which to study atomic PNC. Yb was first proposed as a system for studying PNC by one of Professor Commins former graduate students, David DeMille [18], while he was at Berkeley.

Like Dy, Yb is a rare-earth atom and has seven stable isotopes, including two isotopes with non-zero nuclear spin. However, because the ground state of Yb has a closed $4f$ shell (in addition to a closed $6s$ shell), the low-lying energy levels more closely resemble those found in alkaline-earth atoms such as Ba and Ca. This makes the calculations significantly more reliable than those done in other rare-earth atoms such as Dy and Sm.

In Yb, the weak interaction mixes the even-parity $5d6s\,^3D_1$ state with the odd-parity $6s6p\,^1P_1$ state (see Fig. 3). The mixing between these two states is expected to be large due to relatively small energy separation between the two states ($\approx 600\,cm^{-1}$) and favorable configurations among the states (the $^1P_1$ state is not a pure $6s6p$ configuration and contains $\approx 15\%$ $5d6p$; allowing mixing between the $6s$ electron in the $^3D_1$ state and the $6p$ electron in the $^1P_1$ state).

This mixing leads to a small electric-dipole ($E1$) transition amplitude between the $6s^2\,^1S_0$ ground state and the $5d6s\,^3D_1$ state. The size of this PNC-induced transition amplitude was estimated in DeMille's original proposal to be $\approx 10^{-9}ea_0$ (≈ 100 times larger than in Cs) [18]. This estimate has been confirmed by more elaborate calculations of M.G. Kozlov, S. Porsev, and Yu. Rakhlina, [19] and B. P. Das [20].

In addition to the large enhancement of the PNC effect, the transition also has a highly suppressed magnetic-dipole ($M1$) amplitude ($\approx 10^{-4}\mu_B$ [21]), and a moderately sized Stark-induced amplitude ($2.18(33) \times 10^{-8}ea_0/(V/cm)$ [21]) allowing the use of the Stark-interference technique in an atomic beam, which has been successfully employed for cesium [22].

A schematic of the apparatus is shown in figure 4. A dc electric field mixes opposite-parity states and creates a Stark-induced $E1$ transition amplitude between the $^1S_0$ and

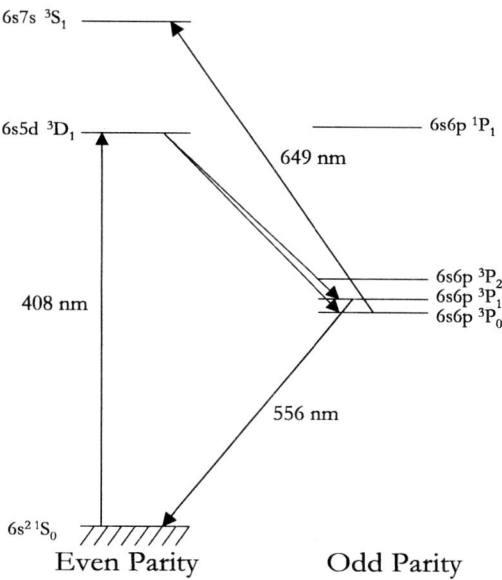

FIGURE 3. Low-lying energy levels for Yb.

$^3D_1$ states (see Fig. 3). This amplitude interferes with the PNC-induced $E1$ transition amplitude. A magnetic field is applied to separate the magnetic sublevels of the upper state to prevent the cancellation of the interference terms for the different sublevels. The atoms are excited using resonant laser light at 408 nm which is coupled into a power-build-up cavity. Changing the handedness of the geometry (reversing the electric field or the polarization angle) changes the sign of the interference term, which is proportional to the P-odd rotational invariant

$$(\varepsilon \cdot \mathbf{B}) \, ((\varepsilon \times \mathbf{E}) \cdot \mathbf{B}), \qquad (2)$$

where ε is the light polarization vector. This leads to a change in the number of atoms excited to the upper state. The excited atoms decay predominantly to the $^3P_1$ and the metastable $^3P_0$ states ($\approx 35\%$ and 64%, respectively) as indicated in figure 3. The atoms which decay to the $^3P_1$ state can be detected through fluorescence from the subsequent decay to the ground state at 556 *nm*, while the atoms decaying to the metastable $^3P_0$ state can be probed downstream by exciting them to a higher-lying state and observing the subsequent fluorescence. Thus, it is possible to achieve high-efficiency detection of the number of atoms undergoing the transition.

The geometry described above was first employed by Professor Commins in his PNC experiments in thallium (Tl) (see [2]; those experiments were done in a vapor cell rather than an atomic beam). This geometry differs slightly from that used in the Cs experiment [22]. Calculations done by Professor Commins in the early stages of the experiment

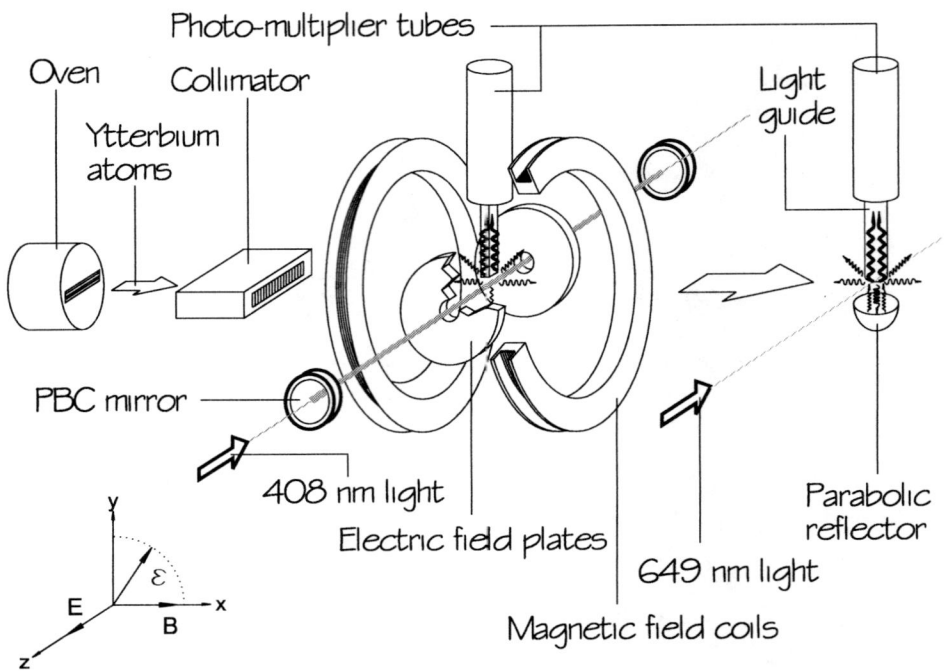

FIGURE 4. Experimental Apparatus for Yb PNC measurement.

suggest that this geometry is less sensitive to systematic effects associated with the nonzero $M1$ transition amplitude between the $6s^2\,{}^1S_0$ and $5d6s\,{}^3D_1$ states.

During the past several years, we have pursued a detailed study of the spectroscopic parameters relevant to performing a PNC experiment in atomic Yb. These preliminary investigations included measurements of the lifetimes of many relevant states, dc-Stark shifts, hyperfine and isotope shifts, and the Stark-induced amplitudes for both the $6s^2\,{}^1S_0 \to 5d6s\,{}^3D_1$ transition and the $6s^2\,{}^1S_0 \to 5d6s\,{}^3D_2$ transition, another candidate for a PNC experiment [23, 21]. Most recently we have completed a measurement of the highly forbidden $M1$ transition amplitude between the $6s^2\,{}^1S_0$ state and the $5d6s\,{}^3D_1$ state using the Stark-interference technique [24]. All of these measurements have confirmed our understanding of the system and lead us to believe a high-precision test of PNC in Yb using the Stark-interference technique in an atomic beam is feasible.

One remaining question for the Stark-interference atomic beam experiment concerns the use of the power-build-up cavity in order to increase the number of atoms making the transition. The ac-Stark shifts resulting from the high-intensity standing wave inside the power-build-up cavity serve to broaden the atomic resonance. At high powers, some of the atoms are shifted out of resonance, decreasing the overall signal. Because the ac-Stark shift depends on the intensity of the light and the signal depends only on its power, it is desirable to have a transverse-cavity-mode spot size which is as large as experimentally feasible. Unfortunately, as the cavity-mode size increases the cavity becomes more sensitive to misalignment and is less stable. Thus there is a practical limit on the size of a single cavity mode. Another approach that we are pursuing is to use a confocal power-build-up cavity. In this case, each transverse mode is degenerate with either the fundamental transverse mode (with a different longitudinal index), or its eigenfrequency falls exactly between two fundamental modes [25]. Thus, instead of coupling into just one transverse mode, it is possible to use a beam with a large spot size and excite many modes at once. The size of the ac-Stark shifts is the final quantity that must be measured before proceeding with the PNC experiment.

In addition to the experiment described above we have investigated the possibility of measuring the PNC effect in a vapor cell, which may allow for higher atomic densities and higher statistical sensitivity [26, 27]. As with the atomic beam experiment the atoms would be excited in the presence of crossed electric and magnetic fields and a change in the transition rate would be detected with a change in the handedness of the fields. The atoms undergoing the transition would be detected by using probe light at 649 nm resonant with a transition from the metastable $6s6p\,^3P_0$ state to a higher-lying $6s7s\,^3S_1$ state. A possible limiting factor in this experiment is the collisional de-excitation of the atoms in the metastable $6s6p\,^3P_0$ state by other Yb atoms as well as buffer gas atoms. We have measured the collisional perturbations of the $6s6p\,^3P_0$ state, including pressure broadening and de-excitation cross sections, due to quenching with possible buffer gases (He, Ne) and with Yb atoms [26]. These experiments suggest that indeed it may be possible to perform a Yb PNC experiment in a vapor cell with greater statistical sensitivity than is possible with an atomic beam. However, there are many questions yet to be answered concerning the feasibility of such a measurement such as collisional broadening of the transition, ionization of the Yb atoms in the high-intensity light fields required for the experiment, and collisionally assisted transition rates.

Another possibility for a PNC experiment is an optical-rotation experiment in a vapor cell [28]. In this case the transition studied would be the $M1$ transition between the metastable $6s6p\,^3P_0$ state and the $6s6p\,^1P_1$ state. This transition amplitude is relatively large ($\approx 0.1\,\mu_B$) due to spin-orbit coupling. The PNC mixing described above again creates a $E1$ transition amplitude between states of nominally the same parity. The interference between the $M1$ transition amplitude and the PNC-induced $E1$ transition amplitude leads to optical rotation of linearly polarized light in the absence of external fields. The magnitude of the optical rotation per unit absorption length is about an order of magnitude larger than in the transitions in Tl, Bi, and Pb, where this effect had been measured (this is partially due to a smaller $M1$ amplitude for Yb). Because the lower state is not the ground state, it is possible to measure spurious optical rotation by monitoring optical rotation as a function of time while the population of the $6s6p\,^3P_0$ state decays. While quenching of the $6s6p\,^3P_0$ state in Yb-Yb collisions may be a limiting factor in this experiment, it appears possible that a highly complementary measurement to Stark-interference measurements could be performed this way.

SPECTROSCOPY AND POSSIBLE TESTS OF P,T-VIOLATION IN SAMARIUM

There are several parameters of importance in judging the merits of a system in which an EDM is measured. In addition to the large enhancement factor, it is also necessary to have a long spin-relaxation time (a parameter that enters directly into sensitivity to an EDM). Correspondingly, the majority of studies so far have been performed in atoms or molecules in their ground electronic states (e.g. [29]). The use of excited metastable states (as pioneered by Player and Sandars in 1970 [30]) may offer significant advantages, including the possibility to access the potential enhancement afforded by level degeneracy. The development of this idea for molecules is discussed in a contribution by D. P. DeMille et. al. in these Proceedings.

Another potentially useful system for an EDM experiment with metastable states is atomic samarium (Sm; $Z = 62$). In order to evaluate the feasibility and merit of an EDM search in samarium, an experimental study was undertaken, the results of which were reported in Ref. [31]. A systematic measurement of the lifetimes and tensor polarizabilities of the lowest-lying odd-parity levels was performed. The lifetimes were measured by detecting time-resolved fluorescence following pulsed laser excitation of atoms in an atomic beam; polarizabilities were measured employing the method of Stark-induced quantum beats. An analysis of the data was undertaken to find the best even-parity candidate states for an EDM measurement. For the most favorable candidate state (nominally $4f^6 5d6s$ $^7G_1$ at 15639.80 cm^{-1} which has an opposite-parity "partner" state only ~ 11 cm^{-1} away), the electron EDM enhancement factor was estimated to be $R = \pm 1100 \pm 800 \pm (1300 \text{ to } 1900)$. Here the three factors come from the contributions of various constituent electronic configurations and terms of the nearly degenerate states. Unfortunately, relative signs in the configuration and term decomposition, although they must have been known to the workers whose analysis contributed to the compilation [32], have been apparently lost. This considerable remaining uncertainty in the enhancement factor has put further work on the EDM search in Sm on hold pending further theoretical or experimental input regarding the sign ambiguity. If it turns out that the terms in the enhancement factor are of the same sign, the enhancement factor could exceed that of Tl by as much as an order of magnitude.

The spectroscopic investigations in Sm [31] also led to a somewhat unexpected additional result. Critical analysis of the obtained lifetime and polarizability data along with earlier results in Sm showed quite unambiguously that terms were incorrectly listed in [32] for a class of odd-parity states. The term reassignment cleared some long-standing discrepancies between theory and spectroscopic data in Sm. It also allowed to revise (namely, reduce by a factor of about 40) an estimate of the PNC amplitude in a transition from the ground state of Sm to the $^7G_1$ state originally made by the Oxford group [7] (see also Ref. [33] where experimental possibilities of measuring PNC in this transition were investigated). Our revised estimate showed that while the PNC amplitude in this Sm transition is still much larger than in Cs, it is probably somewhat smaller than that in Yb (see above), and no obvious advantages of Sm were found.

SEARCH FOR PERMUTATION SYMMETRY VIOLATION FOR PHOTONS

The notion that an N-particle wave function describing identical particles should be either symmetric (in the case of integer-spin particles), or antisymmetric (in the case of half-integer-spin particles) with respect to permutation of any two particles constitutes one of the fundamental pillars of our current understanding of Nature. This permutation symmetry postulate (PSP) and the spin-statistics connection (SSC) are of similar importance as, and closely related to the CPT theorem, an important cornerstone of modern physics. In spite of the very general assumptions underlying the proof of the spin-statistics theorem, the argument leading to it is far from being straightforward, and involves some subtle assumptions (see E. Wichmann's contribution in the current Proceedings). An attempt of an intuitive explanation was given by Feynman in his 1986 Dirac memorial lecture [34], which subsequently was vigorously dismissed by a number of authors (see [35] for a review of theory and experiments related to PSP and SSC). It is precisely the fundamental nature of PSP and SSC that makes it worthwhile to evaluate how well it can be checked experimentally. The discoveries of P- and CP-violation give us examples of how it can be fruitful to perform tests of a seemingly solid physical law.

While sensitive searches for violations of PSP and SSC were performed for the electrons and some composite bosons, direct experimental data for important fundamental bosons – the photons – turn out to be surprisingly scarce (see [36] for a detailed review).

Atomic spectroscopy offers a possibility of testing quantum statistics for photons via the use of a powerful (but not widely known) selection rule for two-photon transitions [37, 23]: while two-photon transitions between an F=0 and F'=1 state (where F, F' are total angular momenta) are generally allowed for *non-degenerate* photons, the transition is strictly forbidden for *degenerate* photons. This rule is closely related to the well-known Landau-Yang theorem in particle physics [38, 39] which states that a vector particle cannot decay into two photons. Bose-Einstein statistics is at the core of the proof of this theorem.

The idea of the experiment is to apply a strong laser field to atoms in the ground F=0 state, and look for forbidden excitation events to a two-photon resonant, high-lying F'=1 state. The sensitivity of the experiment can be calibrated by using two laser fields of appropriate polarizations whose photon energies add up to the two-photon transition energy, but which are non-degenerate.

The first experiment based on this approach was described in Ref. [40], and reported a limit on the relative fraction (ν) of anti-symmetric photon pairs present in the laser field of $\nu \leq 10^{-7}$. This experiment used barium atoms in a vapor cell and pulsed lasers and was limited by their final bandwidth.

Currently, we are pursuing a new version of the experiment that is using barium atomic beam and a narrow-band cw laser [41]. This is expected to reach the level of sensitivity of $\nu \lesssim 10^{-11}$ with eventual further improvements by several orders of magnitude.

Simultaneously, we are also performing auxiliary spectroscopic measurements with Ba using the pulsed lasers and the atomic beam apparatus of Ref. [31]. These measurements are aimed at choosing the optimal F=0→F'=1 transition and detection scheme [41], and a better understanding of the configuration and term composition of the levels involved. They include determination of lifetimes and branching ratios of even-parity excited states, and measurements of their tensor electric polarizabilities. In addition, we will look for auto-ionizing resonances at energies corresponding to three-photon absorption from the ground state, the presence of which could, in principle, degrade the sensitivity of the forbidden two-photon transition Bose-Einstein statistics test [40, 36].

ACKNOWLEDGEMENTS

We are grateful to Prof. D. P. DeMille of Yale (a former Commins' student), collaboration with whom has to a large extent defined the directions of this work. C. J. Bowers, D. Clyde, G. D. Chern, and B. DeBoo have contributed to various parts of this research. Recently, Dr. Gabriela Stoessel and undergraduate students L. Zimmerman and S. Anjum have also participated in our efforts. Research on ytterbium PNC and Bose-Einstein statistics tests for photons has been supported by NSF (grant PHY-9877046 and CAREER grant PHY-9733479). Research on dysprosium and samarium was supported by D.B.'s start-up funds, and by the UC Berkeley Committee on Research. D.B. and S.J.F. also wish to acknowledge partial support from the U.S. Department of Energy, Office of Science, under Contract No. DE-AC03-76SF00098 through the LBNL Nuclear Science Division.

REFERENCES

1. P. H. Bucksbaum, E. D. Commins, and L. R. Hunter, Phys. Rev. D, Part. Fields **24**(5), 1134 (1981).
2. P. S. Drell and E. D. Commins, Phys. Rev. A, Gen. Phys. **32**(4), 2196 (1985).
3. E. D. Commins (1988), p. 1093, Ninth Vavilov Conference on Nonlinear Optics Novosibirsk, USSR 16-18 June 1987; Kvantovaya Elektron. Mosk. (USSR).
4. I. B. Khriplovich, *Parity nonconservation in atomic phenomena* (Gordon and Breach Science Publishers, Philadelphia, 1991).
5. M. Zolotorev and D. Budker, Phys. Rev. Lett. **78**(25), 4717 (1997).
6. V. A. Dzuba, V. V. Flambaum, and I. B. Khriplovich, Z. Phys. D, At. Mol. Clusters (West Germany) **1**(3), 243 (1986).
7. A. Gongora and P. G. H. Sandars, J. Phys. B, At. Mol. Phys. (UK) **19**(8), L291 (1986).
8. L. M. Barkov, M. S. Zolotorev, and D. A. Melik-Pashaev, Sov. Journ. Quant. Electron. **18**(6), 710 (1988).
9. D. Budker, E. D. Commins, D. DeMille, and M. S. Zolotorev, Opt. Lett. **16**(19), 1514 (1991).
10. D. Budker, D. DeMille, E. D. Commins, and M. S. Zolotorev, Phys. Rev. Lett. **70**(20), 3019 (1993).
11. D. Budker, D. Demille, E. D. Commins, and M. S. Zolotorev, Phys Rev A **50**(1), 132 (1994).
12. A. T. Nguyen, D. Budker, D. DeMille, and M. Zolotorev, Phys. Rev. A, At. Mol. Opt. Phys. **56**(5), 3453 (1997).
13. V. A. Dzuba, V. V. Flambaum, and M. G. Kozlov, Phys. Rev. A, At. Mol. Opt. Phys. **50**(5), 3812 (1994).
14. D. Budker, in *Physics Beyond the Standard Model, proceedings of the Fifth Intrnational WEIN Symposium*, edited by P. Herczeg, C. M. Hoffman, and H. V. Klapdor-Kleingrothaus (World Scientific, 1999), p. 418.
15. A. T. Nguyen, G. D. Chern, D. Budker, and M. Zolotorev, Phys. Rev. A, At. Mol. Opt. Phys. **63**(1), 013406/1 (2001).

16. A. T. Nguyen, D. E. Brown, D. Budker, D. DeMille, D. F. Kimball, and M. Zolotorev, in *Parity Violation in Atoms and Polarized Electron Scattering*, edited by B. Frois and M. A. Bouchiat (World Scientific, 1999), vol. 457, p. 295.
17. A. Abragam, *The principles of nuclear magnetism*, International series of monographs on physics (Oxford, Oxfordshire) (Claredon Press, Oxford, 1962).
18. D. DeMille, Phys. Rev. Lett. **74**(21), 4165 (1995).
19. S. G. Porsev, Y. G. Rakhlina, and M. G. Kozlov, Pis'ma Zh. Eksp. Teor. Fiz. (Russia) **61**(6), 449 (1995).
20. B. P. Das, Phys. Rev. A, At. Mol. Opt. Phys. **56**(2), 1635 (1997).
21. C. J. Bowers, D. Budker, S. J. Freedman, G. Gwinner, J. E. Stalnaker, and D. DeMille, Phys. Rev. A, At. Mol. Opt. Phys. **59**(5), 3513 (1999).
22. C. S. Wood, S. C. Bennett, J. L. Roberts, D. Cho, and C. E. Wieman, Can. J. Phys. (Canada) **77**(1), 7 (1999).
23. C. J. Bowers, D. Budker, E. D. Commins, D. DeMille, S. J. Freedman, A. T. Nguyen, S. Q. Shang, and M. Zolotorev, Phys. Rev. A, At. Mol. Opt. Phys. **53**(5), 3103 (1996).
24. J. E. Stalnaker, V. V. Yashchuk, D. Budker, and S. J. Freedman, *Manuscript in preparation* (2001).
25. A. E. Siegman, *Lasers* (University Science Books, Mill Valley, 1986).
26. D. F. Kimball, D. Clyde, L. D. Budker, D. DeMille, S. J. Freedman, S. Rochester, J. E. Stalnaker, and M. Zolotorev, Phys. Rev. A, At. Mol. Opt. Phys. **60**(2), 1103 (1999).
27. B. DeBoo, D. F. Kimball, C. H. Li, and D. Budker, J. Opt. Soc. Am. B, Opt. Phys. **18**(5), 639 (2001).
28. D. F. Kimball, Phys Rev A **63**(5), 052113 (2001).
29. E. D. Commins, Ad. At. Mol. Opt. Phys. **40**, 1 (1999).
30. M. A. Player and P. G. H. Sandars, J. Phys. B, At. Mol. Phys. (UK) **3**(12), 1620 (1970).
31. S. Rochester, C. J. Bowers, D. Budker, D. DeMille, and M. Zolotorev, Phys. Rev. A, At. Mol. Opt. Phys. **59**(5), 3480 (1999).
32. W. C. Martin, R. Zalubas, and L. Hagan, *Atomic energy levels–the rare-earth elements* (U.S. Dept. of Commerce National Bureau of Standards; U.S. Govt. Print. Off., Washington, 1978).
33. I. O. G. Davies, P. E. G. Baird, P. G. H. Sandars, and T. D. Wolfenden, J. Phys. B, At. Mol. Opt. Phys. (UK) **22**(5), 741 (1989).
34. R. P. Feynman and S. Weinberg, *Elementary particles and the laws of physics : the 1986 Dirac memorial lectures* (Cambridge University Press, Cambridge ; New York, 1987).
35. R. C. Hilborn and G. M. Tino, *Spin-statistics connection and commutation relations : experimental tests and theoretical implications : Anacapri, Capri Island, Italy, 31 May–3 June 2000*, AIP Conference Proceedings, 545 (American Institute of Physics, Melville, N.Y., 2000).
36. D. DeMille, D. Budker, N. Derr, and E. Deveney (2000), p. 227, Spin-Statistics Connection and Commutation Relations. Experimental Tests and Theoretical Implications Anacapri, Italy 31 May-3 June 2000; AIP Conf. Proc. , 545.
37. K. D. Bonin and T. J. McIlrath, J. Opt. Soc. Am. B, Opt. Phys. **1**(1), 52 (1984).
38. L. D. Landau, Dokl. Akad. Nauk SSSR **60**, 207 (1948).
39. C. N. Yang, Physical Review **77**, 242 (1950).
40. D. DeMille, D. Budker, N. Derr, and E. Deveney, Phys. Rev. Lett. **83**(20), 3978 (1999).
41. D. Brown, D. Budker, and D. P. DeMille (2000), p. 281, Spin-Statistics Connection and Commutation Relations. Experimental Tests and Theoretical Implications Anacapri, Italy 31 May-3 June 2000; AIP Conf. Proc. , 545.

SEARCHES FOR NEW PARTICLES AND INTERACTIONS

Hunting the Fifth Force on the Snake River[1]

Wm. R. Bennett, Jr.

Department of Applied Physics, Yale University, New Haven, CT. 06530-1968
Present Address: 158 Rose Lane, Haverford, PA. 19041-1618

Abstract. A modulated-source Eötvös experiment was performed at the Little Goose Lock on the Snake River during the summers of 1988 and 1990. Although results of the first experiment were published by the author in 1989, the results from a more sophisticated version of the experiment performed in 1990 are published here for the first time. The 1990 experiment involved a freely oscillating toroidal Cu-Pb pendulum suspended from a 5-µm tungsten fiber having a period of 790 sec and a torsion constant of 0.00035 erg/rad. The lock contained ≈ 1.7 x 10^8 kg of water which could be filled or drained within about ten minutes. Results from 40 2-hr runs (during half of which the pendulum was rotated 180°) showed no differential acceleration on the Cu-Pb masses from the water with a limit of ±2.4×10^{-9} cm/sec^2. Angular deflection measurements were based on a time-interval method developed previously by the author. Fiber drift was roughly constant during runs and always in the counterclockwise direction looking down on the apparatus, a result which may have arisen through interaction of the Coriolis force with vertical vibration of the pendulum. The principal error in measurement was from small discontinuous changes in the fiber drift-rate. The fiber motion suggested the presence of a small amount of second harmonic which was removed from the data by digital filtering. Direct measurement showed that pendulum tilt was a negligible source of systematic error. A 2-σ limit was set on the "isospin coupling constant" of $\alpha_0 = \pm 0.001$ at λ=100-m.

INTRODUCTION

It seemed especially appropriate to give a talk on the present topic at the Symposium honoring Eugene Commins because I had originally hoped to collaborate with him on this very experiment. Alas, he couldn't tear himself away from his other experiments at Berkeley, and I had to do the work by myself. However, he was represented at the Little Goose Lock site by his sister, Frances Commins Bennett, who also happens to be my wife. Eugene contributed to the project by sending her lots of reading material during our stay in that rather desolate part of eastern Washington.

---

[1] The data reported here were taken at the Little Goose Lock and Dam on the Snake River in Eastern Washington during the summers of 1988 and 1990. The magnitude of the project and several serious illnesses prevented me from publishing the final results earlier. However, there is ample precedent: Eötvös, himself, died eleven years after taking his famous data of 1908 without having published it.

BACKGROUND OF THE EXPERIMENT

Baron Roland von Eötvös[2] made a series of measurements between 1889 and 1908 using a torsion balance with which he compared the gravitational acceleration toward the Earth experienced by various materials.[1] He concluded in that work that there was no significant material-dependent difference in the acceleration within ≈ 5 parts in 10^9. Although a paper based on that work got him the Benecke Prize from the University of Göttingen in 1909, the paper went unpublished. Eötvös died in 1919, the year that Eddington measured the bending of light from a distant star during a solar eclipse. Eddington's measurement was among the earliest experimental confirmations of Einstein's theory of general relativity.

Although the Eötvös experiments provided strong support for the "Principle of Equivalence", Einstein himself apparently did not learn of that work until much later; he wrote in 1934 that he had had no serious doubts about the constancy of gravitational acceleration even "without knowledge of the admirable experiments of Eötvös." However, three years after the 1919 eclipse, two students of Eötvös, Desiderius Pekár and Eugen Fekete, thought it would be timely and appropriate to publish their mentor's work.[2] A summary of results from that reference is given in Table 1.

Underground gravitational measurements by Stacey and others [3] during the 1980's and 1990's exhibited anomalies which could be described by adding a Yukawa potential to the Newtonian gravitational potential. In this model, the interaction potential between two masses m_1 and m_2 a distance r apart becomes,

$$V(r) = -(G\, m_1 m_2/r)[1+\alpha \exp(-r/\lambda)] = V_N(r) + \Delta V(r), \qquad (1)$$

where it was initially thought that $\alpha \approx -0.01$ and λ would be in the range from 100 to 1000 m.

TABLE 1. Results for the difference from Pt of acceleration toward the Earth of various materials expressed as a fraction of g. (From Eötvös et al.[2])

MATERIAL	DIFFERENCE FROM PLATINUM
Magnalium	$(+0.004 \pm 0.001) \times 10^{-6}$
Schlangenholz	$(-0.001 \pm 0.002) \times 10^{-6}$
Kupfer	$(+0.004 \pm 0.002) \times 10^{-6}$
Wasser	$(-0.006 \pm 0.002) \times 10^{-6}$
Kristall, Kupfersulfat	$(-0.003 \pm 0.002) \times 10^{-6}$
Kupfersultatlösung	$(-0.001 \pm 0.002) \times 10^{-6}$
Asbest	$(+0.001 \pm 0.002) \times 10^{-6}$
Talg	$(-0.002 \pm 0.002) \times 10^{-6}$

---

[2] "Eötvös" is pronounced "utvush" with the "u" as in "bush."

In January of 1986, Fischbach et al [4] took the problem a step further by pointing out a material-dependent correlation within the quoted error of the 1922 paper by Eötvös et al, which they thought might explain the geophysical results.

They suggested a massive hyperfield whose quanta would have mass $\approx 10^{-9}$ eV and proportional to $1/\lambda$. Exchange of these hyperons would give rise to ΔV, α being related to the unit of hypercharge. They proposed that the coupling coefficient in Eq. (1) would be of the form

$$\alpha = \alpha_0 \, C_1 C_2 \tag{2}$$

where the "charges" were proportional to baryon number and normalized to the atomic weight,

$$C_i = (N + Z)/\mu_i. \tag{3}$$

By this definition, the additional term in Eq. (1) becomes dependent on the total number of particles, rather than on the masses involved. Hence, the gravitational acceleration would vary with the material and this assumption would rule out the Principle of Equivalence.

Fischbach et al did not actually quote a value for α_0 in Eq. (2), although they implied their model would be compatible with the value of $\alpha \approx -0.01$ obtained by Stacey et al with $\lambda = 200 \pm 50$ m. There were various things seriously wrong with both the geophysical "bore-hole" measurements and with details of the interpretation of the Eötvös data by Fischbach et al. Among other objections [5], Chu and Dicke noted that the systematic trend shown within the error of the data in Table 1 could easily be explained by air resistance effects at the atmospheric pressure used in the original Eötvös torsion balance.[6] However, our main concern here is with subsequent developments stimulated by the paper of Fischbach et al.

Although the Fischbach interpretation of the original Eötvös data seemed doubtful, it stimulated a number of contradictory "fifth force" experiments, two of which gave positive results of about the magnitude predicted. One was by Thieberger, [7] who placed an almost-completely submerged hollow copper sphere in water adjacent to the cliffs at Palisades New Jersey. He found that the sphere always moved slowly away from the cliffs and interpreted the result as a kind of horizontal Archimedes Principle. Assuming baryon coupling, he obtained $\alpha_0 \approx -0.012$ for $\lambda \approx 100$-m.

Next followed a series of torsion balance experiments based on topological features (sloping hillsides, cliff sides, geological faults and mountain sides) to detect a "fifth-force" effect. These placed decreasingly smaller limits on the size of the baryon coupling coefficient. In particular, the 1-σ limits ($-.0004 < \alpha_0 < +.0003$ at $\lambda = 100$-m) in the early work by Stubbs et al [8] appeared adequate to rule out the baryon coupling interpretation of the Thieberger result and the anomalous geophysical data. This conclusion was supported by further measurements of Adelberger *et al* [9], Fitch *et al* [10] and Cowsik *et al* [11].

However, Boynton *et al* [12] reported a positive effect. Using an Al-Be toroidal pendulum, they measured the period of oscillation with the mass-dipole placed normal and then parallel to the side of a mountain at Index, Washington. They reported a difference between the two period measurements which was several times the statistical error and which implied a value of $\alpha_0 \approx -0.01$ at $\lambda \approx 100$-m. They interpreted their results using isospin coupling for which the normalized "charge" was given by

$$C_i = (N - Z)_i / \mu_i \tag{4}$$

in contrast to Eq. (3).

In these torsion experiments, one "sees" the difference in coupling coefficients

$$\Delta\alpha = \alpha_0 (C_1 - C_2) C_{\text{Source}} \tag{5}$$

where $C_{1,2}$ correspond to the "charges" of the two materials in the pendulum. Generally, with experiments based on topological features, the charge of the source is not well-known. As long as one restricts the model to baryon coupling, the nature of the source doesn't matter very much because then $C_{\text{Source}} = (N + Z)/\mu \approx 1$ for almost anything. However, with isospin coupling, $C_{\text{Source}} \approx (N - Z)/\mu$ and can vary from large negative values at the low end of the Periodic Table to large positive values at the other end. Here we are concerned with the difference $\Delta\alpha$ in coupling constants that arises from copper-water and lead-water interactions, where

$$\Delta\alpha = \alpha_0 (C_{\text{Cu}} - C_{\text{Pb}}) C_{\text{H2O}}$$

$$= \begin{cases} +0.0009\alpha_0 \text{ (baryon coupling)} \\ +0.0134\alpha_0 \text{ (isospin coupling)} \end{cases}$$

and the values have been averaged over normal isotopic abundance.

Boynton *et al* noted that the presence of reasonable amounts of ground water around the physics lab at Seattle could completely cancel the effect of the source in the surrounding hillside assumed from geological analysis of the terrain. Boynton and colleagues also introduced the notion of mixed coupling in which the ith charge is given by

$$C_i = [\beta(N + Z)_i + (1 - \beta)(N - Z)_i]/\mu_i \tag{6}$$

and noted that the prior results (except for the geophysical ones) were compatible for a value of $\alpha_0 = -0.009 \pm 0.005$ for $\lambda = 100$-m, and a value of $\beta = -0.005 \pm 0.002$, which was very close to pure isospin coupling.

THE PROTOTYPE EXPERIMENT (1988)

It was at that point that I became interested in the problem. It occurred to me that many of the disadvantages of the previous Eötvös experiments could be avoided by setting up a torsion balance next to a shipping lock. That approach offered several strong advantages:
1) A massive source could be turned on and off in a short time without moving the apparatus.
2) The source (water) would be clearly identified.
3) The source distribution would be well-known.
4) Because of the large negative "hypercharge" for water, the experiment would be very sensitive to isospin coupling. (See Table 2.)

I was attending a laser conference in Los Angeles when I thought of this possibility. I was so excited about the experiment that I called my wife that night and dictated a rough proposal to her for the NSF over the telephone. (Fran is an excellent editor and typist. Among other manuscripts, she worked on Eugene's PhD thesis at Columbia - not to mention my own, those of Arno Penzios and several other contemporary graduate students.) When I got back east, I sent the proposal off to the NSF to see if they would be interested in funding an experiment of that type. Instead of funding it, they lost it! I decided to go ahead and do a prototype experiment anyway, but on a very modest scale.

The Lock Site

During a careful study of major canal locks in the western hemisphere, ranging from the Panama Canal to the St. Lawrence Seaway, I came across the ones situated on the Columbia and Snake Rivers in Oregon and Washington. These had by far the highest "lifts" and shortest fill and drain times of any I encountered. Of these, the geometry of the Little Goose Lock on the Snake River was the most attractive because of its unique side wall construction. The walls are supported by concrete buttresses separated by about 30 ft. Hence, apparatus can be moved to within about 11 m of the water. The buttresses and overhead concrete fish ladder also provided excellent shielding from the wind. This particular lock is about 80 miles by

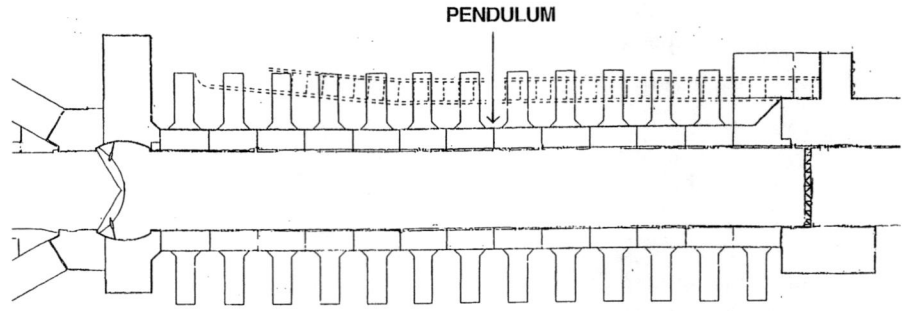

FIGURE 1. Cross-sectional drawing of the Little Goose Lock.

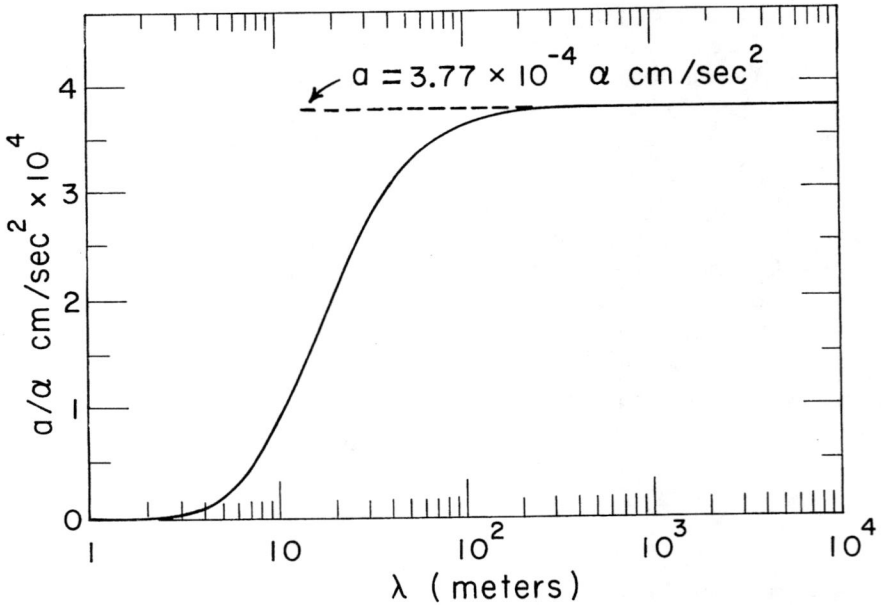

FIGURE 2. Lateral acceleration from filling the lock at pendulum location for varying α and λ (from Bennett, 1989. [13])

road ENE of Pasco, Washington, where the Columbia River joins the Snake. (It is also only ≈ 60 miles downwind from the Hanford plutonium plant.)

The Little Goose Lock is about the size of two-football fields placed end-to-end: it is 219 m long and 25.6 m wide. The lift during measurements was about 30.0 m (≈ 100 ft), during which some 43 million gallons of water (≈ 1.7×10^8 kg) could be filled or drained within ten minutes. A cross-sectional drawing of the Lock is shown in Fig. 1. The pendulum was placed at the center on the west side, 11 m from the water and was about 9 m above the lowest level of the water when the lock was empty.

The simple rectangular geometry made it easy to compute the gravitational potentials and their gradients and to determine the expected value of α from the Yukawa potential in Eq. (1) from measured values of the differential acceleration of the pendulum. That was done to compute Fig. 2, which constitutes a primary calibration for the experiment.

I had initially hoped to use a pendulum made from "spent" uranium and polyethylene because of their especially desirable charges for isospin coupling (see Table 2) and had obtained a large cylinder of uranium for that purpose through the help of D. A. Bromley. Alas, it was confiscated by the Yale Health Physicist before I could have it machined. For reasons of mechanical stability, I also was forced to give up the idea of using polyethylene. Copper and lead were the most reasonable materials left on the list. For practical reasons in precision machining, the pendulum was chosen to be 1-in (2.54-cm) in diameter. That small size also reduced systematic errors that could arise from potential gradients.

A schematic diagram of the apparatus used in the prototype experiment is shown in Fig. 3. The details of the experiment were given in an earlier publication by the author [13] and only a brief summary will be presented here. The detection apparatus was mounted on 3500-lb granite surface plate, which was bolted inside a truck (see Fig. 4) and which provided a flat (± 0.15 mil) level surface of 3 ft x 5 ft cross section. The truck was jacked off the ground at the lock site. A 1-in diameter split disc Cu-Pb torsion pendulum (Fig. 5) was suspended on a 10-μm gold-plated tungsten fiber that contained a mirror mounted in the plane of the mass dipole direction and was placed inside a large stainless steel vacuum housing kept at a pressure below 0.1 Torr throughout the experiment by a fore pump resting on the ground outside the truck. The angular position of the normal to the mirror was monitored with a Model D897 Davidson Optronics Telescope having a limiting resolution of 0.01 arcsec and a viewing range of ±5 arcsec. The total pendulum mass was 17.01 g, including a 23 x 17 -mm^2 mirror. The Pb and Cu halves were each 8.11 gm and machined within 1 mil tolerances. The Cu half was lightly tinned and the two halves were then soldered together. The pendulum torsion constant was 0.00484 erg/rad and determined from the measured period (330.7 sec). The sensitivity of the assembly to a possible Fifth Force acting differently on the

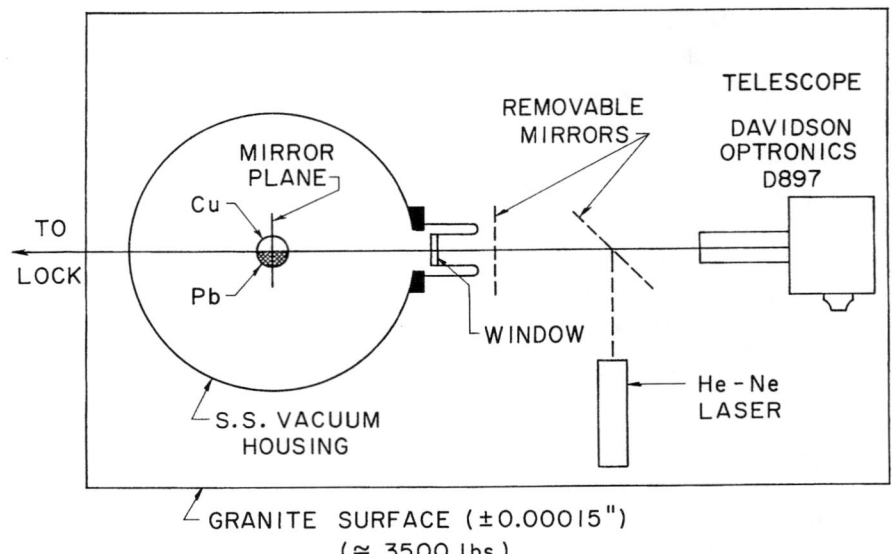

FIGURE 3. Schematic diagram of apparatus used in the prototype experiment. (from Bennett, [13])

FIGURE 4. Truck Location.

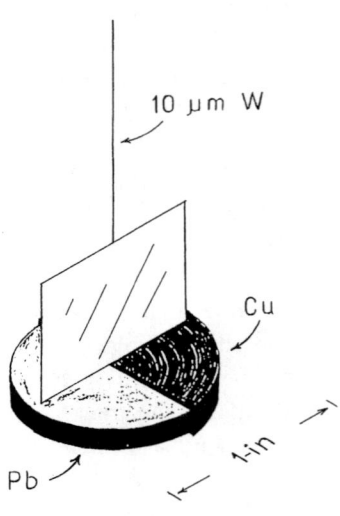

FIGURE 5. Split-disc Pb-Cu pendulum.

Table 2. Hypercharge of suitable materials for isospin coupling.

Material	Baryon $(N + Z)/\mu$	Isospin $(N - Z)/\mu$
Water (H$_2$O)	1.00008	-0.11087
Polyethylene (CH$_2$)$_n$	0.998818	-0.14186
Copper (Cu)	1.00101	+0.08819
Lead (Pb)	1.00012	+0.20866
Uranium (U)	0.99979	+0.22677

copper and on the lead was such that a mirror deflection of 1-arcsec from the equilibrium position of the mass dipole parallel to the long dimension of the lock corresponded to a differential acceleration of $\Delta a = 0.5375 \times 10^{-8}$ cm/sec^2. Values of $\Delta\alpha$ were then determined from the data in Fig. 2.

Method of Angle Measurement

I had initially hoped to be able to get the pendulum to settle down in stable equilibrium with the mass dipole parallel to the lock. That expectation quickly proved to be naïve. However, the constant small-angle oscillation of the pendulum turned out to be a blessing in disguise for it suggested a very powerful way of measuring the equilibrium angle based on time-interval determinations. By deliberately setting the pendulum into oscillation with a large amplitude, the natural frequency of the pendulum can be used as a "carrier" of the angular deviation. The basis of the method is illustrated in Fig. 6. The high resolution telescope was used to determine both the time between zero crossings of the pendulum mirror normal with the telescope axis and the slope at those crossings. The magnitude of the angular displacement is given precisely by the cotangent relation shown in Fig. 6 and in Eq. (7). A small-angle approximation, $\vartheta \approx m\ (t_1-t_2)/4$, is useful for estimating pendulum drift during runs, but the full cotangent equation was always used in the final analysis of the data,

$$\vartheta = |m|(t_1 + t_2)\mathrm{Cot}[2\pi t_1/(t_1 + t_2)]/2\pi \qquad (7)$$

where $|m|$ is the magnitude of the slope at the zero crossing, t_1 and t_2 are the time intervals defined in Fig. 6, and $t_1 + t_2 \approx T$ is the period of the pendulum. The actual measured values for T vary somewhat due to noise and slow decay of the oscillation with time. In addition, drift of the fiber equilibrium position will show up in the measured values. Some other phenomena associated with the angular measurement will be discussed later.

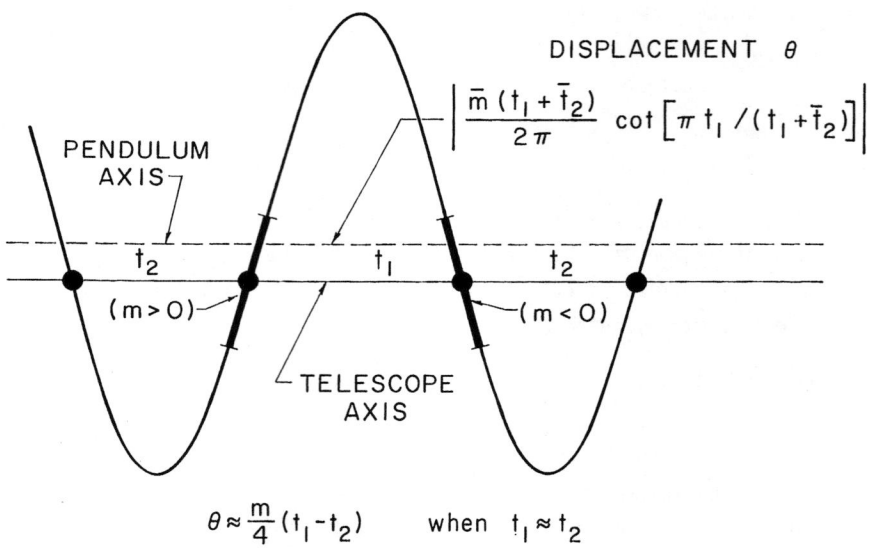

FIGURE 6. Method of angular measurement.

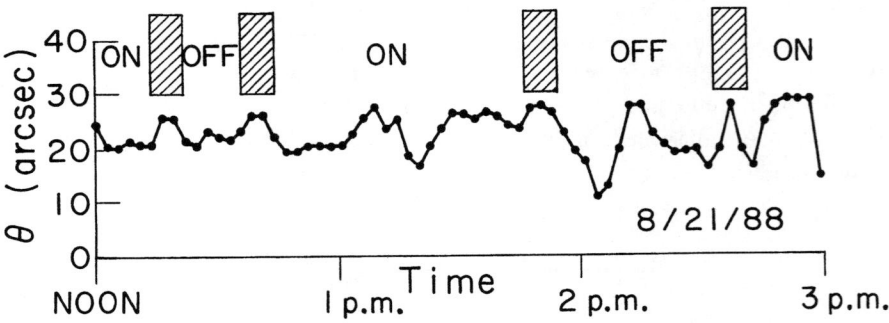

FIGURE 7. Results from the prototype experiment. Mean angular deviation ϑ of the mass dipole from parallelism to the lock, averaged over four successive lock transitions during one 3-h period. Shaded areas denote fill and drain periods. A positive angular displacement means copper rotated towards the water. ON denotes lock is full. (from Bennett, [13].)

Table 3. Results from the prototype experiment. Positive deflection and acceleration are towards water. The error quoted is the weighted standard deviation from the weighted mean. (from Bennett, [13].)

Number of Measurements	Deflection (arcsec)	Cu-Pb Acceleration (10^{-8} cm/ sec^2)
12	-0.74 ± 1.72	-0.40 ± 0.99
12	-1.60 ± 1.31	-0.86 ± 0.70
28	+4.23 ± 5.09	+2.27 ± 2.74
14	+3.52 ± 6.51	+1.89 ± 3.50
12	+10.15 ± 5.81	+5.46 ± 3.12
12	+4.66 ± 5.08	+2.50 ± 2.73
26	-2.24 ± 6.65	-1.20 ± 3.57
Weighted Mean	-0.47 ± 0.96	-0.25 ± 0.52

Results of the Prototype Experiment

The best data taken in the prototype experiment are reproduced in Fig. 7 in which the mean angular deviation of the mass dipole from parallel to the lock is shown in half-period increments over one 3-hr run during which there were four lockages. As noisy as the data appear, they were sufficient to rule out the Boynton conjecture. A summary of corresponding numerical results is shown in Table 3. From the data in Table 3, it was concluded that Δa_{Cu-Pb} = (-0.25 ± 0.52) x 10^{-8} cm/sec^2 as a result of filling the lock. Using the data in Fig. 2, one can then determine limits on $\Delta \alpha$ as a function of λ. For example, for λ=100m, $\Delta \alpha$ = (-0.7±1.4) x 10^{-5} and for pure isospin coupling the limits on the intrinsic coupling coefficient become α_0 = (-0.52 ± 1.04) x 10^{-3}.

THE FINAL EXPERIMENT (1990)

Sometime after I had published my Physical Review Letter of January, 1989, I received another communication from the NSF. They had at last found my proposal and were going to fund it after all -- but *not* retroactively. There was some question in my mind whether I should bother doing the experiment over again because additional measurements had been published by Cowsik *et al* [14] and Adelberger *et al* [15] which appeared to set still more stringent limits on the existence of the Fifth Force. Still, there were a number of ways in which my prototype experiment could be improved, and it seemed useful to do more with that approach to the problem. The method at least would have different sources of systematic error from those found in any other work. Although there still wasn't enough money to hire a graduate student (the going rate at Yale would have taken 3/5 of the grant) and still have enough money for the experiment, we would at least be able to rent an RV ("Recreational Vehicle") to stay in at the lock. (During the prototype experiment, Fran and I lived in a station wagon next to the truck; fortunately, there was a public lavatory in the nearby Fish Viewing Station.)

Improvements in the Final Experiment

The first thing I did was to "FAX" a 20-ton, 2-ft thick concrete slab to the lock to serve as a base for the truck. (Of course, I really FAXed scale drawings of the slab and its location to the construction company that poured the concrete.) The only condition placed by the officials at the lock was that I had to have the slab removed after the experiment was completed. The slab was 16-ft long by 8-ft wide with the long dimension perpendicular to the lock. It was to be positioned 89 in. from the V-shaped indentation of the lock wall at the center of the west side and to have a clearance at the corners of at least 30 in. from the emerging concrete buttresses in order to avoid direct contact with the lock vibrations. I was relieved to find that the slab had been poured in precisely the right place when we got back to the lock several months later. The truck would be raised off the concrete slab by six jack stands, four of which were directly under the corners of the stand supporting the 3500-lb granite table. The other two jack stands were to balance the front end of the truck.

Pendulum Design and Fiber Preparation

For practical reasons, the 1-in outside diameter of the first pendulum was retained. However, I wanted to reduce the fiber diameter d by a factor of two to 5 μm to improve the sensitivity of the experiment. The strength of the fiber increases with d^2, whereas the torsion constant increases as d^4. The overall sensitivity, which depends on the ratio of the mass of the pendulum to the torsion constant therefore varies as $1/d^2$. Hence, a limiting improvement of about a factor of four in sensitivity could be expected by reducing the fiber diameter from 10 to 5 μm. Eventually, the mass of the mirror assembly would prevent obtaining that limit in practice.

An improvement in sensitivity was also obtained by using an annular ring rather than a disc. In order to satisfy the requirements on both mass and geometry, the Yale shop removed the center from the original Pb-Cu pendulum used in the prototype experiment. The final result is shown in Fig. 8 where a light magnesium spider supports the outer Pb-Cu ring and a Mg post supports the mirror. The total machining errors in departure from cylindrical symmetry were less than 0.1 mil. The

Table 4. 5-μm Pendulum Properties.

Material	Mass (gms)	Moment of Inertia (gm-cm^2)
Pb	1.730	2.4032
Cu	1.730	2.4032
Mirror + Mg Mount	1.560	0.6671
Total Values	5.020	5.4735

pendulum was made up as shown in Table 4. Several test fibers broke at ≈ 7.0 to 7.5 gms, hence the pendulum weight was about 70% of the probable breaking strength. The total moment of inertia of the pendulum about the fiber axis was $I = 5.473$ gm-cm^2 and the average measured period T when suspended on the 5-μm fiber was 790 secs (≈ 13.2 min), a time slightly longer than it took to fill or drain the lock. The torsion constant K was determined from the average pendulum oscillation period through the relation $K = (2\pi/T)^2 I = 0.00035$ erg/rad.

A major advantage of an axially symmetric torsion pendulum is that lateral gravitational gradients cancel out identically. The latter may be seen by expanding the transverse accelerations on each of the two masses in a double Taylor series in the two transverse rectangular coordinates. In the present case, when the mass dipole is parallel to the lock, all of the terms in the series drop out except for one, yielding

$$L = (a_1 - a_2)(R_b^3 - R_a^3) 2h\rho/3 \tag{8}$$

where L is the total torque about the vertical axis. Here, a_1 and a_2 are the lateral accelerations normal to the lock on the two different materials, R_a and R_b are the common inner and outer radii of the pendulum and the quantity $h\rho$ is defined by the relation $h_1\rho_1 = h_2\rho_2 \equiv h\rho$ where h_1 and h_2 are the heights of the two different test materials with densities ρ_1 and ρ_2. When the mass dipole of the pendulum is rotated normal to the lock, the total torque, of course, vanishes. Setting $L = K\vartheta$ for the present case, it is seen that a deflection of $\vartheta = 1$ arcsec from the parallel condition in the experiment would correspond to a differential acceleration on the two masses of

$$\Delta a = 1.31 \times 10^{-9} \text{ cm/sec}^2. \tag{9}$$

Hence, the new pendulum suspended on a 5-μm fiber actually was about a factor of 4 more sensitive than the pendulum used in the prototype experiment with a 10-μm fiber. Finally, from the data in Fig. (2), it may be seen that this same deflection of 1 arcsec would correspond to an intrinsic isospin coupling coefficient of

$$\alpha_o = 3.64 \times 10^{-6} / 0.0134 \approx 2.7 \times 10^{-4} \tag{10}$$

for $\lambda = 100$ m.

Working with 5-μm-diameter tungsten fibers was a major difficulty. The wire used had been manufactured and gold plated by the Philips Company for the Yale particle physics group and was kindly provided by Jack Sandweiss. The fiber came tightly wound on a small-diameter spool and seemed to remember its former radius of curvature for days after it had been unwound. The fiber, whose diameter was some ten times the wavelength of visible light, was hard to see without magnification. I used a pair of binocular lenses mounted on a headband similar to those worn by dentists. In order to see the fiber clearly, I used a 1000-W photoflood lamp shining in from the side while the fiber floated in the air over a black background. It was important not to touch the fiber with anything over the length to be used in the experiment. The slightest touch invariably caused it to break at that point when under tension. The method used to handle the fiber is illustrated in Fig. 9. A pair of bent long-nose pliers is shown holding the center pin from a male BNC connector. (The rubber band kept the pliers closed during the procedure.) The BNC pins came gold-plated and the shop drilled a 10-mil hole through the axis of each pin, which holes were counter sunk on the end where the fiber was to enter.

Attached to the pin in Fig. 9 is a piece of small-diameter Tygon tubing whose other end is fit over the needle of a hypodermic syringe. The approach used was to move the pliers around on the work table until the open end of the pin was near the end of the floating fiber. At that point, pulling the plunger of the syringe out quickly would suck the fiber through the hole. The Tygon tubing was then removed carefully and the fiber soldered to the pin at the end where it protruded. Once a pin was attached to one end of the fiber it became easier to manage and the whole operation was then repeated at the other end, while adjusting the active length of the fiber to be 13 in.

I made up seven such fibers with pins soldered to each end. These were "aged" for many weeks in New Haven by suspending a 5.5-gm weight on each. It was hoped this aging process would help to minimize drift during the experiment. (I learned later at the lock site that a much more effective "aging process" consisted of giving the fiber a torsional "kick" by rotating the top end back and forth by $\approx \pm 1$ degrees over a period of ≈ 12 sec every two hours.) The fiber pins could be attached interchangeably by friction fit to the hole in the top of the pendulum mirror post shown in Fig. 8. The pin at the other end was attached to a female BNC connector mounted on the rotational adjustment assembly. The mirror was aluminized on both sides so that the pendulum could be rotated by 180° to check for systematic errors.

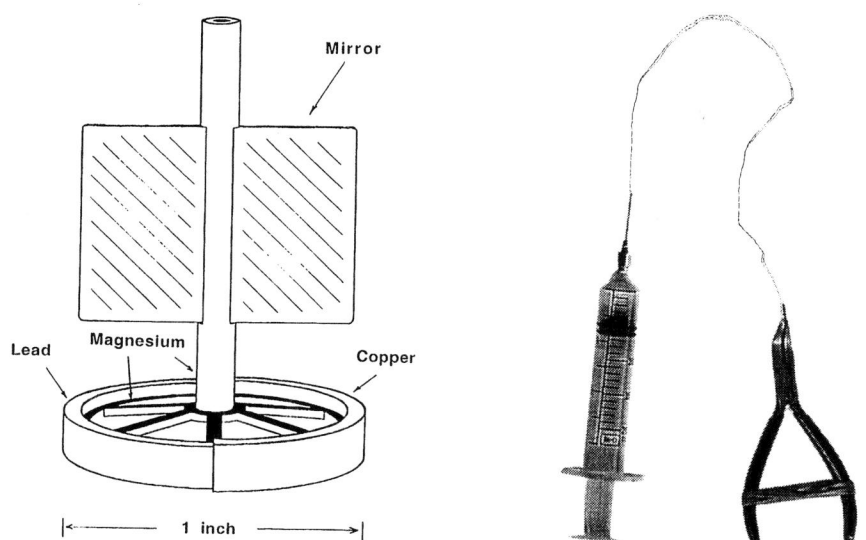

FIGURE 8. Toroidal Pb-Cu pendulum.

FIGURE 9. Handling 5-μm fiber.

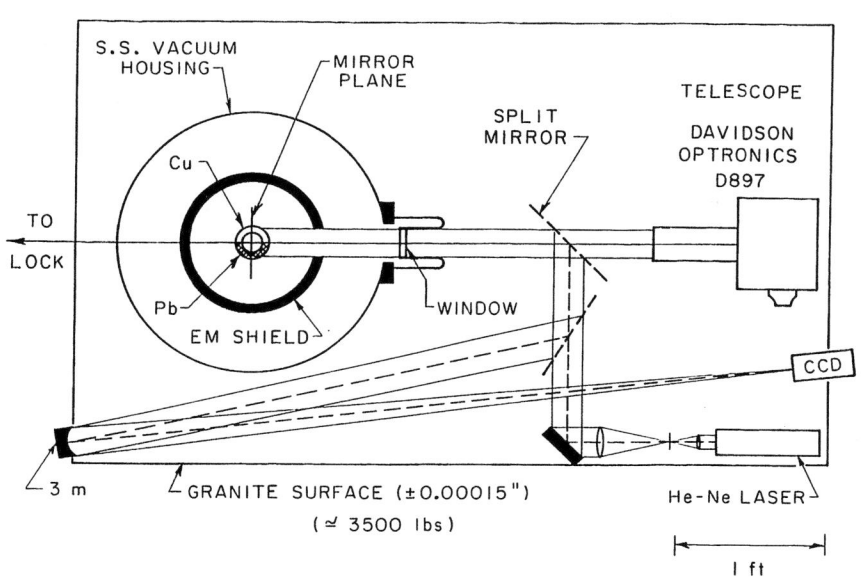

FIGURE 10. Schematic diagram of apparatus used in final experiment.

The first of the seven fibers was sacrificed in demonstrating that it could not withstand the vibration during a lockage without further isolation of the surface plate. (See below.) All the remaining 5-μm data were taken using fiber No. 2. (See below.) (Fibers No. 2 through 7 bit the dust at the lock site later when I attempted to suspend an awkwardly large pendulum made from polyethelene and lead.)

Apparatus used in the Final Experiment

As shown in Fig. 10, a number of things were added to the apparatus of the prototype experiment. A 1/4-in thick, grounded aluminum cylinder whose inner surface was coated with gold was placed around the pendulum to improve electrostatic shielding. Everything within that cylinder (including the pendulum) was coated with gold. Three layers of μ-metal shielding were wrapped around the aluminum cylinder for magnetic shielding. Although not obvious from the drawing, a precision micro-stepping motor made by the Daedal Company was placed in the vacuum housing on top of the stand from which the pendulum fiber was suspended. The motor was also wrapped in μ-metal shielding. This motor permitted rotating the fiber mounting in steps as small as a quarter of an arcsecond and at varying rates. It also permitted rotating the pendulum 180° without entering the truck. The motor was controlled from an instrument panel located 75 feet away in the RV rented for the experiment. A laser telescope (lower right hand corner) was added to monitor both the rotation and tilt of the pendulum mirror. The reflected laser beam was focused on a high resolution CCD video camera whose output was fed into a video "keyer." The "split mirror" was originally intended to be a large optical flat coated to transmit the blue-green light from the autocollimating telescope while reflecting the red beam from the He-Ne laser. That approach was to prevent the two light signals from interfering in the telescope detection circuit. However, after three large and expensive optical flats were ruined in attempts by a commercial company to provide the correct coating, that approach was abandoned. A split mirror was used for a large number of runs to check ground tilt and provide an independent check of the telescope calibration. After it became clear that ground tilt during lockages was not a significant source of systematic error, the split mirror was removed to provide maximum resolution of the autocollimating telescope.

The entire 3500-lb surface plate was suspended by four automobile-suspension coil springs placed at the corners. The plate could be clamped rigidly to the frame for transportation and while adjusting the apparatus. With the clamp removed, the entire apparatus could oscillate up and down at a resonant frequency of 0.3 Hz. Since the main ground vibration was at about 16 Hz, the suspension provided good vibration isolation. After the pendulum was initially suspended and the vacuum chamber pumped down, no one went near the truck. The truck itself was covered with a 4-in layer of Styrofoam insulation held on by ropes and covered by a large tarpaulin.

Data Recording

A total of five video cameras was used to record data during the experiment, each having a resolution of about 400 lines. The arrangement is shown schematically in Fig. 11. Two cameras (#1 and #2) were located in the Lockmaster's control booth at the Southwest corner of the lock where one camera focused on the depth gauge and the other viewed the lock itself. Signals from those two cameras were multiplexed using VHF Channels 3 and 4 to prevent the need to run two separate cables over the 600-ft distance to the RV. It was important to know when large boats were in the lock and when each lockage started. A microphone on one of those cameras also monitored sounds in the booth, ranging from the whirring noise of the motors opening and closing drain valves to the conversations between the lockmaster and the tugboat captains. I was also in radio communication part of the time with the lockmasters. Camera No. 3 monitored the analog output meter (running from -4 to +4 divisions) from the high-resolution telescope in the truck, while Camera #4 showed various meters including vacuum and temperature gauges and a Hall-effect gaussmeter. The outputs from those cameras were fed into a "Quad Splitter", which combined the four images into one video frame. The output of the Quad Splitter went to a "Video Keyer" which superimposed the spot from the laser telescope. The central horizontal and vertical lines in the composite image from the Quad Splitter provided x- and y-axes for the spot location. Finally, the signal from an SMPTE Video Clock (basically, a video frame counter) which provided time measurements every 60th of a second was added to the combined video image. The SMPTE video time code is designed so that "30 frames/sec" actually amount to 29.97002617 frames/sec.[16] Hence, the clock ran slow by about 1 part in 1000, which was of no great importance in the experiment. The microphone signal from the control booth was recorded on one of the audio channels. The composite video output was fed into a pair of SVHS video recorders (each with a resolution of about 400 horizontal lines) and viewed with a high-resolution monitor in the RV. The camera aimed at the autocollimator output also entered the date on each video record.

Without all of that video technology, it would have been impossible to keep track of everything that was going on and also have a permanent record of the data for later analysis. For example, by viewing the tapes in "fast forward", it was easy to see tug boats and barges bouncing off the lock walls -- a major previously

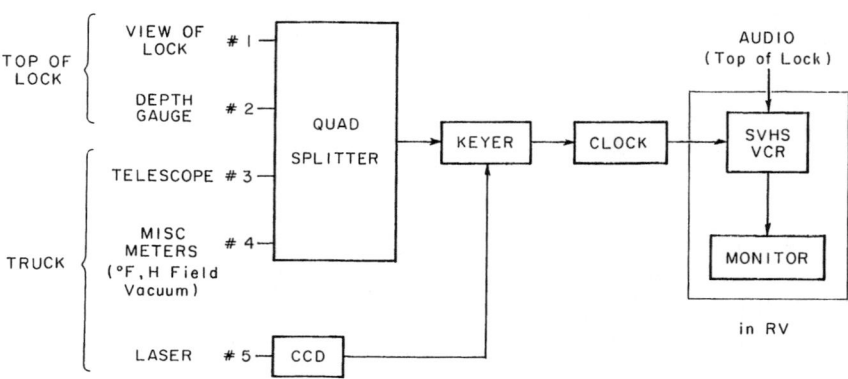

FIGURE 11. Video method used to record data.

FIGURE 12. Sample video frame of recorded data. Clockwise from upper right: Telescope output meter; view of lock; SMPTE clock; depth gauge; laser telescope spot; thermometer & gaussmeter.

unexplained source of vibrational noise in the prototype experiment. It was necessary to take data continuously 24-hours-a-day for many weeks in succession. There was seldom more than a half-hour warning before a lockage occurred and they were just as likely to take place in the middle of the night as in the daytime. For that reason, two VCR's using two-hour SVHS tape cassettes were used. One video frame of data recorded in this manner is illustrated in Fig. 12. The only problem with this approach was that there were some 200 two-hour video tapes to analyze after returning home from the lock. In practice, it is hard to do that analysis by hand in less time than it took to record the data in the first place. Eleven years after the data were taken, viewing those tapes was just like being at the lock and doing the experiment all over again. In the process of analyzing the tapes, I gained much insight regarding the sources of noise and systematic error in the experiment.

Vertical Gradient

It had seemed desirable to perform a rough check of the sensitivity of the apparatus to the vertical gradient in the lateral gravitational force from the water. Numerical calculations showed that this gradient should be $\partial a_y/\partial z \approx 0.4 \times 10^{-7}/\sec^2$ at the pendulum location. For this test, the split-cylinder pendulum shown in Fig. 13 was made entirely from copper and was suspended from a 10-μm fiber. That pendulum had a 1-in diameter and a mass of 18.43 gms (including the mirror which is not shown in the figure). The total moment of inertia was $I = 75.89$ gm-cm^2. The period was 483 sec and the torsion constant was 0.0128 erg/rad. It was comparable in those properties to the pendulum used in the prototype experiment (Fig. 5), except that it was deliberately made to be sensitive to the vertical gradient. The torque produced on this pendulum from filling the lock should have resulted in a telescope deflection of ≈ 4.4 arcsec upon filling the lock. Although there was indication of a deflection of about that amount, it was close to the noise level in single runs. It would have required quite a bit of data averaging to provide an accurate measurement of the gradient, and it was decided to spend the remaining time available looking for the "fifth force" instead. The data taken with the pendulum in Fig. 13 did provide assurance that vertical gradients would be of no importance in the final results. In addition, data were obtained using the laser telescope during those runs to measure ground-level tilt during lockages.

FIGURE 13. Split-cylinder Cu pendulum used on a 10-μm fiber to check the sensitivity of the apparatus to the vertical gradient of the normal gravitational force. (The mirror assembly is not shown, but was the one used on the toroidal pendulum in Fig. 8.)

Systematic Error from Ground Tilt

D. R. Long [17] noted that some change in ground tilt at the pendulum location would be expected when such a huge mass of water is added or drained from the lock and could produce a significant systematic error through cross-coupling effects in the auto-collimating telescope. As previously described by the author,[18] direct measurement of cross-coupling in the telescope showed that a 1 arcsec instrumental shift would require a tilt in the ground plane of about 0.9 arcmin. In the final experiment, direct measurement of pendulum tilt using the laser telescope in Fig. 12 during lockages showed that the maximum change in tilt from filling the lock was ≈0.3 arcmin, corresponding to a maximum cross-coupling shift of ≈0.33 arcsec. This effect may have introduced the small negative shift in the angular difference during lockages that is shown in later Tables 5 and 6 which amounted to about -0.025 meter divisions (≈0.15 arcsec.) After showing the effect was negligible within the expected error of the experiment, the split mirror in Fig. 10 was removed so that the full aperture of the high-resolution telescope would be exposed to the pendulum mirror.

Fiber Drift

The final data were taken in some 40 runs typically lasting about 2 hrs each. Before each run, the pendulum was given a torsional "kick start" consisting of a triangular rotational velocity pulse going linearly from 0° to 1° and back again to 0° over a period of about 12 seconds. The object was to start the pendulum rotating with enough energy to last for a full two-hour run without displacing it from its equilibrium position. A slight angular correction for drift from the preceding run was generally added to the new "kick-start" pulse in this process. (The damping factor of the new pendulum on the 5-μm fiber was about 5 times faster than that for the original pendulum on a 10-μm fiber.) In this way, the initial conditions and drift during successive runs became roughly similar after a day or two. The start pulse and correction were applied from the RV control panel 75-ft away using the Daedal micro-stepping motor from which the pendulum fiber was supported.

It was easiest to keep track of the drift over long times by plotting those corrections. That has been done in Fig. 14 parts a) and b) over the two-week marathon data-taking session that provided the results for the final experiment. In these figures, the number of steps for each adjustment of the micro-stepping motor is plotted as a function of time, where 25,000 steps corresponded to a 2° rotation. The basic idea was to wait until the pendulum had settled down before making a serious measurement. The dots in the figure correspond to separate runs in which a drift correction was made. Run numbers for potentially useful lockages were indicated above each curve. In both cases the mass dipole was kept closely parallel to the lock, except that a -180° rotation of the pendulum made over ≈ 2.5 hrs preceded the data shown in Fig. 14 b).

The results for the first four days (Fig. 14a) were almost dismal enough to make one pack up and go home. It then seemed very unlikely that the pendulum would *ever* settle down. But, quite remarkably, the drift reduced sharply during the last three days of the first week and there were occasional periods when it even became as low as ≈1 meter division per hour. Of course, those periods had to coincide with a lockage without tugboats and barges to be usable in the experiment. As noted before, the sudden reduction in drift probably arose from the repeated "kick-start" technique applied at two-hour intervals. Data taken during the nearly horizontal periods are summarized in later tables.

The large slowly-varying drift probably arose from the wire trying to return to its previous tightly-coiled state on its original spool. The sharp discontinuities in the drift may arise from crystal dislocations stimulated by vibration, especially during lockages. Looking at the drift on a finer and finer scale during runs showed that there were often small discontinuities in the drift rate amounting to a few meter divisions per hour. Sometimes those discontinuities coincided with the drain or fill periods in a lockage, but at other times they did not. During the runs reported here, the principal error was due to slight changes in the rate of drift which in itself was always negative (counterclockwise looking at the rotation from above), was not correlated with the time of day, and was randomly distributed about an average value ≈ -1 meter division per pendulum half-period with a standard deviation of about 0.5 divisions. The counterclockwise characteristic of the drift suggests that it might have resulted from the Coriolis force from the Earth's rotation interacting

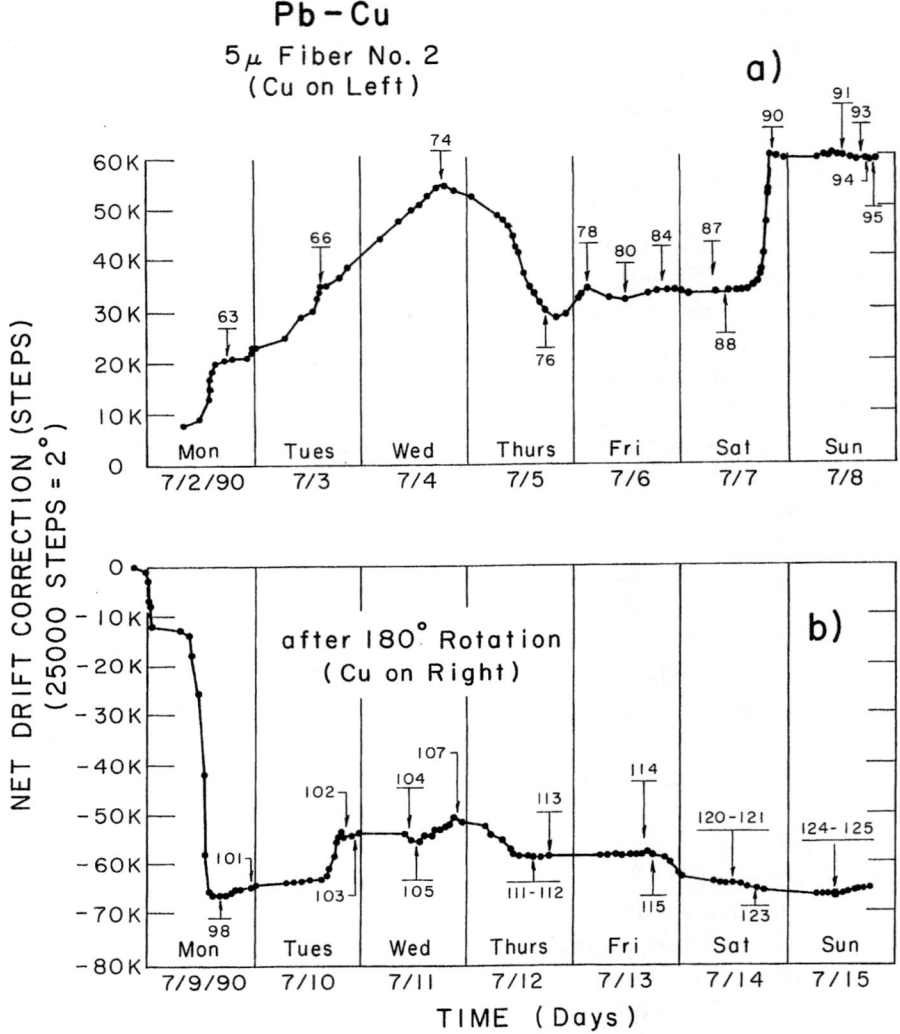

FIGURE 14. Drift corrections during the period when the final data were taken. The numbers indicate potentially useful runs.

with vertical vibrational motion of the pendulum induced by ground vibrations.[3] A similar counterclockwise drift ascribed to ground vibrations was reported in the Eötvös experiment of Roll et al. [19]

Digital Filtering and Second Harmonic Oscillation

In addition to the drift, a modulation of varying amplitude at precisely the pendulum frequency was generally encountered in the data. This effect is illustrated for an extreme case in Fig.15a. Regardless of its source, it is easily eliminated by digital filtering through an algorithm of the type

$$\text{For } i = 2 \text{ to } N-1, \text{ let } \vartheta'_i = [(\vartheta_{i-1} + \vartheta_{i+1})/2 + \vartheta_i]/2 \tag{11}$$

where N is the number of points in the run. One loses the first and last points in the process, but with great reduction in this coherent source of noise. (See Fig. 15b.) Obviously, we are not interested in structure within a single pendulum period. If a Fifth Force were actually present, it would show up over an extended number of periods before and after the lockage.

I believe this coherent modulation was due to a small amount of second harmonic introduced in the pendulum motion: The method of angle measurement used is sensitive to variations in the slope at the zero crossings (see Fig. 6); the presence of a second-harmonic component would modulate the slope at zero-crossings at precisely the pendulum frequency; and there were generally two components present in the exponential decay of the measured slopes, suggesting the presence of two coupled oscillating modes. (See Fig. 16.)

If we suppose that a fractional amplitude ε of second harmonic was introduced by the "kick-start" excitation process in the initial oscillation and didn't change (e.g., through nonlinear effects), the deviation angle would vary as

$$\vartheta = \exp(-\mu_1 t)\sin\omega t + \varepsilon\exp(-\mu_2 t)\sin 2\omega t \tag{12}$$

where $\omega = 2\pi/T$ and where μ_1 and μ_2 are the decay rates of the two modes. The slope, $m = d\vartheta/dt$, at the zero-crossings ($\omega t = n\pi$) would then vary as

$$d\vartheta/dt = (2\pi/T)\exp(-n\mu_1 T/2) \{(-1)^n + 2\varepsilon\exp[+(\mu_1 - \mu_2)n T/2]\} \tag{13}$$

where n=0,1,2,3.... The first term in the wiggly brackets oscillates over the range ± 1

---

[3] The reader may wish to check this interpretation by repeating the experiment in the southern hemisphere.

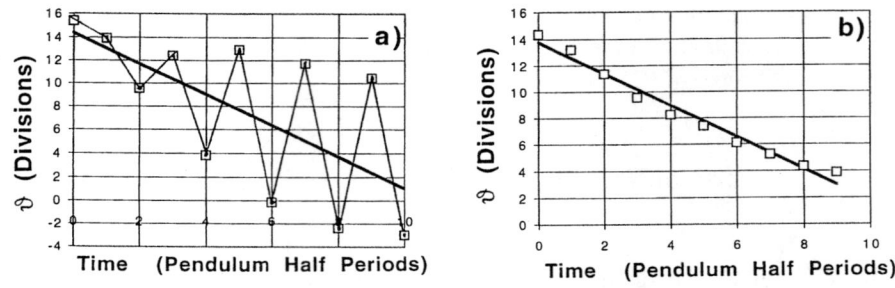

FIGURE 15. a) Amplitude modulation at the pendulum period in the most extreme case encountered (Run 75.) b) The same data after applying the digital-filtering algorithm discussed in the text.

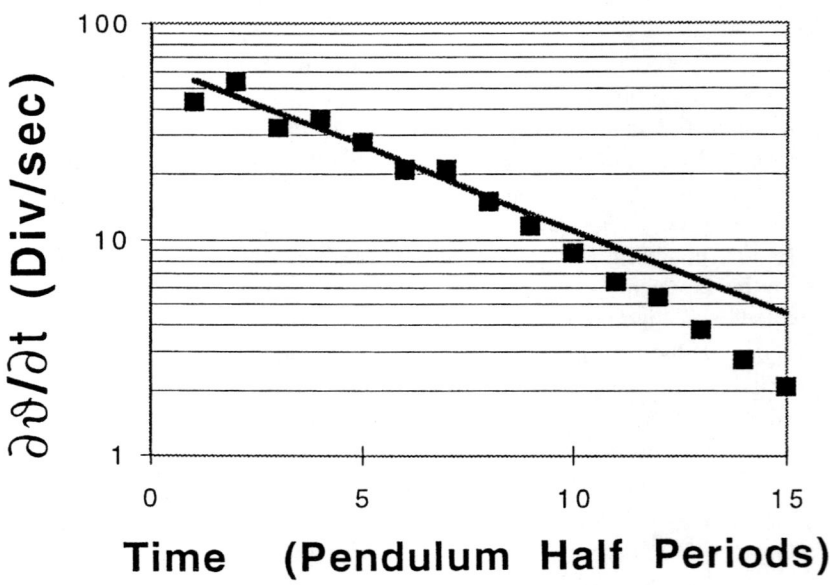

FIGURE 16. Double exponential decay in the slope measurements encountered in Run 76. The fluctuations about the straight line at short times are due to the same type of coherent modulation at the pendulum frequency shown in Fig. 15a).

and for $\mu_1 > \mu_2$ (a relationship only occasionally found in the data), the second term would build up as the overall signal damped out -- an effect that *was* generally observed throughout the experiment. The size of the fractional term ε varied from very large to almost nonexistent during runs. Although the torsion-pendulum equation alone would not support second harmonic oscillation in the absence of nonlinear terms,[4] the vertical vibrations (especially during lockages) would certainly couple in transverse violin-string modes, half of which have nodes at the mid-point of the fiber. It seems likely that those nodes would also serve as a node to permit a torsional mode at twice the normal pendulum frequency. The observed coherent modulation usually grew after a lockage. From the data such as that in Fig. 16, the high-Q torsional mode loses energy by coupling to a higher-loss mode (probably the 2nd-harmonic mode) after the lockage. In any case, the digital filtering algorithm used [Eq. (11)] eliminates the effect of this coherent noise source on the data and would not interfere with the observation of a "fifth force."

Magnetic Fields

Variations from the Earth's magnetic field were encountered during locakages of large barges and tug boats. These fields were monitored with a Hall-effect gaussmeter. Typically, the largest changes amounted to about ±13 mG. Because the pendulum assembly was wrapped in three layers of μ-metal shielding and data were *only* taken when there were no large boats or barges in the lock, magnetic field effects may be safely ignored. The three layers of μ-metal shielding wrapped around the Al cylinder in Fig. 10 reduced the external magnetic field near the apparatus ($\approx$ 0.6 G) by a factor of 15 at the pendulum location.

Temperature Variation

The temperature variation in the truck during lockages was generally less than $\approx$ 0.1 or 0.2 °F, although large diurnal variations did occur. Because the truck was well-insulated and shielded from the Sun, the peak in temperature was shifted from the daylight hours to about 2 am. A small electric heater with a thermostat set at 80°F was left running in the back of the truck, but was off most of the time during June and July of 1990. The temperature in the truck varied slowly from day-to-day

---

[4] For example, if $K = K_0[1+6\varepsilon\cos(\omega t)]$ where $\varepsilon \ll 1$, then $\vartheta \approx \sin(\omega t) + \varepsilon \sin(2\omega t) + \text{Order}(\varepsilon^2)$ where $\omega^2 = K_0/I$ and I is the moment of inertia.

and ranged from about 85°F at the start of the two-week run (July 2, 1990) to a maximum of 103°F (July 12, 1990) and down to 95°F on the last day (July 15, 1990.)

Final Experimental Results

The results obtained for a large number of lockages are shown in Tables 5 and 6, where the Run numbers are shown in Fig. 14. On Sunday, July 8, 1990 at 11:28 pm and before Run 96, I started a -180° rotation of the pendulum at a constant rate of $\approx 1°$ per minute over ≈ 2.5 hr period. That sharply increased the drift to $\approx -5°$ over the next 12-hr period. (See Fig. 14b.) After that, the system recovered.

In all cases, the data were filtered using the digital algorithm given in Eq. (11). A symmetrically-located, equal number of points was used on each side of each lockage and an unweighted least-squares fit was made of that data to a straight line. The resultant linear drift was typically ≤ 10 meter divisions/hr over each run. This drift was subtracted from the data and an average value and standard deviation were then computed for the difference in angular deviation due to the lockage. No data were used when large tug boats and barges were in the lock, although several runs were taken in the presence of very small motor boats. Tug boats frequently pushed as many as five barges which took up the entire cross-sectional area of the lock. The procedure in those cases was to use only the data substantially before and substantially after the boats were in the lock. The drift was always determined for the entire period of the data and an unweighted least-squares fit was made to a straight line for that period. Because there were generally two or three pendulum half-periods before the large boats came in and went out, the "before" and "after" points for the analysis were substantially separated and those runs had the largest noise levels.

The weighted-mean net deflection of the pendulum during a lockage over some 20 runs in each of the two orientations was less than 0.01 meter divisions with a weighted standard deviation of less than 0.5 divisions. The 180° rotation did show a small net, negative value well-within the standard deviation, which probably had its origin in ground tilt. No net effect greater than the standard deviation was encountered in any of the 40 runs. That, of course, implies that the noise level was not perfectly gaussian and that the error in the overall measurement is likely to be less than one standard deviation; i.e., for purely random noise, the data should only fall within one standard deviation of the mean about 68% of the time.

TABLE 5. Summary of all 40 runs analyzed from the groups in Fig. 14a and Fig. 14b. Angles are given in telescope meter divisions. The average (-0.018 Div) was about equal to the weighted mean given in the table. (Also see Table 6.)

Run Number	Lockage Direction	$\Delta\vartheta_{Full-Empty}$ (Divisions)	Error (σ) (Divisions)
66	Full → Empty	-0.17	± 2.11
71	Full → Empty	-0.45	± 0.64
73	Full ← Empty	+0.10	± 0.98
74-1	Full → Empty	+0.76	± 0.80
74-2	Full ← Empty	+0.50	± 0.79
75	Full → Empty	-0.023	± 0.84
76	Full → Empty	+0.075	± 2.08
77	Full ← Empty	-0.26	± 0.65
78	Full → Empty	+0.98	± 1.25
80-1	Full → Empty	-0.04	± 2.10
80-2	Full ← Empty	+0.02	± 0.10
82-1	Full → Empty	-0.002	± 1.28
82-2	Full ← Empty	-0.02	± 0.42
84	Full → Empty	+0.08	± 0.88
85	Full ← Empty	-0.06	± 0.78
87-1	Full → Empty	+0.17	± 0.42
87-2	Full ← Empty	-0.065	± 0.17
88	Full ← Empty	-0.27	± 0.33
90	Full → Empty	+0.16	± 1.08
91	Full ← Empty	+0.06	± 0.45
93	Full → Empty	-0.19	± 0.54
94-1	Full ← Empty	-0.15	± 1.12
94-2	Full → Empty	-0.008	± 0.29
97	Full ← Empty	-0.014	± 0.43
98	Full → Empty	-0.59	± 2.40
107	Full ← Empty	-0.52	± 1.38
108	Full → Empty	-0.020	± 0.113
111	Full ← Empty	-0.049	± 0.768
112	Full → Empty	-0.12	± 1.16
114	Full ← Empty	-0.46	± 1.82
115	Full → Empty	-0.23	± 1.58
116	Full → Empty	+0.017	± 0.960
118	Full → Empty	-0.20	± 0.74
120-1	Full ← Empty	-0.006	± 0.684
120-2	Full → Empty	-0.008	± 0.115
121	Full → Empty	+0.56	± 2.64
122	Full ← Empty	-0.06	± 0.78
123	Full ← Empty	-0.007	± 0.834
124-1	Full ← Empty	-0.163	± 0.472
124-2	Full ← Empty	-0.012	± 0.721
Total Weighted Mean and error:		-0.0185	± 0.365

TABLE 6. Summary of the ten least-noisy runs from each group in Fig. 14. The weighted means and weighted standard deviations are given for each group separately and for the entire group of twenty lowest-noise runs. Angles are again given in telescope meter divisions. A net positive deflection in each case meant that the mass on the right (facing the lock) rotated towards the lock. There was a consistent shift of the weighted mean within the standard deviation in both groups by about -0.03 meter divisions ($\approx$-0.18 arcsec) that was probably due to ground tilt when the lock was filled from cross-coupling in the auto-collimating telescope. This suggests that the overall statistical error in the measurements is probably somewhat less than σ.

Run Number	Lockage Direction	$\Delta\vartheta_{\text{Full-Empty}}$ (Divisions)	Error (1 σ) (Divisions)
a) Pb on the right facing the lock:			
71	Full $\rightarrow$ Empty	-0.45	± 0.64
77	Full $\leftarrow$ Empty	-0.26	± 0.65
80-2	Full $\leftarrow$ Empty	+0.02	± 0.10
82-2	Full $\leftarrow$ Empty	-0.02	± 0.42
87-1	Full $\rightarrow$ Empty	+0.17	± 0.42
87-2	Full $\leftarrow$ Empty	-0.065	± 0.17
88	Full $\leftarrow$ Empty	-0.27	± 0.33
91	Full $\leftarrow$ Empty	+0.06	± 0.45
93	Full $\rightarrow$ Empty	-0.19	± 0.54
94-2	Full $\rightarrow$ Empty	-0.008	± 0.29
	Weighted	Mean: -0.031	Weighted σ = 0.441
b) Cu on the right facing the lock:			
97	Full $\leftarrow$ Empty	-0.014	± 0.43
108	Full $\rightarrow$ Empty	-0.020	± 0.113
111	Full $\leftarrow$ Empty	-0.049	± 0.768
118	Full $\rightarrow$ Empty	-0.20	± 0.74
120-1	Full $\leftarrow$ Empty	-0.006	± 0.684
120-2	Full $\rightarrow$ Empty	-0.008	± 0.115
122	Full $\leftarrow$ Empty	-0.06	± 0.78
123	Full $\leftarrow$ Empty	-0.007	± 0.834
124-1	Full $\leftarrow$ Empty	-0.163	± 0.472
124-2	Full $\leftarrow$ Empty	-0.012	± 0.721
	Weighted	Mean: -0.020	Weighted σ = 0.155
a) + b)	Total Weighted	Mean: -0.026	Weighted σ = 0.331

Telescope Calibration

The analog meter on the telescope was only approximately calibrated at the start of the experiment, and it seemed best to record all of the angular measurements in meter divisions with the object of doing a precise calibration at the end of the experiment. When the experiment was completed, the Daedal rotation unit was placed on the granite table with the pendulum resting on it and carefully aligned with the telescope axis. Rotating the pendulum by calibrated amounts showed that

$$1 \text{ Telescope Division } \approx 6 \text{ arcsec.} \tag{14}$$

The Daedal microstepping motor itself was quite remarkable. Although it had some backlash, the resetting error when an angular position was reversed and approached from the same direction was less than 3 arcsec. (There were ≈ 3.47 steps per arcsec.)

Errors

The errors in individual angular measurement are easiest to estimate from the small-angle approximation to Eq. (7),

$$\vartheta \approx m\ (t_1 - t_2)/4. \tag{15}$$

(Also see Fig. 6.) The half-period time measurements were all made to within 1 part in 70,000. Hence, the error in determining (t_1-t_2) was ≈ 1 part in 35,000 (aside from the correctable and unimportant absolute error in the SMPTE clock, which ran slow by about 1 part in 1,000.) The most error was in the slope determinations at zero crossings, which were largest at early times in each run, where $m \approx 50$ meter divisions per sec. The slope was determined from the time required for the telescope meter to deflect by about 6 meter divisions. Such early time-interval measurements were good to within ≈ 1 part in 60. Allowing for a meter nonlinearity of $\approx 2\%$, values of m determined at the start of a run had an uncertainty of $\approx 3\%$. The first measurements in a run were seldom used because the lockages typically occurred at least a half-hour into the run. As the run progressed and the pendulum velocity at zero crossings slowed down, the accuracy of the slope measurements approached the accuracy of the time-interval difference measurements quoted above, except for limiting errors $\approx 2\%$ from meter nonlinearity at the extreme ends of the runs. A correction for the slope change due to decay of the pendulum motion was not made.

However, that effect and the meter nonlinearity were about the same for all runs. Hence, individual relative values of ϑ were good within a few percent. (The uncertainty in absolute values using the calibration in Eq. (14) was about 10%.)

The principal source of error in the measurements was discontinuous change in the drift rate during runs. The drift rate was typically very constant on either side of the discontinuity, yielding a "V-shaped" appearance to the differences from an overall fit to one straight line throughout the entire run. Computed standard deviations from a straight line fit on either side of the "V" were often an order of magnitude smaller than those quoted in Tables 5 and 6. The drift discontinuities usually occurred during the Fill or Drain periods in the lockage. Here, it was especially important to use an equal number of points in the final analysis that were symmetrically located about the lockage. Not to do so could generate the illusion of a "fifth force" merely from changes in drift rate. Without a quantitative understanding of the causes of the drift and changes in its rate, a more detailed analysis of the data could not be made. Because the drift during runs was always counterclockwise as seen from above the apparatus, it seems likely that the drift arose from vibrational motion of the pendulum interacting with the Coriolis force due to the Earth's rotation. Changes in the nature of the vibration during lockages would then explain the drift-rate discontinuities.

Comparison with other Results

Combining the results for the 20 lowest noise-runs shown in Table 6 with the telescope calibration in Eq. (14) and the inherent sensitivity of the pendulum given by Eqs. (9) and (10), it is apparent that no "fifth-force" effect was observed within the limits of error (≈ 1.8 arcsecs) in angular measurement. No differential acceleration acting on the two masses by the water was observed in excess of $\Delta a \approx \pm 2.4 \times 10^{-9}$ cm/sec^2. By coincidence that limit is in the middle of the error range of the original Eötvös results quoted in Table 1. One is tempted to say that little has changed in the last hundred years. However, the detailed analysis of the original Eötvös experiment given by Dicke in 1961 leaves one wondering how the Baron could ever have claimed such a low limit of error in the first place. Among other things, his original apparatus was actually *designed* to detect vertical gravitational gradients and would have been extremely sensitive to horizontal gradients. According to Dicke,[1] Eötvös quoted probable errors of about one 200th of the smallest scale division on his telescope, picked the center of the spot within a 40th of the full width of the diffraction pattern, and the presence of his own body would have created a deflection at least 200 times his quoted error. One has to have done one of these experiments oneself to appreciate the magnitude of that feat! In view of

Table 7. Comparison of different author's notation

Author	Coefficient of ΔV	Mixed-Coupling of charges
Fischbach et al [4]	α	---
Boynton et al [12]	$\xi = -\alpha$	$\beta(N+Z)/\mu + (1-\beta)(N-Z)/\mu$
Bennett [13]	$\alpha = \alpha_0 C_1 C_2$	$\beta(N+Z)/\mu + (1-\beta)(N-Z)/\mu$
Cowsik et al (1988) [11]	$\alpha_{12} = q_1 q_2 \xi = \alpha_0 C_1 C_2$	$\beta(N+Z)/\mu + (1-\beta)(N-Z)/\mu$
Cowsik et al (1990) [14]	$\alpha = q_1 q_2 \xi = \alpha_0 C_1 C_2$ $\beta = 1/(1 + \tan\vartheta)$	$(N+Z)\cos\vartheta/\mu + (N-Z)\sin\vartheta/\mu$
Adelberger et al (1990) [15]	α_5 q_5 q_5' $\alpha_0 = \alpha_5 \cos\vartheta_5$ $\beta = -\tan\vartheta_5 /2$	$(N+Z)\cos\vartheta_5/\mu + Z\sin\vartheta_5/\mu$

Dicke's conclusions, the notion of examining systematic trends within his quoted limit of error (see Table 1) seems absurd.

From the lock calibration in Fig. 2, it may be shown that the present experiment yielded a 2-σ limit for the intrinsic isospin coupling coefficient of $\alpha_0 \approx \pm 0.001$ at $\lambda \approx 100$ m. The sensitivity, of course, falls off at shorter values of λ and saturates above $\lambda \approx 200$ m. Comparisons with other work are hard to make because the results are often quoted using mixed-coupling models that are sometimes rather obscure and are almost as numerous as the authors reporting them. (See Table 7.) One would think that the easiest comparisons to make would be with the limiting differential acceleration observed between two test masses against a given source. But many authors omit reporting that data entirely. (See Table 8.) The recent emphasis on coupling schemes to describe limits on a possible effect that has no real theoretical basis is somewhat perplexing. The fundamental aspect of the problem really is whether or not a material-dependent difference in gravitational acceleration exists. The answer to that question seems clearly to be, "No", within very stringent limits in the cases studied.

A comparison of the present 2-σ limit for isospin coupling with those of a number of other authors is shown in Fig. 17. One striking thing about the data is the way it moved from the initial results of Boynton *et al* to converge within narrower and narrower limits on zero. The three bottom entries in Fig. 17 resulted from three entirely different approaches to the problem over a wide range of experimental dimensions. However, each of those three used lead as either a test mass or a source.

TABLE 8. Comparison of 1-σ limits on differential acceleration.

Reference	$\Delta a \times 10^{10}$ cm/sec^2	Test Masses	Source
Thieberger [20]	850 ± 260	Cu-H$_2$O	Cliff
Fitch et al [10]	30 ± 49	Cu-CH$_2$	Sloping Terrain
Bennett [13]	25 ± 52	Cu-Pb	H$_2$O
Bennett[a]	2 ± 22	Cu-Pb	H$_2$O
Adelberger *et al* [15]	-0.15 ± 2.6	Be-Al	Pb

[a]Bennett, Wm. R., Jr. (present work; data of 1990)

The limits set by Adelberger *et al* (which are for their coupling angle of $\vartheta_5 = -63°$ -- hence, only approximately isospin coupling) hold as well at small values of the range for the Yukawa potential (but for $\lambda \gg 1$ m) and are especially stringent.

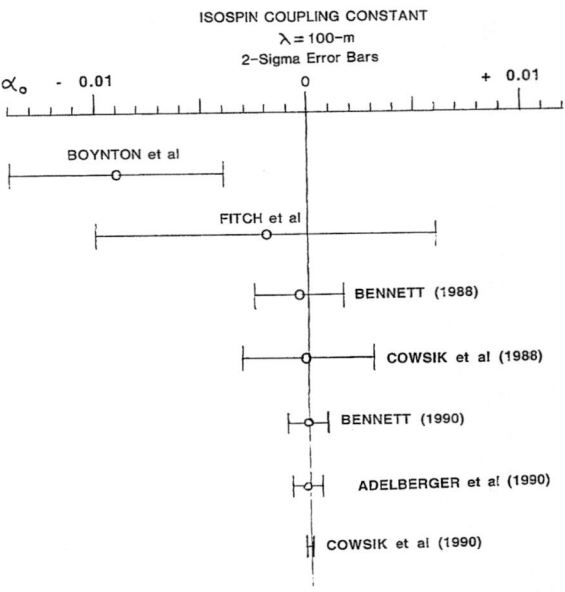

FIGURE 17. Comparison of different determinations of the intrinsic coupling coefficient α_0 for isospin coupling. The data shown are from the following sources: Boynton *et al* [12]; Fitch *et al* [10]; Bennett (1988) [13]; Cowsik *et al*, (1988) [11]; Bennett (1990), the present paper; Adelberger *et al* (1990), Ref. 15; and Cowsik *et al* (1990), [14].

ACKNOWLEDGMENTS

I am especially indebted to my wife, Frances Commins Bennett, for her help in carrying out the present experiments. She should really have been a co-author of this paper, but declined that invitation because her formal field is literature. I am also indebted to Paul Winborg, Wayne John, Roy Eakin, and Steve Featherston of the U. S. Army Corps of Engineers for permission to do the experiment and for the cooperation and hospitality shown by everyone at the Little Goose Lock and Dam. Lockmasters Wally Heubner, Charles Bartlett, Jeff Hawkins, and Ron Jewell were especially helpful. At Yale University, Skip Gemmel, Gil Vogel and other members of the physics department shop provided expert machining help. Richard Downing did the precision machining on the pendulum for the final experiment. Jayne Miller,

Dr. Peter Kindlmann and Dr. Christy Holland provided important back-up help at Yale while I was at the Dam. I would also like to acknowledge helpful discussions with Dr. Robert Krotkov and Dr. Janice Button-Shafer of the University of Massachusetts at Amherst, Dr. Jack Sandweiss of Yale University, and Dr. Eugene Commins of the University of California at Berkeley. The author is also indebted to Drs. Robert Apfel, Robert Wheeler, and Werner Wolf for their encouragement and for helping to arrange a modest grant from the Mayer Fund at Yale.

The Mayer Fund at Yale purchased the high-resolution Davidson telescope used; an NSF Grant in Gravitational Physics PHY-8903682 provided operating support for the final experiment; the remaining equipment support was provided by the late Vi M. Bennett and the author. I am also indebted to Dr. Jean Bennett, Wm. Robert Bennett and Nancy Bennett for the loan of their high-resolution video cameras used in the final experiment.

REFERENCES

1. See, for example, Dicke, R. H., "The Eötvös Experiment", *Scientific American* **205**, No. 6, pp. 84-94 (Dec., 1961).
2. Eötvös, R. v., Pekar, D., and Fekete, E. *Annalen der Physik* (Leipzig) **68**, 11-66 (1922).
3. See the review by Stacey, F. D., Tuck, G. J., Moore, G. I., Holding, S. C., Goodwin, B. D., and Zhou, Y. R. ,*Rev. Modern Phys.* **59**, 157 (1997).
4. Fischbach, E., Sudarsky, D., Szafer, A., Talmadge, C., and Aronson, S. H. , *Phys. Rev. Letters* **56**, 3 (1986)
5. See, Thodburg, H. H., *Phys. Rev. Letters* **56**, 2423 (1986).
6. Chu, S. Y., and Dicke, R. H., *Phys. Rev. Letters* **57**, 1823 (1986).
7. Thieberger, P., *Phys. Rev. Letters* **58**, 1066 (1987).
8. Stubbs, C. W., Adelberger, E. G., Raab, F. J., Gundlach, J. H., Heckel, B. R., McMurry, K. D., Swanson, H. E., and Watanabe, R., *Phys. Rev. Letters* **58**, 1070 (1987).
9. Adelberger, E. G., Stubbs, C. W., Rogers, W. R. F., Raab, F. J., Heckel, B. R., Gundlach, J. H., Swanson, H. E., and Watanabe, R., *Phys. Rev. Letters* **60**, 1801 (1987).
10. Fitch, V. L., Isaila, M. V., and Palmer, M. A., *Phys. Rev. Letters* **60**, 1801 (1988).
11. Cowsik, R., Krishnan, N., Tandon, S. N., and Unnihrishnan, C. S., *Phys. Rev. Letters* **61**, 2179 (1988).
12. Boynton, P. E., Crosby, D., Ekstrom, P., and Szumiko, A., *Phys. Rev. Letters* **59**, 1385 (1987).
13. Bennett, Wm. R. ,Jr., *Phys. Rev. Letters* **62**, 365-368 (1989).
14. Cowsik, R., Krishnan, N., Tandon, S. N., and Unnihrishnan, C. S., *Phys. Rev. Letters* **64**, 336 (1990).
15. Adelberger, E. G., Stubbs, C. W., Heckel, B. R., Su, Y., Swanson, H. E., Smith, G. and Gundlach, J. H., *Physical Review D*, **42**, pp. 3267-3291 (1990).
16. See, *Time Code Handbook* (Cipher Digital, Frederick, MD); also, Lehrman, P. D., "Decoding SMPTE" in *Electronic Musician* **7**, pp. 62-70 (1991).
17. Long, D. R., *Phys. Rev. Letters* **63**, 809 (1989).
18. Bennett, Wm. R. ,Jr., *Phys. Rev. Letters* **63**, 810 (1989).
19. Roll, P. G., Krotkov, R., and Dicke, R. H., *Annals of Physics* **26**, 442-517 (1964); p. 489.
20. Thieberger, P., Phys. Rev. **58**, 1066 (1987) (Used in present Table 7.)

Remarks On Search Methods For Stable, Massive, Elementary Particles

Martin L. Perl

Stanford Linear Accelerator Center, Stanford University
Stanford, California 94309, U. S. A.

Abstract. This paper was presented at the 69th birthday celebration of Professor Eugene Commins, honoring his research achievements. These remarks are about the experimental techniques used in the search for new stable, massive particles, particles at least as massive as the electron. A variety of experimental methods such as accelerator experiments, cosmic ray studies, searches for halo particles in the galaxy and searches for exotic particles in bulk matter are described. A summary is presented of the measured limits on the existence of new stable, massive particle.

1. INTRODUCTION

1.1 Stable Elementary Particles

I gave this paper at the 69th birthday celebration of Professor Eugene Commins, honoring his research achievements. This paper is about the experimental techniques used in the search for stable, *elementary,* particles at least as massive as the electron, 0.5 MeV/c^2. The particles may be neutral, may have unit charge or may have fractional charge. They may interact through the strong, electromagnetic, or weak force, or through some unknown force, as long as the force allows direct or indirect detection of the particle.

Particle stability is a loose criterion. For example, I discuss particles whose lifetimes are of the order of the lifetime of the universe so that searches for particles produced in the early universe can be meaningful. However, I also describe searches at accelerators for new particles with lifetimes sufficiently long to be directly detected, that is, longer than about 10^{-10} s Thus, a stable particle is operationally defined as one that can be observed in isolation and has sufficient lifetime to be detected by the search techniques discussed in this paper. At present the class of so defined stable particles has just two known members: the electron - a basic elementary particle - and the proton - a composite elementary particle.

The reader may wonder about the status in my stability classification system of particles such as the neutron, the muon and the π meson. They are not stable but my stability classification system is operational in the sense that there would be tremendous interest in the discovery of a *new* particle with such a lifetime. What about the large class of stable nuclei, ^{4}He for example? Their stability is based upon the intrinsic stability of the proton and on the stability of a neutron inside some nuclei, I don't think there is anything to learn in this area about the deeper nature of elementary particle stability.

These remarks are based on a recent long review paper [1] by my colleagues and myself on experimental, observational and theoretical aspects of searches for stable and massive elementary particles. I have abstracted from that review a summary of the measured limits on the existence of new stable, massive particle.

1.2 Dark Matter

Many present searches for massive elementary particles are motivated by the desire to identify the elementary particle or particles that compose what we call dark matter. But in this paper I consider more general search interests. An unknown, stable, massive particle may exist in numbers much too small to explain the amount of dark matter in the universe, but the existence of such a particle would be of great importance.

1.3 Massive Particle Production In The Early Universe

In models for massive particle production in the early universe, it is usually assumed the massive particle X and its antiparticle existed in thermodynamic equilibrium in the very early universe. Then, as the universe cools and expands, the X particles 'freeze-out' of equilibrium and become *thermal relics*, their final density depending upon their mass. Conventional cosmological theory leads to an upper limit on the mass of thermal relics so produced. Griest and Kamionkowski [2] give an upper limit of about 10^6 GeV/c^2.

There are proposals for getting around this thermal-relic mass upper limit. For example. Kolb *et al.* [3] have proposed processes by which particles with masses in the range of 10^{12} to 10^{16} GeV/c^2 might be produced in sufficient quantity to explain the abundance of dark matter. If one is searching for massive particles with less than dark matter abundance, these non-thermal relic processes become even more attractive.

1.4 Massive Particle Searches Using Colliders

Searches for stable, massive particles at high energy colliders have two obvious advantages. First, the search mass range is known and, if a production cross section is assumed, the existence of a hypothetical particle may be directly tested. Second, the stability requirement is less restrictive compared to other search methods, usually a lifetime greater than 10^{-10} s is sufficient for direct detection of the particle.

The limitations of collider searches are obvious. The mass range is limited by the available energy. The significance of a null collider search is limited by our knowledge of the production cross section. Thus a search for free quarks cannot use conventional quantum chromodynamics (with its confinement hypothesis) to calculate the expected production cross section.

1.5 Massive Particle Searches In Cosmic Rays And Halo Particles

There are two known and observed classes of particles impinging on the earth: primary cosmic rays coming from outside the earth, mostly proton and nucleons; and the secondary particles produced by interactions in the atmosphere such as muons and pions. The primary cosmic ray class is of more interest particularly because of the detection of very high energy particles, those with energies greater than 10^{19} eV. Historically the class of secondary particles has been of great importance because this class was the source of the discoveries of many particles. But, with the steadily increasing energy and intensity of colliders the secondary class is of decreasing interest.

The third class of particles impinging on the earth consists of the assumed *halo particles* of the galaxy. Interest in this class has, of course, been greatly stimulated by the search for the constituents of dark matter. These halo particles are assumed to have been present along with ordinary matter in the early history of the galaxy. But their large mass prevented their losing sufficient enough energy so that they could settle along with ordinary matter into the disc of the galaxy.

The mass boundary between when particles are trapped in ordinary matter and when they remain in the halo of the galaxy depends upon the interactions of the particle. The calculation of the mass boundary is not precise because of the different models used for particle interactions as the galaxy cools. Thus for a positive, unit-charge, massive particle, De Rújula *et al.*[4] give the boundary as above 2×10^4 GeV/c^2 for these massive particles to remain in the halo. On the other hand, Dimopoulos *et al.* [5] give a much higher mass boundary, 10^8 GeV/c^2.

The accepted model for the formation of the galaxy leads to halo particles having velocities of the order of 10^{-3} to 10^{-4} in units of $\beta = v/c$. Hence some of the later discussions in this paper on the observed limits for the flux of halo particles are concerned with velocities in this range.

1.6 Searches For Stable, Massive Particles In Ordinary Matter

Stable, massive particles might be introduced into ordinary matter through two different processes. In one process occurring in the course of the cooling of the galaxy, massive particles might condense with baryonic matter to form stars, planets and smaller bodies such as asteroids. This process will occur if the particles are not too massive or if in their interaction with ordinary matter they have sufficiently large *dE/dx*.

The other process that could introduce massive particles into ordinary matter would take place once stars and planets are formed. Some types of massive cosmic ray particles or halo particles impinging on the these bodies could lose sufficient energy to become trapped in the ordinary matter of these bodies.

2. SEARCHES FOR INTEGER CHARGE, STABLE, MASSIVE PARTICLES

2.1 General Considerations

In this section I discuss search methods for integer charge massive particles, $X^{\pm}$, and summarize the experimental limits on their abundance. The long term interest in such searches has been to find trace amount of such particles. But in the first years of the last decade, De Rújula *et al.*[4] and Dimopoulos *et al.* [5] suggested that charged particles could have sufficient mass and abundance to be the major constituent of dark matter. These proposed massive particles were called *CHAMPS*. This led to increased interest in searches for charged massive stable particles. However searches in the last decade, some of which are described in Secs. 2.3-2.5, have ruled out the possibility that CHAMPS could be a *major* constituent of dark matter.

2.2 Collider Searches For Integer Charge, Stable, Massive Particles

The pair production of integer charge particles in electron-positron annihilation, Eq. 1, provides the most straightforward way to search for such stable massive particles.

$$e^+ + e^- \rightarrow X^+ + X^- \qquad (1)$$

The beam energy and the measured momentum provide a determination of the mass of the sought particle and the pair production cross section is known. At the LEP2 electron positron collider searches were conducted up to about 100 GeV/c^2 per beam. For example, in the Acciarri *et al.* [6] L3 experiment at LEP the search upper limit was 93.5 GeV/c^2 and no massive particles were found. Similar null searches were carried out by other experimenters at LEP [1]. Thus up to about 100 GeV/c^2, the existence is excluded of new, massive spin ½ particles with the conventional electromagnetic interaction. As summarized in the Ref. 1, searches using Eq. 1 have been used to exclude other types of massive particles,

Searches for integer charge, stable, massive particles at the Tevatron reach to larger masses than searches at LEP2. Abe *et al.* [7] looked for pair production of particles with masses up to the order of 500 GeV/c^2, Fig. 1. No evidence for such particles was found. However, proton-antiproton production cross section calculations do not have the certainty of electron-positron electromagnetic pair production cross sections, hence the significance of the null results depends upon the model used for the production cross section,

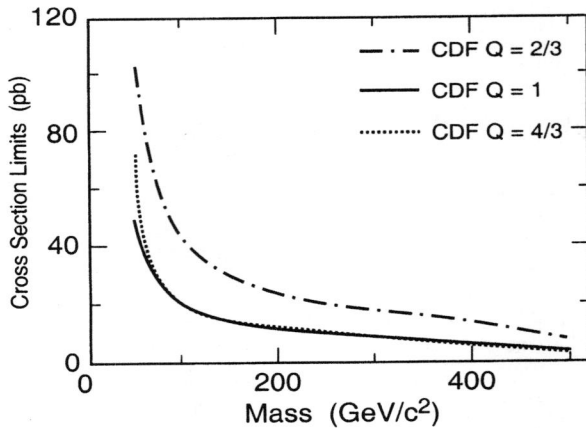

Figure 1. The cross section upper limits (95% C.L.) for the pair production of stable, charged particles with charges 1, 2/3, and 4/3 from Abe *et al.* [7] using the Tevatron.

2.3 Balloon and Satellite Searches For Integer Charge, Stable, Massive Particles Impinging On the Earth's Surface

Although no satellite or balloon experiments were designed to explicitly search for integer charged impinging on the earth's surface, limits on their abundances have been obtained [5,8,9] using data from experiments that were originally intended for cosmic ray studies. The limits, shown on the left side of Fig. 2, exclude the possibility that dark matter is entirely composed of integer charged particles with $10^2 < M_X < 10^7$ GeV/c². M_X is the mass of X.

2.4 Terrestrial Detector Searches For Integer Charge, Stable, Massive Particles Impinging On the Earth's Surface

There have been two sensitive, terrestrial searches for integer charge massive particles impinging on the earth [10,11]. The search by Barish *et al.* [10] for monopoles falling onto the earth's surface, can be used to place limits on X particles in the mass range $M_X > 10^8$ GeV/c². Below this mass, a particle with velocity $\beta \sim 10^{-3}$ at the top of the atmosphere ranges out before hitting the ground. Their 90% C.L. upper flux limit for particles with velocity $5 \times 10^{-4} < \beta < 2.7 \times 10^{-3}$ is:

$$\phi < 4.7 \times 10^{-12} \text{ cm}^{-2}\text{sr}^{-1}\text{s}^{-1} \qquad (2)$$

The other search for is from the underground MACRO experiment, Ambrosio *et al.* [11], also primarily designed to search for magnetic monopoles. The null monopole search results can be used [1] to derive a limit on the flux of integer charge X particles with sufficient energy to reach the experiment. The 3300 m water equivalent above the detector places a lower limit on the mass of X. For example, if X has a typical halo particle velocity, $\beta \sim 10^{-3}$, X must have a minimum mass of $M_X > 10^{11}$ GeV/c^2 in order to reach the MACRO detector. The 90% C.L. upper flux limit for particles with velocity $10^{-4} < \beta < 10^{-3}$ is:

$$\phi < 4 \times 10^{-14} \text{ cm}^{-2}\text{sr}^{-1}\text{s}^{-1} \tag{3}$$

For faster particles with $1.2 \times 10^{-3} < \beta < 10^{-1}$

$$\phi < 2.6 \times 10^{-16} \text{ cm}^{-2}\text{sr}^{-1}\text{s}^{-1} \tag{4}$$

The limits from Eqs. 2-4 are shown in Fig, 2

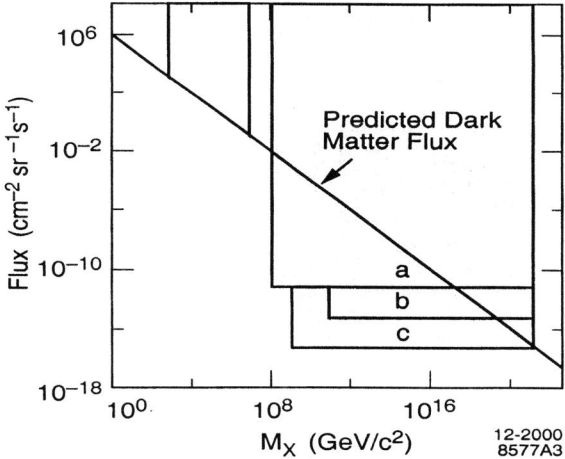

Figure 2. Observed upper limits on the flux of integer charged massive particles versus the particle mass, M_X. The diagonal line gives the predicted dark matter flux as a function of M_X. The excluded, shaded region on the left comes from the observations discussed in Sec. 2.3. The excluded, shaded regions on the right came from Eqs. 2-4.

2.5 Searches For Integer Charge, Stable Particles In Terrestrial Water

It is an attractive idea that a stable, massive, positively charged particle, X^+, falling onto and through oceans and lakes, will form the heavy water molecule HXO. Figure

3 gives the upper limits on the concentration of HXO in H_2O as found in three different searches [12-14]. No evidence was found for an X^+ in the mass ranges given in the figure.

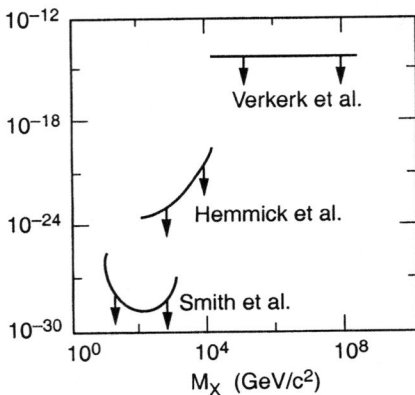

Figure 3. Observed upper limits on the concentration of HXO in H_2O versus the mass of a unit charge particle X. Limits are from Ref. 12-14.

3. DIRECT SEARCHES FOR WEAKLY INTERACTING MASSIVE PARTICLES (WIMPS)

The weakly interacting massive particle (WIMP) is the leading candidate for cold dark mater, hence there have been many searches for WIMPS. The favorite hypothesis for the nature of the WIMP is that it is the lightest supersymmetric particle with mass between 1 GeV/c^2 and 1 TeV/c^2 and with an interaction cross section with ordinary matter of the order of, or less than, the weak interaction cross section. There has been so much published on this subject that I will limit my remarks to a summary of the recent direct searches for WIMPS, I will not discuss indirect searches based on looking for evidence of WIMP annihilation in celestial bodies.

At present there is some conflict between the two most sensitive WIMP search experiments. The DAMA experiment [15] at the Gran Sasso National Laboratory claims to have measured a seasonal modulation of the rate at which WIMPS interact in their detector. They attribute this to a WIMP with mass of about 52 GeV/c^2 and an interaction cross section of about 7×10^{-6} pb.

However, a second experiment, the Cryogenic Dark Matter Search (CDMS) [16,17], finds a null result even though their sensitivity is close to that of DAMA's. The findings of the CDMS experimenters rule out DAMA's signal region at greater than 75% C.L. In addition, the CDMS experimenters find their data to be incompatible with the DAMA modulation signal. The CDMS and DAMA results are shown in Fig. 4. In the next decade this confusing situation will certainly be clarified when an improved CDMS experiment will be installed deep underground and other new WIMP search experiments will begin operation.

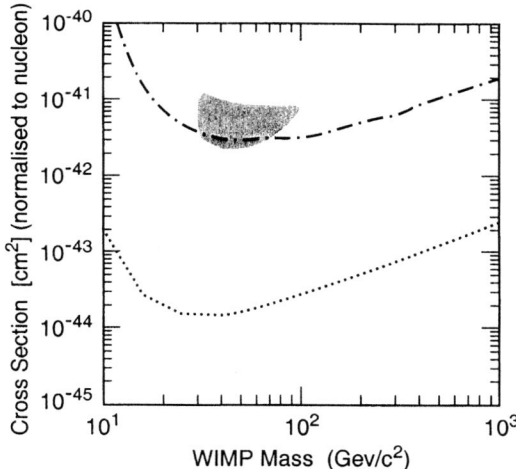

Figure 4. DAMA observation and CDMS limit on the WIMP-nucleon spin-independent cross section as a function of mass. The dark shaded area is the observed signal from the DAMA experiment [15]. The dash-dot curve shows the current upper limit on the cross section found by the CDMS experiment [16,17]. The dot-dot curve shows the *projected* upper limit that may be reached by the CDMS experiment when it is in the Soudan mine. The figure was obtained from the limit plots web tool provided by R. Gaiskell and V. Mandic at http://cdms.berkeley.edu/limitplots.

4. SEARCHES FOR FRACTIONAL CHARGE, STABLE PARTICLES

In this section we ignore the restriction to massive particles. The discovery of a stable, fractional charge particle of any mass would be of great importance. There have been four reviews of searches for fractional charge particles: Jones [18], Lyons [19], Smith [20] and Marinelli and Morpurgo [21]. My remarks emphasize the results of the searches after 1980. All charges are given in units of the magnitude of the electron charge.

4.1 Collider Searches For Fractional Charge, Stable Particles

High energy, high intensity, electron-positron and proton-antiproton colliders are ideal instruments for searches for fractional charge particles. Surprisingly with respect to published results, full use has not been made of these opportunities. For example, most experiments have published search upper limits for particles with 1/3, 2/3, or 4/3 charges without commenting on other fractional charges.

In Z^0 production at an electron-positron collider the Z^0 decay

$$e^+ + e^- \rightarrow Z^0 \rightarrow X^{+q} + X^{-q} \qquad (5)$$

would contribute to the decay width of the Z^0. If the X^q pair is *not* detected because of the fractional charge, as would usually be the case, then this decay would contribute to the *invisible* width of the Z^0. Therefore, if fractional charge particles are pair produced but not detected at the Z^0, their weak interaction coupling must be smaller than that of the neutrino by a factor of order 2 or 3. This condition on the weak interaction coupling constant is thus a limit on the existence of a fractional charge particle with mass less than 45.5 GeV/c^2.

There have also been *inclusive* searches at the Z^0 for fractional charge particles. Buskulic *et al.* [22] of the ALEPH collaboration, searched with null results for particles with charge 1/3, 2/3, or 4/3 in the mass range of 8 to 45 GeV/c^2. Akers *et al.*[23], the OPAL collaboration, reported similar null results.

There have also been searches at higher energy at LEP, again all with null results. For example Ackerstaff *et al,* [24], the Opal collaboration, carried out a search at 130 to 183 GeV for pair produced X particles with charge 2/3. The X mass search range was 45 to about 90 GeV/c^2 and the 95 % C. L. upper limits on the cross section were in the range of 0.05 to 0.2 pb.

Turning to proton-antiproton annihilation, the highest energy search for fractional charge particles has been carried out at the Tevatron by Abe *et al.* [7] at 1.8 TeV Figure 1 shows the 95% C. L. upper limits on the cross section for charge 2/3 and 4/3.

4.2 Searches For Fractional Charge, Stable Particles In Cosmic Rays And Halo Particles

The most comprehensive search to date for fractional charge particles impinging on the earth has been carried out by Ambrosio *et al.* [25] using the MACRO experimental apparatus. They searched for lightly ionizing particles with $.25 < \beta < 1.0$ and were sensitive to charges as small as 1/5. No fractional charge particles were found, the 90% C.L. upper limit on the flux is given in Fig. 5. Recall that the MACRO experiment lies beneath a mountain with a minimum overburden of 3300 m water equivalent. This leads to two limitations on the significance of these flux limits: strongly interacting fractional charge particles would not reach the MACRO detector and non-strongly interacting particles must have sufficient energy to overcome the *dE/dx* loss.

Two other searches for fractional charge particles in cosmic rays and halo particles were carried out in the 1990's; both report upper limits on the flux only for particles with charges of 1/3 and 2/3, Fig. 5. Aglietta *et al.* [26] used the LSD detector in the Mont Blanc tunnel with an overburden of 5000 m of water equivalent. Mori *et al.* [27] used the Kamiokande II detector with an overburden of 2700 m of water equivalent.

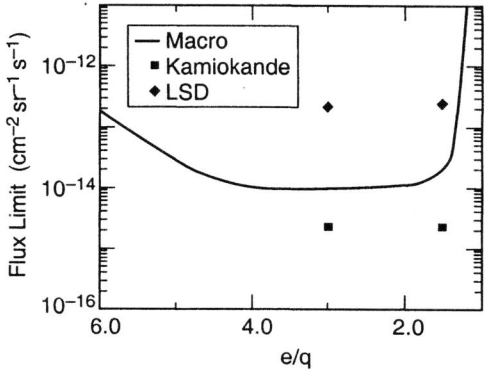

Figure 5. The curve is the observed upper limit on the flux of fractionally charged particles as a function of e/q from the MACRO detector, Ambrosio et al. [25]. The points are upper limits from the Kamiokande II detector, Mori et al. [26] and the LSD detector, Aglietta et al. [27]. q is the fractional charge in units of the electron charge e. Thus $q=1/3$ correspond to $e/q=3$.

4.3 Searches In Ordinary Matter For Fractional Charge, Stable Particles

At present there are two methods for searching for fractional charge, stable particles in ordinary matter- the levitometer method and the Millikan drop method. In the levitometer method, a small object consisting of ordinary matter is magnetically suspended in vacuum using either ferromagnetism [20,21] or superconductivity [28]. The object often, but not necessarily, is a sphere. The object's mass is in the range of 0.03 to 0.1 mg. An electric field is used to drive the object into forced oscillation and the charge on the object is measured by the resonant frequency of the oscillation.

The Millikan drop method goes back to his measurements of the electron charge [29]. In the modern development of the method [30-34] liquid drops with diameters in the range of 5 to 30 μm fall though air in the presence of an oscillating electric field. The drop charge is determined by measuring the drop's terminal velocity caused by the electric field.

Table 1 lists recent sensitive searches for fractional charge particles in ordinary matter. All experimenters reported null results except La Rue et al.[28] who reported finding 1/3 and 2/3 charges in niobium. This 1981 paper produced considerable interest in the possible existence of free quarks, but Smith et al.[35] studied about four times as much niobium and found *no* evidence for fractional charge particles in niobium. The La Rue et al.[28] report is not accepted today.

The sensitivity of the searches in Table 1 may be estimated by noting that there are 6.4×10^{20} nucleons in a milligram. The largest individual search sample is 17.4 mg of silicone oil, Halyo et al.[39]. They report that the concentration of particles with fractional charge more than 0.16e from the nearest integer is less than 4.7×10^{-22} particles per nucleon with 95% confidence.

TABLE 1. Searches for fractional charge particles in ordinary matter. All experimenters reported null results except La Rue et al.[28]. See text. There are 6.4×10^{20} nucleons in a milligram.

Method	Experiment	Material	Sample Mass (mg)
superconducting levitometer	La Rue et al.[28]	niobium	1.1
ferromagnetic levitometer	Marinelli et al.[36]	iron	3.7
ferromagnetic levitometer	Smith et al.[35]	niobium	4.9
ferromagnetic levitometer	Jones et al.[37]	meteorite	2.8
liquid drop	Joyce et al.[30]	sea water	0.05
liquid drop	Savage et al.[38]	mercury	2.0
liquid drop	Halyo et al.[39]	silicone oil	17.4

As shown in Table 1, most materials used for fractional charge searches have been chosen for ease of use: niobium for superconducting levitation, iron for ferromagnetic levitation, mercury and oil for the liquid drop method. These are probably not the best choices because fractional charge particles might easily be lost in electrically conducting materials or in refined materials such as oils. Lackner and Zweig[40] have discussed the chemistry of atoms containing fractional charge and have discussed the most suitable materials for fractional charge searches[41]. As summarized by Perl and Lee[32] the most suitable materials appear to be meteoritic material from asteroids, terrestrial minerals that concentrate rare impurities and perhaps material from the moon's surface.

5. FINAL REMARKS

There is no *confirmed* experimental or observational evidence for the existence of massive, stable, elementary particles other than the electron and proton. In particular, in the special case of dark matter there is no *confirmed* evidence for dark matter being composed of massive particles. But there are areas where continued or new searches for massive particles are warranted.

It is to be hoped that there will be new direct searches for massive particles when higher energy colliders go into operation - the proton-antiproton Large Hadron Collider and a very high energy electron-positron linear collider.

The various searches for fluxes of massive particles impinging on the earth, begun as dark matter explorations, should be continued even when the observed upper limit on the flux is too small to explain the expected dark matter density. There are two reasons. First, dark matter might consist of several types of massive particles. Second, the existence of a stable massive particle is of great importance even if it has nothing to do with dark matter. The question is whether the sensitive areas of the detectors can be substantially enlarged so as to increase the sensitivity.

Existing searches in bulk matter for massive particles with fractional or integer electric charge have probed to sensitivities in the range of 10^{-22} to 10^{-28} massive particles per nucleon. Improved search technology would allow substantial improvement of these sensitivities. There is a particular need to find ways to search in bulk matter for integer charge particles with masses greater than 10^8 GeV/c^2.

ACKNOWLEDGMENTS

I am greatly indebted to my colleagues Peter C. Kim, Valerie Halyo, Klaus S. Lackner, Eric R. Lee, Irwin T. Lee and Dinesh Loomba for our joint studies of the search for massive elementary particles.

REFERENCES

1. Perl, M. L. et al., Int. J. Mod. Phys., **A16**, 2137-2164 (2001).
2. Griest. K, and Kamionkowski, M. Phys. Rev. Lett. **64**, 615-618 (1990).
3. Kolb, E. W. et al, in Proc. Int. Conf. Dark Matter in Astro and Particle Phys. ,edited by H.V. Klapdor-Kleingrothanus and L. Baudis, Inst. of Phys., Bristol, 1998, pp. 592-614.
4. De Rújula, A, et al., Nucl. Phys. **B333**, 173-194 (1990).
5. Dimopoulos, S. et al., Phys. Rev. **D41**, 2388-2397 (1990).
6. Acciarri, M. et al., Phys. Lett. **B462**, 354-364 (1999).
7. Abe. F. et al., Phys. Rev. **D46**, 1889-1894 (1992).
8. Barwick, S. W. et al., Phys. Rev. Lett. **64**, 2859-2862 (1990).
9. Snowden-Ifft, D. P. et al., Astrophys. J. **364**, L25-L27 (1990).
10. Barish. B. et al., Phys. Rev. **D36**, 2642-2648 (1987).
11. Ambrosio, M. et al., MACRO Coll., hep-ex/0009002.
12. Smith, P. F. et al., Nucl. Phys. **B206**, 333-348 (1982).
13. Hemmick, T. K. et al., Phys. Rev. **D41**, 2074-2080 (1990).
14. Verkerk, P. et al., Phys. Rev. Lett. **68**, 1116-1119 (1992).
15. Bernabei, R. et al., Phys. Lett. **B480**, 23-31 (2000) and references contained therein.
16. Abusaidi, R. et al., Phys. Rev. Lett. **84**, 5699-5703 (2000).
17. Abusaidi, R., et al., Nucl. Instr. and Meth. Phys. Res. **A444**, 345-349 (2000).
18. Jones, L. W., Rev. Mod. Phys. **49**, 717-752 (1977).
19. Lyons, L., Phys. Reports **129**, 225-279 (1985).
20. Smith, P. F., Ann. Rev. Nucl. Part. Sci. **39**, 73-111 (1985).
21. Marinelli, M. and Morpurgo, G., Phys. Reports **85**, 161-258 (1985).
22. Buskulic, D. et al., Phys. Lett. **B303**, 198-208 (1993).
23. Akers, R. et al., Z. Phys. **C67**, 203-212 (1995).
24. Ackerstaff, K. et al., Phys. Lett. **B433**, 195-208 (1998).
25. Ambrosio, M. et al., Phys. Rev. **D62**, 052003-052012 (2000).
26. Aglietta, M. et al., Astroparticle Phys. **2**, 29-34 (1994).
27. Mori, M. et al., Phys. Rev. **D43**, 2843-2846 (1991).
28. LaRue, G. S. et al., Phys. Rev. Lett. **46**, 967-970 (1981).
29. Millikan, R. A., Phys. Rev. **32**, 349-397 (1911).
30. Joyce, D. C. et al., Phys. Rev. Lett. **51**, 731-734 (1983).
31. Hendricks, C. D. et al., Meas. Sci. Technol. **5**, 337-347 (1994).
32. Perl, M. L. and Lee, E. R., Am. J. Phys. **65**, 698-706 (1997).
33. Mar, N. M. et al., Phys. Rev. **D53**, 6017-6032 (1996).
34. Loomba, D. et. a., Rev. Sci. Instr. **71**, 3409-3414 (2000).

35. Smith, P. F. *et al.*, *Phys. Lett.* **B153**, 188-194 (1985).
36. Marinelli, M. and Morpurgo, G., *Phys. Lett.* **B137**, 439-442 (1984).
37. Jones, W. G. *et al.*, *Z. Phys.* **C43**, 349-355 (1989).
38. Savage, M. L. *et al.*, *Phys. Lett.* **B167**, 481-486 (1986).
39. Halyo, V. *et al.*, *Phys. Rev. Lett.* **84**, 2576-1579 (2000).
40. Lackner, K. S. and Zweig, G., *Phys. Rev.* **D28**, 1671-1691 (1983).
41. Lackner, K. S. and Zweig, G., *Novel Results in Particle Physics*, edited by R. S. Panvini, S. Alam, and S. E. Csorna, AIP Conf. Proc. 93, New York, 1982, pp. 1-14.

MORE ADVENTURES IN ATOMIC, MOLECULAR, AND NUCLEAR PHYSICS

From Helium-6 to Krypton-81

Zheng-Tian Lu

Physics Division, Argonne National Laboratory, Argonne, IL 60439, USA
Email: lu@anl.gov   URL: www-mep.phy.anl.gov/atta/

Abstract. A new method of ultrasensitive trace-isotope analysis has been developed based upon the technique of laser manipulation of neutral atoms. It has been used to count individual ^{85}Kr and ^{81}Kr atoms present in a natural krypton sample with isotopic abundances in the range of 10^{-11} and 10^{-13}, respectively. The atom counts are free of contamination from other isotopes, elements, or molecules. The method is applicable to other trace-isotopes that can be efficiently captured with a magneto-optical trap, and has a broad range of potential applications.

COMMINS & RADIOACTIVE ATOMS

Professor Eugene Commins has had a long "friendship" with radioactive atoms, particularly with isotopes of noble gases. This friendship has been an extremely rewarding one, not only for his own scientific quests but also for the generations of students who followed his footsteps. Just as biological genes are passed along a family tree and affect the health and appearance of many generations down the branches, scientific "genes" are inherited through apprentice relationships and affect one's scientific taste, style, and sometimes preferred techniques. The following is a brief sketch of the scientific family branch that I am attached to.

Our story began in 1958 when Gene and his thesis advisor, Polykarp Kusch, measured the magnetic moment of ^{6}He ($t_{1/2}$ = 0.8 s). They did not observe any effects that would have risen if ^{6}He had a magnetic moment, so they concluded that its spin was probably zero [1]. Through this experiment Gene had his first working experience with radioactive atoms. It also marked the beginning of Gene's distinguished career on searching for anomalous effects.

Later at Berkeley, Gene and his students, including Frank Calaprice, did a series of now classic experiments measuring various correlation effects in the β-decays of ^{19}Ne ($t_{1/2}$ = 17 s) [2]. These measurements were used to test fundamental symmetries and to probe the underlying mechanisms of Weak Interaction.

Frank continued the research on ^{19}Ne at Princeton, where he was joined by Stuart Freedman, a younger graduate from Gene's group. Together they measured the angular distribution of positrons emitted in the decay of polarized ^{19}Ne as a test for the 2nd-class Weak Interaction [3].

When laser trapping of atoms was developed, Stuart, then at Argonne, recognized that radioactive atoms trapped in a Magneto-Optical Trap (MOT) provide an exceptionally good source for the type of experiments done so far with ^{19}Ne trapped in a cell. I was a new student in Stuart's group, and was assigned to work on this idea for my thesis [4]. Being at the right place and the right moment, I joined the game of trapping radioactive atoms, and have been in it ever since.

In the following section, I will describe the work of my group at Argonne on the laser trapping of ^{81}Kr ($t_{1/2}$ = 230 kyr) atoms. A more detailed description of this work has been published in reference [5]. It is interesting to note that Gene has worked on ^{6}He, ^{19}Ne and ^{35}Ar ($t_{1/2}$ = 1.8 s). It is my pleasure to continue the family business working on a noble gas isotope down the periodic table.

Table 1. Quests on radioactive noble gases -- a family tradition.

Members	Time	Location	Isotope	Reference
Kusch & Commins	1958	Columbia	He-6	[1]
Commins & Calaprice	1969	Berkeley	Ne-19	[2]
Calaprice & Freedman	1975	Princeton	Ne-19	[3]
Freedman & Lu	1994	Berkeley	Na-21	[4]
Lu & Du	2001	Argonne	Kr-81	[5, 6]

ATOM TRAP TRACE ANALYSIS (ATTA)

Much can be learned from the concentrations of the ubiquitous long-lived radioactive isotopes. W. Libby and coworkers first demonstrated in 1949 that trace analysis of ^{14}C ($t_{1/2}$ = 5.7 kyr, isotopic abundance = 1×10^{-12}) can be used for archaeological dating [7]. Since then, two well established methods, Low-Level Counting and Accelerator Mass Spectrometry [8], have been used to analyze many other trace-isotopes at about the parts-per-trillion level and to extract valuable information encoded in the production, transport, and decay processes of these isotopes. The impact of ultrasensitive trace-isotope analysis has reached a wide range of scientific and technological fields.

We have recently developed a new method, Atom Trap Trace Analysis (ATTA) [5, 6], and utilized it to analyze two rare krypton isotopes, ^{81}Kr ($t_{1/2}$ = 230 kyr, isotopic abundance = 6×10^{-13}) and ^{85}Kr ($t_{1/2}$ = 10.8 yr, isotopic abundance $\sim 1 \times 10^{-11}$). ^{81}Kr is produced in the upper atmosphere by cosmic-ray induced spallation and neutron activation of stable krypton isotopes. It is an ideal tracer for dating ice and groundwater that are older than 100,000 years, which is beyond the range of ^{14}C-dating. ^{85}Kr is a fission product of ^{235}U and ^{239}Pu, and has been used as a general tracer to study air and ocean currents, date shallow groundwater, and monitor nuclear-fuel reprocessing activities. Due to its high mobility, it may be used as a leak sensor to check the seals of nuclear fuel cells and nuclear waste containers.

ATTA is a new laser-based atom-counting method. Our design is based on a type of MOT system [9] that has been used to trap various metastable noble gas atoms. Trapping krypton atoms in the $5s[^3/_2]_2$ metastable level (lifetime ≈ 40 sec) is accomplished by exciting the $5s[^3/_2]_2$ - $5p[^5/_2]_3$ transition. Two repump sidebands are generated via additional AOMs to optically pump the atoms into the F=13/2 level for ^{85}Kr and F=11/2 level for ^{81}Kr where they can be excited by the trapping light. In the analysis, a krypton gas sample is injected into the system through a discharge region, where about 1×10^{-4} of the atoms are excited into the $5s[^3/_2]_2$ level via electron impact excitation. The thermal (300°C) atoms are then transversely cooled, decelerated with the Zeeman slowing technique, and loaded into a MOT. Atoms remain trapped for an average of 1.8 sec as the vacuum is maintained at 2×10^{-8} Torr. This trap system can capture the abundant ^{83}Kr atoms at the rate of 2×10^{8} sec^{-1}. The ratio of the capture rate to the injection rate gives a total capture efficiency of 1×10^{-7}.

With expected capture rates between 10^{-3} sec^{-1} and 10^{-2} sec^{-1} for the rare krypton isotopes, the system must be able to detect a single atom in the trap [10]. In the trap, a single atom scatters resonant photons at a rate of $\sim 10^{7}$ sec^{-1}, of which 1% are collected, spatially filtered to reduce background light, and then focused onto an avalanche photodiode with a photon counting efficiency of 25%. In order to achieve a high capture efficiency and a clean single-atom signal, the setup is switched at 2 Hz between the different parameters optimized for capture and for atom counting. The resulting fluorescence signal of a single atom is 16 kcps (kilo-counts per second) while the background level is 3.4 kcps (Fig. 1).

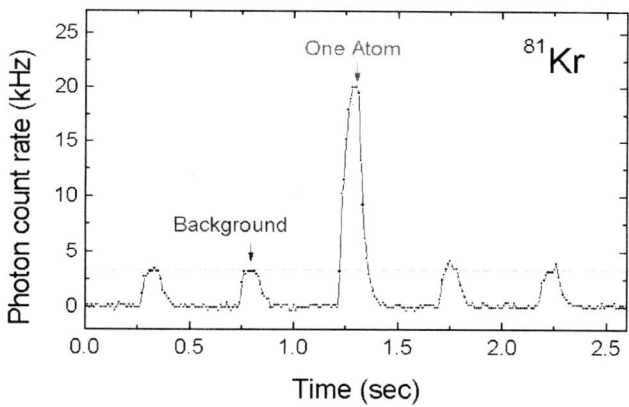

Figure 1. Signal of a single trapped ^{81}Kr atom. The photon counter is only open during the detection phase. Single atom signal ≈ 1600 photon counts, background ≈ 340 photon counts.

We have trapped and counted ^{85}Kr and ^{81}Kr atoms from a natural krypton gas sample. The frequency settings of the trapping laser and the two sidebands are in good agreement with previous spectroscopic measurements obtained using enriched ^{85}Kr gas and enriched ^{81}Kr gas. We have also mapped the atom capture rates versus laser frequency (Fig. 2). Furthermore, repeated tests were performed under conditions in which a ^{85}Kr (^{81}Kr) trap should not work, such as turning off repump sidebands and tuning the laser frequency above resonance. These tests always yielded zero atom counts, which shows that the recorded counts are solely due to laser-trapped ^{85}Kr (^{81}Kr) atoms.

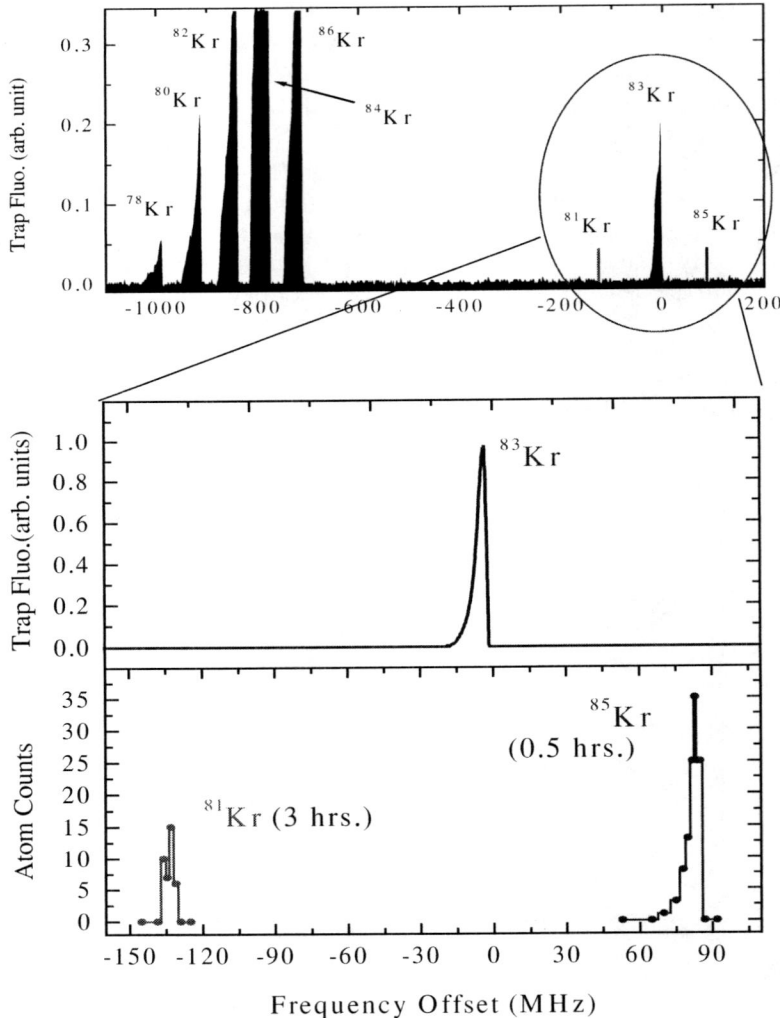

Figure 2. (a) Fluorescence of trapped krypton atoms. Dark bands are the signal of stable isotopes measured with a low-gain photo-diode detector. Line markers mark the positions of the two rare isotopes. (b) Fluorescence of trapped ^{83}Kr atoms versus laser frequency. (c) Number of ^{81}Kr and ^{85}Kr atoms counted versus laser frequency. Each data point represents the number of ^{81}Kr atoms counted in 3 hours, and ^{85}Kr atoms counted in 0.5 hours.

Previous efforts to develop a laser-based technique have encountered serious problems as a result of contamination from nearby abundant isotopes or isobars. ATTA is immune from the contamination for several reasons: fluorescence is only collected in a small region (ϕ 0.5 mm) around the trap center; a trapped atom is cooled to a speed below 1 m/s so that its laser induced fluorescence is virtually Doppler-free; the long observation time (>100 ms) allows the atom to be unambiguously identified (S/N $\approx$ 40); and trapping allows the temporal separation of capture and detection so that both capture efficiency and detection sensitivity can be optimized. Our design also provides additional features, such as chopping off the atomic beam before detecting the trapped atom.

The capture rate of our system depends on the discharge current, laser power, and optical alignment. At one particular setting, we measured capture rates of ^{83}Kr, ^{85}Kr, and ^{81}Kr, which were $(1.5\pm0.3)\times10^8$ sec^{-1}, $(1.9\pm0.3)\times10^{-2}$ sec^{-1}, $(1.3\pm0.4)\times10^{-3}$ sec^{-1} respectively. If we assume the same detection efficiency for all three isotopes, then we get isotopic abundances of $(1.5\pm0.4)\times10^{-11}$ for ^{85}Kr and $(1.0\pm0.4)\times10^{-12}$ for ^{81}Kr, which are in good agreement with previous measurements performed using other methods. The capture efficiencies can be calibrated with enriched samples of known isotopic abundance to correct for any isotope-dependent effects and measure isotopic ratios in unknown samples. For example, in ^{81}Kr-dating, a known amount of ^{85}Kr can be mixed into the sample, thus allowing the ^{81}Kr abundance be extracted by measuring the ratio of ^{81}Kr / ^{85}Kr.

Our system has achieved an overall efficiency of 1×10^{-7}. Use of this system to measure the abundance of ^{85}Kr to within 10% would require 2 hours and a krypton sample of 3 cm^3 STP while measurement of ^{81}Kr to within 10% would require 2 days and a sample of 60 cm^3 STP. This limits the current system to atmospheric applications where large samples of gas are available. Improvements, such as a liquid-nitrogen cooled discharge source and recirculation of krypton gas, are presently under investigation.

This work is conducted in collaboration with K. Bailey, C.Y. Chen, X. Du, Y.M. Li, T.P. O'Connor, and L. Young. This work is supported by the U.S. Department of Energy, Nuclear Physics Division. (contract W-31-109-ENG-38).

REFERENCES

1) E.D. Commins and P. Kusch, Phys. Rev. Lett. **1**, 208 (1958).
2) F.P. Calaprice, E.D. Commins, and D.C. Girvin, Phys. Rev. **D9**, 519 (1974).
3) F.P. Calaprice, S.J. Freedman, W.C. Mead, and H.C. Vantine, Phys. Rev. Lett. **35**, 1566 (1975).
4) Z.-T. Lu et al., Phys. Rev. Lett. **72**, 3791 (1994).
5) C.Y. Chen et al., Science **286**, 1139 (1999).
6) Atomic Physics 17, edited by E. Arimondo, P. DeNatale, and M. Inguscio (AIP 2001).
7) J.R. Arnold and W.F. Libby, Science **110**, 678 (1949).
8) R.A. Muller. Radioisotope dating with a cyclotron. Science **196**, 489 (1977).
9) E.L. Raab et al. Phys. Rev. Lett. **59**, 2631 (1987).
10) Z. Hu and H.J. Kimble, Opt. Lett. **19**, 1888 (1994).

Quantum Control

Philip H. Bucksbaum

Physics Department, University of Michigan
Ann Arbor, MI 48109-1120

Abstract. Tunable lasers, sources of intense narrow-band coherent radiation, enabled the parity violation experiments in atoms in the 1970's. Ultrafast lasers are source of *broadband* coherent radiation. This paper reviews recent progress in experiments to control quantum dynamics in condensed phase and gas phase systems, using shaped ultrafast radiation. Many of the same techniques that have led to laser pulses in the 10-100 fsec range can also be applied to the control of quantum systems with similar dynamical time scales

INTRODUCTION

The following limericks were written during data collection runs on the thallium parity violation experiment in 1977 or 1978:

> Commins et al. at UC
> Thought they saw a small asymmetry
> After sorting out statics
> And false systematics
> They said it was due to the Z.

> The scientists at Novosibirsk
> Said "Comrades, we discovered it firsk!
> At first we were sad
> 'Cause our noise was so bad
> But compared to yours it look no wirsk!"

> Fortson up north in Seattle
> Says, "My setup's developed a rattle!
> It comes and it goes.
> A Z-boson? Who knows?
> For weak force, it sure puts up a battle."

> And now clear from Oxford we hear
> That a change in results may be near.
> While the boss was out speaking
> The students were tweaking.
> Now sometimes strange signals appear.

This *festschrift* contribution in honor of the career of Gene Commins is an opportunity to look for connections between those parity violation measurements we all worked on together, and broader and more applied areas of laser science. The 1970s was a decade of great excitement in spectroscopy because of the invention of the tunable laser. As I arrived at Berkeley, Gene's student Steve Chu was designing and building flashlamp pumped dye lasers from parts machined in the shop or scavenged from the Livermore junk pile. The parity violation experiments were an opportunity to apply precision laser spectroscopy to a fundamental problem in physics. Under Gene's mentorship, we learned the joy of this challenge.

The term "quantum control" does not describe a single field of physics, any more than do the terms "laser spectroscopy" or "precision measurements." Rather, quantum control is a method of using laser fields to study different kinds of physical problems. Part of the fun is finding new frontiers where these techniques can make an impact. Examples of these frontier fields are quantum information science and coherent chemistry.

Quantum control is based on the notion that a well-controlled time-varying intense optical field can transform any pure quantum state of a system into a different state with different properties. In some problems, the target state, initial state, and system Hamiltonian are all known, so that the challenge is to get from the start to the finish efficiently. Methods such as Optimal Control Theory have been developed for these situations. Other times, the target state and system Hamiltonian are not known, but some system properties are known. An example is breaking a specific bond in a complex molecule. In this case we cannot prescribe the optimal control field in advance, but the solution can be found through a systematic search. One of our goals has been the construction of a "learning machine," that is, an automated experimental apparatus that can use feedback signals from a quantum system to help optimize control strategies for a particular application. The learning machine consists of a search algorithm, together with programmable experimental inputs and readouts.

FUNDAMENTALS OF CONTROL

The ideas behind quantum control are illustrated by an examination of the Hamiltonian for a system in the presence of external fields:

$$(H_{System} + H_{External} + H_{Control})\Psi = i\hbar \partial \Psi / \partial t \qquad (1)$$

Here $H_{Systerm}$ is the time-independent Hamiltonian of a quantum system, such as an atom or molecule. $H_{External}$ is the interaction of the quantum system with external fields. These could be due to collisions, or couplings to other modes in the case of a complex molecule, and also include coupling to the vacuum modes of the radiation field that cause fluorescent decay. $H_{Control}$ is the coupling that we add in order to control the system, such as the field of a laser. Ψ is the Schrödinger wave function, and quantum control is the process of directing Ψ into a desired target state.

The evolution of the wave function is governed by the combination of internal and external contributions. In the absence of $H_{External}$, the system has stationary states, the energy eigenstates, and any Ψ can be constructed from a suitable linear combination of these. In real systems $H_{External}$ is never totally absent, however; it leads to loss of the quantum coherences ("T_2"processes), and ultimately to relaxation of the excited state amplitudes ("T_1" processes).

In most situations in laser spectroscopy, the control Hamiltonian is a small perturbation compared to the external Hamiltonian. In that case only a tiny fraction of probability amplitude can be transferred away from the initial state before relaxation sets in. Quantum control is not possible, since the loss of phase information means loss of control. Control, then, requires strong fields. In the notation of equation (1), we require $H_{Control} > H_{External}$.

Another useful way to look at the control problem is the competition of different time scales in the problem. The control Hamiltonian transfers coherent population among states of the system, a process characterized by the Rabi frequency Ω, which is proportional to $H_{Control}$. The amount of probability amplitude transferred to the excited state goes like $\Omega\tau$, where τ is the laser pulse duration. Significant population transfers should be much more rapid than the decoherence time, i.e. $\Omega > T_2^{-1}$, and this is equivalent to the strong field control condition expressed above.

Some control experiments can relax this requirement by making use of a "launch state." This is an eigenstate of H_{System} prepared at the beginning of the experiment, which which is relatively immune to $H_{External}$, and thus provides the reference phase for further excitation of a wave packet. Pulsed excitation with a weak field (i.e. where $\Omega\tau < 1$) can still produce controllable wave packets; however, if we have $\tau > T_2$, then all coherence is lost. This then leads to the two conditions for quantum control: strong fields and ultrafast pulses. Many techniques now exist to produce such optical fields and to shape them. Quantum control has grown out of these advances in technology.

SHAPING ULTRAFAST OPTICAL PULSES

Ultrafast optical pulses are produced by mode-locked broadband lasers. The mode-locking mechanism is usually a nonlinear focusing element such as a Kerr lens in the cavity, which makes the laser optical cavity most stable if the circulating light is a single very short pulse. If the gain medium is very broadband, the pulse may have only a few optical cycles. We use a Kerr-Lens modelocked titanium sapphire oscillator[1]. The output is amplified in a 10 Hz regenerative chirped pulsed amplifier[2]. The output pulse is approximately 100 fsec long, with a central wavelength of about 790 nm.

The light pulse is then shaped by filtering its spectral components in a zero-dispersion pulse stretcher, consisting of a 1:1 telescope between two diffraction gratings.[3] The heart of this pulse shaper is a programmable acousto-optic modulator (AOM) made of Tellurium Dioxide *(TeO2)*, which forms a Fourier filter.[4] The acoustic wave creates a transient transmission grating which diffracts the optical wave at the Bragg angle. Since the acoustic wave is essentially frozen as the optical pulse travels through the crystal, the complex amplitude of the acoustic wave traveling through the crystal in the transverse direction $A(t)\cos\omega_c t = A(y/v_s)\cos\omega_c t$ is mapped onto the optical field $E(\omega)$ as it passes through the AOM. The shaped beam then has the form

$$E_{shaped}(\omega) = E_{input}(\omega) \times a(\omega) \times e^{i\varphi(\omega)t} \quad (2)$$

where $a(\omega)e^{i\varphi(\omega)} = A[y(\omega)/v_s]$.

The shaped pulses can be measured using spectral interferometry. In this technique, the pulse to be measured is combined with an unshaped reference pulse, and then analyzed in a spectrometer. The spectrally resolved interference between signal and reference is a direct measure of the spectral phase function.

SCULPTED RYDBERG WAVE PACKETS

This section describes our development of wave packet "sculpting" and the quantum interference techniques used to view these sculptures.[5] We use Rydberg atoms to explore ideas in strong field quantum control and quantum information problems. Rydberg atoms are a testing ground for quantum state preparation and measurement.[6]

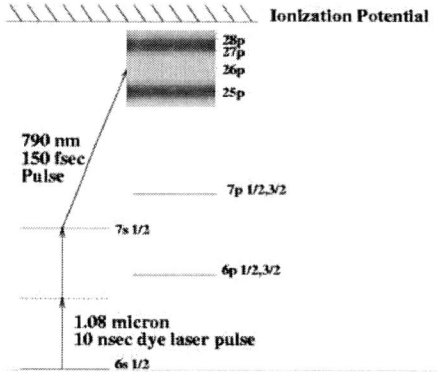

Figure 1. Excitation scheme for producing sculpted Rydberg wave packets in atomic Cs.

Rydberg wave packets are produced in a Cs atomic beam by first exciting ground state atoms to the 7s state using a two-photon transition at 1.08 μm. Sculpted laser pulses then excite the atoms to Rydberg states. This produces coherent superpositions of $p_{1/2}$ and $p_{3/2}$ states with principal quantum numbers between $n=24$ and $n=35$. In most applications we are only interested in tracking the shape over the first few picoseconds, so the spin-orbit interaction (on the order of 100 ps for the states considered here) can be neglected. This means that the wave packet can be described without spin, in the nlm basis. The value of the magnetic quantum number m then depends on the relative orientation of the quantization axis and the laser polarization that induces transitions from the 7s state.

Decoherence and dephasing due to Doppler effect and collisions can be reduced by using an atomic beam. Residual dephasing mechanisms are stray electric fields, and some residual first-order Doppler shifts if the beam is not perpendicular to the laser. These contribute to coherent dephasing times on the order of 10^{-7}s. Decay of the state amplitudes can occur due to excitation by black-body radiation, or fluorescence decay, with associated T_1 times on the order of 10^{-5}s. These times set the limits for producing and measuring wave packets.

The cesium Rydberg wave packet $\Psi(\mathbf{x},t)$ is determined by the amplitude and phase of the constituent eigenstates:

$$\Psi(\mathbf{x},t) = \sum_{np} a_n e^{i\Phi_n} u_{npm}(\mathbf{x}) e^{-i\omega_n t} + a_0 \varphi_s e^{-i\omega_s t}, \tag{3}$$

where ϕ_s is the launch state. For weak excitation, we can use Fermi's Golden rule to calculate the coefficients:

$$a_n e^{i\Phi_n} \propto E(\omega_{np} - \omega_{7s})\langle npm|z|7s\rangle \propto E(\omega_{np} - \omega_{7s})/n^3 \tag{4}$$

This expression relates the quantum state amplitude and phase to the amplitude and phase of the sculpted optical field.

Amplitude Measurement

Quantum amplitudes of Rydberg wave packets are measured using two techniques. The simplest method for atomic beams is the technique of state-selective field ionization[7]. If a uniform electric field is applied to the Rydberg wave packets, field ionization becomes possible. However, the field ionization probability is not a monotonically increasing function of field, because the energy structure of the wave packet in the field is discretized into constituent Stark eigenstates. Each energy eigenstate with energy E has its own characteristic critical ionization field F. For alkali atoms, this is roughly given by $|F| = (1/4)E^{-2}$ in atomic units.

When a wave packet experiences the critical field for one of its constituent states, the ionization occurs with probability proportional to the square of the amplitude for that state. Thus, the atomic beam ionization distribution maps the squared state amplitudes.

In some situations, state selective field ionization is inconvenient or impossible. For example, wave packets in a dense gas or a liquid may relax so rapidly that the ramped field technique takes too long, or the ionized electrons cannot be extracted. In that case, it is often possible to use quantum wave packet interferometry to analyze the wave packet amplitudes. Quantum interferometry uses a Michelson interferometer to split the optical pulse that excites the wave packet into two parts with a variable time delay τ. Each pulse in the pair excites an identical wave packet in the atom, and the two wave packets coherently interfere, depending on the time delay:

$$\Psi(\mathbf{x},t,\tau) = e^{i\omega_{gs}\tau} \sum_n a_n u_n(\mathbf{x}) e^{-i\omega_n t} (1 + e^{-i\omega_n \tau}) \tag{5}$$

The total wave function depends only on the squared amplitudes and the time delay:

$$\langle\Psi|\Psi\rangle \propto \sum_n |a_n|^2 \cos\omega_n\tau. \qquad (6)$$

A Fourier analysis of the ionization current $|\langle\psi|\psi\rangle|^2$ will yield the amplitudes.

Phase Measurement

The phase of the state coefficients a_i can be determined using a holographic technique similar to spectral interferometry. In this measurement, the atoms are excited to a second, *reference* wave packet superposition with a time delay τ with respect to the sculpted packet. The reference excitation has real amplitudes, produced by a transform limited optical field.

$$\Psi_{ref}(\mathbf{x},t,\tau) = e^{i\omega_{gs}\tau}\sum_n b_n u_n(\mathbf{x})e^{-i\omega_n t} \qquad (7)$$

The probability for state i depends on the relative phase ϕ_i between signal and reference

$$P_i = |a_i|^2 + |b_i|^2 + 2|a_i||b_i|\cos[(\omega_i - \omega_{gs})\tau - \phi_i]. \qquad (8)$$

Since this delay time τ is not stable with respect to the optical frequency $\omega_i-\omega_{gs}$, the phases are extracted by averaging several laser shots and constructing the correlation function

$$r_{ij} = \frac{\langle P_i P_j\rangle - \langle P_i\rangle\langle P_j\rangle}{(\Delta P_i)(\Delta P_j)} = \cos[(\omega_i - \omega_j) - (\phi_i - \phi_j)]. \qquad (9)$$

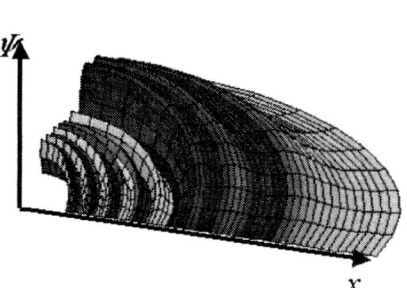
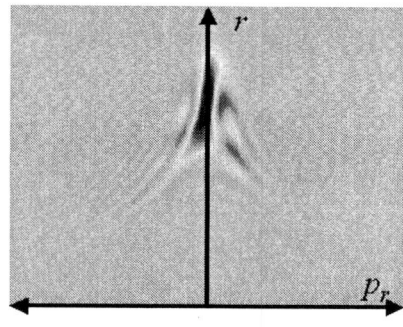

Figure 2. Two views of a sculpted wave packet. Left: Schrödinger wave function. The amplitude is shown as a function of x and z for a *p*-state wave packet oriented along x. phase is represented by the shading of the wave packet. Right: Wigner representation of the same wave packet in the (r,p_r) phase space plane.

QUANTUM CONTROL APPLICATIONS

Data Storage and Retrieval

Quantum control techniques can be used to to load, hide, retrieve, and manipulate information in wave packets. Consider, for example, the storage of a simple binary number, such as 000001000. A wave packet could encode this number in various ways. For example, one could load an n-state quantum wave packet with an n-bit number according to the prescription that at a specified time t, the phase is real and positive for binary 0, but real and negative for binary 1. Since every state in the wave packet has equal amplitude, ordinary spectroscopic techniques such as state-selective field ionization cannot reveal which state stores the binary 1. This bit is hidden from view. If the same data were stored in a classical binary register with n locations, one would have to search each location to find the marked bit. The search would take, on average, $n/2$ steps; however, the rules of quantum state manipulation provide some simple methods for revealing the marked bit.

Quantum holographic techniques provide a simple means to find the marked bit. To find the hidden information, the atom is excited by a search pulse which produces a second wave packet with all quantum phases relatively real. This, in fact, is the very same "reference" pulse that was used to measure the phases in the sculpted wave in the previous section. The superposition leads to destructive interference of any states where a binary 0 was stored, but constructive addition to any states with a binary 1. The combined wave packet then has the same information as before, but now it is encoded into quantum amplitudes rather than phases. These can then be read out using state-selective field ionization.[8]

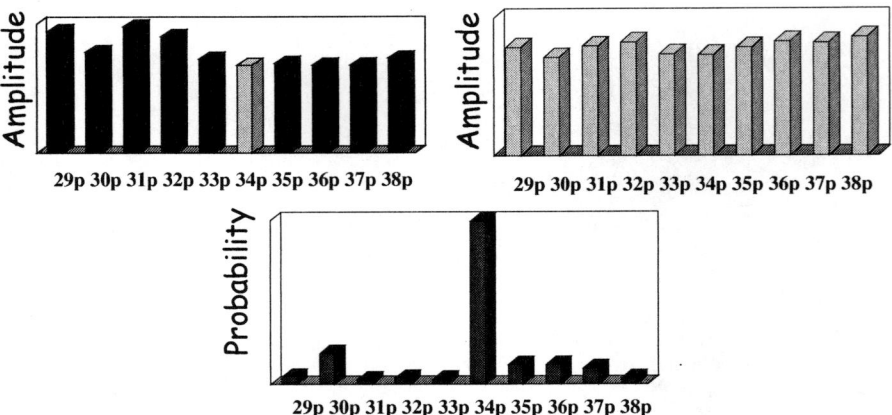

Figure 3. (top left) Bar graph representing a Rydberg data register for the binary number 0000010000. The binary bit is encoded into the phase of the n-state, so that binary 0 states are real and positive (dark phase), while binary 1 states are real and negative (light phase). (Top right) The decoding wave packet has equal amplitudes, and all negative real phases. (Bottom) The superposition of both wave packets amplifies the binary 1 bits, while destructive interference destroys the binary 0 bits.

This is a simple example of a general class of search algorithms that make use of the properties of superposition and quantum interference, which were introduced by Grover.[9] Our particular implementation of a Grover-style search algorithm has some unresolved difficulties if the register is too large. The simple form of the decoding pulse only works in the perturbation theory limit, where almost all of the probability amplitude in the wave packet resides in the "launch" state (7s for our work in Cs.) If thre are too many states in the Rydberg wave packet, the launch state will become depleted. The data retrieval is still possible, but now the unitary transformation that amplifies the "1" bits and suppresses the "0" bits must depend on the total pulse energy. In other words, we move into *the strong field regime*.

Feedback and Learning Control

The techniques of wave packet sculpting are "feed-forward," since the construction and measurement processes make no use make no use of the measured results. A "feedback" loop may be used to adjust the pulse shaper in response to information gathered by wave packet holography. This type of feedback is more properly called "learning," since the readjustment of the apparatus does not occur during a measurement, but rather follows the data acquisition, so that corrections can be made to future experiments.[10] An automated form of learning control on wave packets in Rydberg states can control the shape of the wave packet.[11]

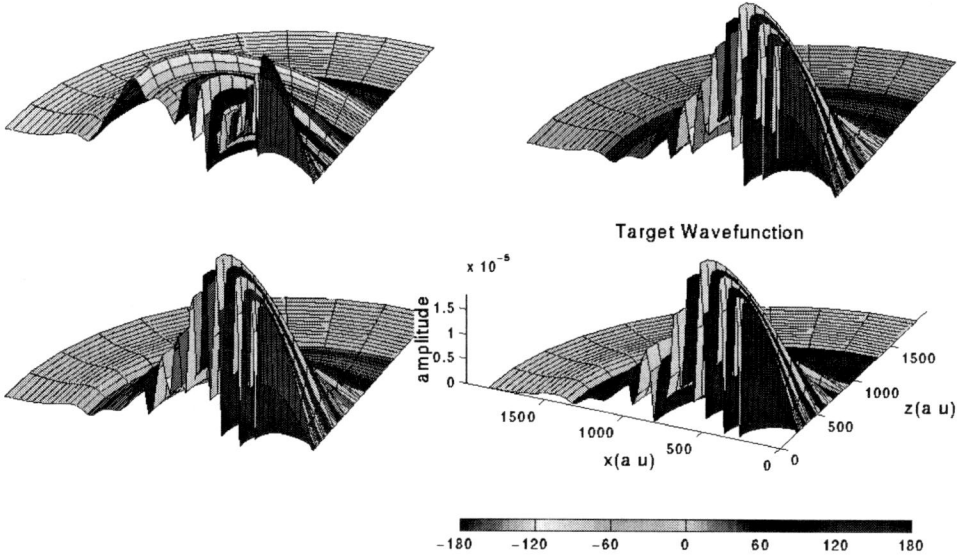

Figure 4. Learning feedback control of a Rydberg wave packet. The desired target wave packet is shown on the lower right. The initial guess produced by the pulse shaper is on the upper left. After two iterations, shown on the upper right and lower left, the experiment converges to the target shape.

This simple feedback control is only effective for systems with a known Hamiltonian and eigenstate spectrum, and for laser interactions in the weak-field limit.

The more general situation is more complicated, but also more useful. Nonlinear optical interactions are used to generate new coherent light sources, and also to probe dynamics. In principle, a strong driving field could be shaped to enhance the desired nonlinear interaction, so that pulse shaping could become a general tool for nonlinear dynamics. Often, the Hamiltonian for any complex quantum system is not known well enough, and with out it one cannot predict likely excitation pathways or derive the optimal pulse shapes. Furthermore, dissipation mechanisms limit the coherence time severely. Learning algorithms for strong field control may aid these problems. In learning control, the experiment runs itself by means of an intelligent feedback loop. It tries various pulse shapes, assesses their success in achieving the desired target excitation, and uses the knowledge gained in this way to improve the pulse shapes on subsequent experiments, all without the intervention of the researcher.

We have been studying learning control in several different nonlinear systems, using a search strategy known as the genetic algorithm (GA).[12] "Genetic" in this case means that the algorithm creates new pulses through a non-local approach based on splicing together traits of successful "parent" laser pulses, rather than by following a fitness gradient function, as in other evolutionary methods.

In our GA implementation, each individual corresponds to a pulse shape, which is encoded as a string of floating point numbers (the individual genome) specifying the phase and amplitude at the various frequency components of the laser pulse. In the first generation, the population consists of sixty individual pulse shapes, chosen at random. We studied quantum dynamics where the target measurement depended on the shape of the driving laser field. The experiment was performed for each pulse shape in turn, and the results assigned a numerical fitness value based on the result. This fitness value determines the chances that a particular pulse shape is selected to reproduce. For example, "roulette wheel" selection, an individual's reproduction probability is proportional to its fitness.

Selective Excitation of Quantum Modes in Molecular Liquids

Our most successful use of the GA in quantum systems thus far has been in selective excitation of C-H and C-D stretch modes in organic molecules in liquid phase. Control has been demonstrated in methanol, ethanol, benzene, and in mixtures of regular and deuterated benzene (C_6D_6). Liquids pose a special challenge to strong field physics, because of rapid relaxation, and also because many nonlinear processes influence the light and make it difficult to define an observable that is related to the desired fitness.

In the liquid phase different experiments were performed, including control of self-phase modulation in carbon tetrachloride and excitation of Raman modes in methanol and benzene. Physical insight is obtained not only through the pulse shapes that the GA finds, but also by how the GA arrived at those solutions.

An example of the kind of control that is possible is shown in the stimulated Raman spectra of methanol excited by GA-shaped pulses. By changing the relative phase between the various spectral components in the pulse, we can selectively excite either the symmetric or asymmetric modes. Further work on state-selective excitation in molecular liquids is underway..

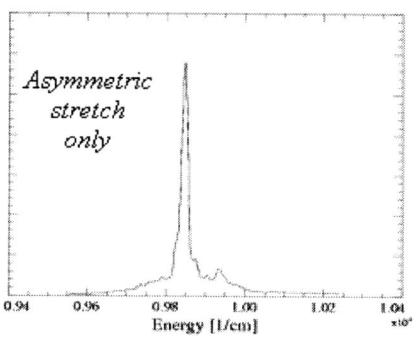 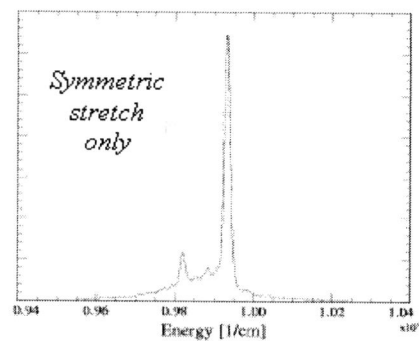

Figure 8. Selective Stokes peaks in the forward scattered light with a methanol sample. The driving laser pulse had a central frequency of 12,700 cm^{-1}.

ACKNOWLEDGMENTS

My students bear no responsibility for any mistakes or omissions in this contribution, but they deserve full credit for the work I have described here. I would particularly like to acknowledge Tom Weinacht, Jae Ahn, and Brett Pearson. The National Science Foundation provided the funding for this work.

REFERENCES

1. C. Spielman, P. F. Curley, T. Brabec, and F. Krausz, IEEE J. Quant Electron. QE-30 1100 (1994), and references therein.
2. D Strickland and G Mourou. Compression of ampli…ed chirped optical pulses. Opt. Commun., **55**(6):447–449, October 1985.
3. J. X. Tull, M. A. Dugan and W. S. Warren, Adv. Opt. Mag. Resonance 20, 1 (1990); A. M. Weiner D. E. Leird J. S. Patel, and J. R. Wullert , J. of Quantum Electronics **28** (1992)..
4. Tull, op.cit. 3.
5. T.C. Weinacht, J. Ahn, and P.H. Bucksbaum, Phys. Rev. Letters **80**, 5508 (1998).
6. T. Gallagher, Rydberg Atoms. Cambridge, Massachusetts: Cambridge Press, 1995.
7. Gallagher, op. cit. 6.
8. J. Ahn, T. C. Weinacht, and P. H. Bucksbaum, Science **287**, 463 (2000)
9. L.K. Grover, Phys. Rev. Lett. **79**, 325 (1997); **79**, 4709 (1997).
10. H. Rabitz and R. Judson, Phys. Rev. Lett. **68**, 1500 (1992).
11. T. C. Weinacht, J. Ahn, and P. H. Bucksbaum, Nature **397**, 233 (1999).
12. Genetic algorithm: Davis, L. Ed. *Handbook of Genetic Algorithms*, Van Norstrand Reinhold: New York, (1991).

The Creation and Measurement of Chiral Coherences

Robert A. Harris* and Jamie D. Walls

Department of Chemistry, University of California, Berkeley, CA. 94720-1460
*(raharris@uclink4.berkeley.edu)

Abstract. The effect of parity nonconservation (PNC) on the coherent tunneling dynamics of chiral molecules is discussed. The measurement of the components of an arbitrary chiral density matrix, as well as the creation of special chiral coherences, is examined.

The realization that PNC inevitably leads to chirality in atoms was experimentally verified by Commins and his group, and others, in the 1970s [1,2,3]. The most significant consequence of PNC in molecules is that the degeneracy of the double well representation of molecules which support chirality may be broken [4]. Hence, the metastable energy levels of left-handed molecules may differ from of their right handed mirror images.

Although there exists a number of proposals for measuring these energy differences, no successful experiment exists [5]. Indeed, only recently an upper bound of 13 Hz was found in the difference between particular hyperfine components of vibration-rotation transitions of L-CHFClBr and R-CHFClBr [6,7]. This bound is an improvement of two orders of magnitude over those of earlier investigations [6].

Two of the proposals to measure parity violations are based on the fact that $|L\rangle$ and $|R\rangle$ are not stationary states [8,9]. Thus, the time dependence of $|L\rangle$ and $|R\rangle$ would be due to both tunneling and parity violations. By taking the eigenstates of the Pauli matrix, σ_z, to be,

$$\sigma_z |L\rangle = +|L\rangle, \tag{1a}$$

$$\sigma_z |R\rangle = -|R\rangle, \tag{1b}$$

we may write a Hamiltonian which incorporates both tunneling and parity violations as [8],

$$H = \delta\sigma_x + \varepsilon\sigma_z, \tag{2}$$

where 2δ is the tunnel splitting and ε is the asymmetry induced by parity violations.

In nature, molecules are usually found in $|L\rangle$ or $|R\rangle$. Hence, if no other states are accessible, tunneling is necessary for measuring parity violations. With both tunneling and parity violations occurring, the oscillations between $|L\rangle$ and $|R\rangle$ would be asymmetric [8]. If the molecule was initially in $|L\rangle$, it could never reach $|R\rangle$. Measurement of the asymmetry, which is $O\left(\frac{\varepsilon}{\delta}\right)^2$, would provide a direct measure of parity violations in chiral molecules.

This method, however, is fraught with difficulties, the major one being that tunneling in chiral molecules has never been observed. However, these difficulties might be circumvented by the rapid advancement of single molecule traps, which would help to remove many decoherence mechanisms.

On the other hand, in the absence of tunneling, the measurement of ε requires a state which is a superposition of $|L\rangle$ and $|R\rangle$. An eigenstate of σ_x could be prepared by exciting coherently to an excited electronic state whose vibrational eigenstates are eigenstates of parity, or through using a nonlinear pulse excitation to a higher state of the ground electronic state [9]. Suppose the resulting state at t=0 is $|+\rangle$, where

$$|+\rangle = \frac{1}{\sqrt{2}}(|L\rangle + |R\rangle). \tag{3}$$

After a time t, $|+\rangle$ would evolve into

$$|t\rangle = \frac{1}{\sqrt{2}}\left(|L\rangle + e^{i2\varepsilon t}|R\rangle\right) \tag{4}$$

The probability of finding the system in the state of negative parity, $|-\rangle$, would be,

$$P(-,t) = \sin^2 2\varepsilon t. \tag{5}$$

As the states of opposite parity in the excited electronic state are assumed to be well separated, the transition, $|-\rangle \to |+, excited\rangle$, is expected to show an initial time dependence of t^2. This transition could then be measured by time-dependent absorption spectroscopy [9].

The question naturally arises: how can one prepare and measure an arbitrary superposition of $|L\rangle$ and $|R\rangle$? In 1994 this question was addressed for the first time [10]. Its approach involved the preparation and measurement of an arbitrary density matrix of $|L\rangle$ and $|R\rangle$ using an excited electronic state which was not characterized by a double well potential. Using phase locked pulses of radiation, arbitrary superpositions could be created [10,11,12].

In this note it is discussed how different components of the density matrix can actually be measured. It is also shown how the eigenstates of the Pauli matrices may be prepared, and their significance discussed.

As usual, the two state density matrix may be written as,

$$\rho = \frac{1}{2}\left(1 + \vec{P}\cdot\vec{\sigma}\right) \tag{6}$$

In the σ_z representation, σ_x is the parity operator,

$$\sigma_x|L\rangle = |R\rangle. \tag{7}$$

When parity is conserved, the eigenstates of σ_x are the eigenstates of the double well Hamiltonian,

$$\sigma_x|\pm\rangle = \pm|\pm\rangle, \tag{8}$$

where,

$$|\pm\rangle = \frac{1}{\sqrt{2}}\left(|L\rangle \pm |R\rangle\right). \tag{9}$$

The eigenstates and eigenvalues of σ_y are,

$$\sigma_y|1,2\rangle = \pm|1,2\rangle, \tag{10}$$

where,

$$|1,2\rangle = \frac{1}{\sqrt{2}}\left(|L\rangle \pm i|R\rangle\right). \tag{11}$$

These states are chiral because

$$P|1,2\rangle = |2,1\rangle. \tag{12}$$

However, they can also be interconverted by time reversal,

$$T|1,2\rangle = |2,1\rangle. \tag{13}$$

This phenomenon is known as "false chirality" [13]. The simplest such example of such states are the eigenstates of momentum in one dimension. It also may be shown that σ_y is proportional to the momentum operator in the two state limit [14].

When there is no tunneling, the states $|1\rangle$ and $|2\rangle$, the eigenstates of σ_y, may be prepared by a simple generalization of the second method, described above, to measure ε. Suppose we prepare the state $|+\rangle$ at t=0. In the presence of an arbitrary chiral potential, $v\sigma_z$, $|+\rangle$ develops into,

$$|t\rangle = \frac{1}{\sqrt{2}}\left(|L\rangle + e^{2ivt}|R\rangle\right). \tag{14}$$

Hence, when $2vt = \frac{\pi}{2}$, $|t\rangle = |1\rangle$. When $2vt = \frac{3\pi}{2}$, we have $|t\rangle = |2\rangle$

We note that if parity violations are to be determined through coherent chiral dynamics, then all "chiral impurities" will contribute. These impurities will create states such as those given by Eqn (14). It is of importance to remove their effect from any measurements to determine parity violations.

We now return to the measurement and significance of the polarization vector, $\bar{P}$. From Eqns [7] through [11], it can readily be seen that $\bar{P}$ may be written in terms of the following probability differences,

$$P_z = P_L - P_R, \tag{15a}$$

$$P_x = P_+ - P_-, \tag{15b}$$

and

$$P_y = P_1 - P_2. \tag{15c}$$

To date only p_z has been measured. However, it was demonstrated that, in the absence of any tunneling, an excited electronic state could be used for the measurement of p_x and p_y [10,11]. Absorption induced by a pair of phase locked pulses followed by measurement of the resulting fluorescence is sufficient to determine p_x. However, it is simpler to measure the difference in absorption cross sections to the + and - vibrational states in the electronic excited electronic state.

Furthermore, p_y can be directly measured by electric field optical activity or electric field circular dichroism [15].

In the presence of tunneling, the components of an arbitrary density matrix may be measured by using pulses of a chiral potential. This is the type of experiment described herein; the point being that σ_z is always the operator being measured. However, such measurements may be converted into measurements of σ_x and σ_y indirectly, by appropriate pulse sequences, a method ubiquitous in NMR and optical spectroscopy.

The Hamiltonian in this case is

$$H(t) = \delta \sigma_x + v(t) \sigma_z. \tag{16}$$

where v(t) is a time dependent chiral potential. The measured quantity is always

$$\langle \sigma_z(t) \rangle = p_L(t) - p_R(t), \tag{17}$$

which is of the form,

$$p_L(t) - p_R(t) = Tr \sigma_z U(t) \rho(0) U^+(t). \tag{18}$$

$U(t)$ is the unitary operator generated by $H(t)$. By using the invariance of the trace we have,

$$p_L(t) - p_R(t) = Tr U^+(t) \sigma_z U(t) \rho(0). \tag{19}$$

If $U(t) \equiv U_x(t)$ where $U_x(t)$, is defined to have the property,

$$\sigma_x = U_x^+(t) \sigma_z U_x(t), \tag{20}$$

Then,

$$p_L(t) - p_R(t) = p_+ - p_- = p_x \tag{21}$$

Similarly, if $U(t) \equiv U_y(t)$, then we will have

$$p_L(t) - p_R(t) = p_1 - p_2 = p_y. \tag{22}$$

Thus, the complete density matrix may be determined. Explicitly, when v(t)=0, the propagator under the above Hamiltonian can be written as

$$U_y\left(t=\frac{2\pi}{4\delta}\right) = e^{-i\delta\left(\frac{2\pi}{4\delta}\right)\sigma_x}. \tag{23}$$

In order to obtain $U_x(t)$, the molecule is allowed to tunnel for a time, $t=\frac{\pi}{2\delta}$. Assuming that tunneling is slow compared to the time dependence of external fields, a constant z field, v, is applied for a time, $\tau=\frac{\pi}{2v}$, where $v >> \delta$.
Thus,

$$U_x(t+\tau) = e^{-i\frac{\pi}{2}\sigma_z} U_y\left(\frac{\pi}{2\delta}\right). \tag{24}$$

It is in this way that all the components of the density matrix can be determined.

In conclusion, we have briefly discussed the role parity violations may play in the dynamics of a chiral system. These dynamics necessarily involve superpositions of left and right localized states characterized by a double well potential. We then examined the components of an arbitrary density matrix. We suggested how to measure the three components of the density matrix. Finally, we suggested a method of constructing the eigenstates of σ_y, the states of maximal false chirality.

ACKNOWLEDGMENTS

R. A. Harris wishes to thank Jeff Cina and Leo Stodolsky for their invaluable contributions to the work described here. He also thanks Laurence Barron for many profitable discussions. Finally, for R.A.H. the friendship of Eugene Commins -- of over thirty-five years -- has been an enormous joy.

REFERENCES

1. See for example, E.D. Commins, *Physica Scripta* **T46**, 92 (1993).
2. Budker, D., in *Fifth International Symposium*, edited by P. Herczeg et al., World Scientific, 1999, p. 418.

3. Khriplovich, I.B., *Parity Nonconservation in Atomic Phenomena*, Gordon & Breach, Philadelphia, PA, 1991.
4. See R. Berger and M. Quack, *J.Chem. Phys.* **112,** 3148 (2000), for the latest in calculations and possessing a very complete set of references.
5. E.g. Kompanets, O.N., et al., *Opt. Commun.* **19**, 414 (1976); and Arimondo, E., et al., *Opt. Commun.* **23**, 369 (1977).
6. Daussy, Ch., et al., *Phys. Rev. Lett.* **83**, 1554 (1999).
7. Quack, M., and Shohner, J., Phys. Rev. Lett. **84**, 3807 (2000).
8 Harris, R.A., and Stodolsky, L., *Phys. Lett* **18B**, 313 (1978).
9. Quack, M, *Chem. Phys. Lett.* **132**, 1471 (1986).
10. Cina, J.A., and.Harris, R.A., *J.Chem. Phys.* **100,** 2531 (1994).
11. Cina, J.A. and Harris, R.A. *Science* **267**, 832 (1994).
12. Duarto-Zamorano, R.P., and Romero-Rochin, V., *J.Chem. Phys.* **114**, 9276 (2001).
13. Barron, L, *Chem. Phys. Lett.* **123**, 423 (1986).
14. Silbey, R., and Harris, R.A., *J.Phys. Chem.* **93**, 7062 (1984).
15. Harris, R.A., *Chem. Phys. Lett.* **223**, 250 (1993).

CASCADE: A New Efficient and Position Sensitive Detector for Thermal Neutrons on Large Areas

M. Klein, H. Abele, D. Fiolka, Chr. J. Schmidt

Physikalisches Institut der Universität Heidelberg, Philosophenweg 12, 69120 Heidelberg, Germany
Fax: 06221/475733, e-mail: martin.klein@physi.uni-heidelberg.de

Abstract. We present the CASCADE-detector, a new concept for efficient and position sensitive detection of thermal neutrons on large areas. The detector concept is based on solid neutron converter layers in a common gas detector system, which guarantees insensitivity to γ-rays. High detection efficiency is reached by cascading several solid neutron converter layers each coated onto a charge transparent carrier substrate. The GEM-foils are employed as such a transparent substrate. In this design position information, the locus of neutron conversion, can undistortedly be imaged through a cascade of several converter layers onto a readout structure underneath. The detector works with ordinary counting gases under normal pressure. Lightweight, easy to handle and in particular large area detectors can be constructed at low costs. A constant flow of fresh, purging counting gas avoids aging effects, which guarantees long term stability as well as long lifetime. The use of GEM-foils provides a high dynamic range from single neutron counting up to high count rates of 10^7 n/cm^2 s and sub-µs absolute time resolution. This feature is required for modern spallation source instrumentation and opens the door towards new ambitious TOF applications. The concept of the detector, together with first results from a prototype device are presented.

1. DESIGN CONSIDERATIONS

The principal change from a gaseous to a solid neutron converter allows to employ any cheap counting gas under normal pressure for charge amplification and detection. Convenient ambient pressure can be used because detection efficiency is now independent from gas pressure. Furthermore, lightweight, easy to handle and in particular large area detectors can be constructed. They can even be arranged in an array configuration with very limited loss to blind areas. Further, semi-spherical constructions, ideally suited for scattering applications are now easily feasible. In order to maintain the negligibly low gamma sensitivity of gas detectors, their enormous advantage over other detector concepts, a solid converter of ^{6}Li or, alternatively ^{10}B is used. Both, for their low Z, produce hardly any photoelectrons on γ-impact. ^{10}B has, as opposed to ^{6}Li, a much higher cross-section of absorption for thermal neutrons. The primer advantage of B, however, is its chemical inertness, so that the material can be exposed to roughly any harsh environment. The charged conversion products (α-particle and ^{7}Li-nuclei) of the neutron capture reaction in ^{10}B produce very high ionization densities in adequate counting gases even at atmospheric pressure. The center of gravity of the charges produced is limited to a range of 1-3 mm around the point of conversion. This results in a spatial resolution of 2-6 mm, by far sufficient for many applications. Cleaning by constant throughput of fresh counting gas avoids aging effects, which guarantees long term stability and long lifetime of the detector.

1.1 The Problem of Insufficient Detection Efficiency

The principal challenge when using solid converters is the detection of the charged particles emerging from the conversion process. Most of these charged products are absorbed within the solid converter itself so that they cannot deposit their energy in the detection medium, e.g. a surrounding counting gas. Only if the conversion products manage to leave the bulk of the converter they can be detected. So apart from neutron capture cross section the ratio of surface to volume of the solid converter determines the overall detection efficiency for neutrons. The advantage of high density of nuclei in a solid converter is jeopardized by the insufficient escape probability of the charged fragments from the bulk of the solid converter.

Up until now it was impossible to use several thin layers of solid converter material in a cascade in order to enhance detection efficiency, as the subsequent layer was impenetrable for the ionization signal, a cloud of electrons and ions. Extraction in other directions inevitably spoils the information on position. An ideal carrier substrate for the converter material would have to be completely transparent for the charges produced.

1.2 Charge-Transparent, Solid Neutron Converting Layers, the GEM

Since 1997 a substrate is available that has this property of charge transparency, the GEM [1]. The GEM was developed by F. Sauli at CERN in 1997 as a Gas Electron Multiplier. GEMs are Kapton foils (50-100 µm thick), that are coated on both sides by a thin layer of copper (5-15 µm thick). Most importantly, these foils are covered with a regular pattern of through going holes, 50 µm in diameter and at a lattice spacing of about 150 µm. Both layers can be individually contacted electrically. If the potentials on these electrodes are chosen adequately, homogeneous field lines far above the hole structure will be channeled right through these holes and the GEM. The fields are formed in such a way, that charges from above the GEM will be channeled and imaged undistorted to below the GEM. In an intuitive manner, these GEMs can thus be cascaded one behind the other. Application of a thin coat of a solid neutron converter material onto such a GEM provides the charge-transparent, solid neutron converter layer.

With charge transparent layers of a solid neutron converter available, the complete technology of standard gas detectors can readily be employed towards the construction of a neutron detector. The transparency for charges allows to cascade several such layers one behind the other collecting single layer detection efficiencies. In such a cascade, an overall detection efficiency for an impinging neutron can be obtained that challenges the standard single channel ^{3}He counter tube.

2. THE CASCADE-DETECTOR

The CASCADE neutron detector is a detector constructed from a cascade of several GEM-foils that are coated with a layer of neutron converter material such as ^{10}B. This detector is operated at ambient pressure. The robust detector can easily be built to cover large areas with a lightweight housing (1-2 kg for 10 000 channels and a surface of 30x30 cm^2).

The planar stack of the converter foils can be read out by means of a simple readout structure. This structure can be adapted to meet specific needs such as e.g. pure azimuth resolution and integration over polar angle or simply 2D resolution. With a stack of several (5-10) converter sheets, detection efficiencies of about 50% for incoming thermal neutrons (1.8 Å) can be achieved.

The use of GEM-foils allows high count rates up to 10^7 n/cm^2 s [2] and because of the low-Z of ^{10}B, because of the high energy of the charged conversion products and together with the property of being a gas detector the detector is free of γ-underground. Furthermore, the detector provides sub-µs absolute time resolution, opening the door towards new TOF applications.

3. FIRST EXPERIMENTAL RESULTS

By now, first completely functional, natural boron coated GEM-foils (sensitive area 4 x 4 cm^2) have been successfully employed in a cascaded setup, proving their functionality as cascadable, charge-transparent substrate for the neutron converter coating. In a four, singly coated layer cascade, spatial resolution was found to be 3 mm. Position information was undistortedly imaged through the four layer cascade onto a charge-detection structure (e.g. MSGC [3]). Detection efficiency was determined to be 4% in agreement with theoretical expectations. The detector has now successfully been scaled up to a sensitive surface area of 8 x 8 cm^2. A two-dimensional readout was additionally installed so that first imaging experiments with data presented here could be carried out.

4. FINAL REMARKS AND OUTLOOK

The further expansion of the sensitive area to a size of 30 x 30 cm^2 is prospected. This larger detector will be equipped with GEM-foils coated with isotopically pure ^{10}B, which will further enhance detection efficiency by a factor of fife. The final goal is the construction of a full sized (30 x 30 cm^2) detector of 50% detection efficiency for incoming thermal (1.8 Å) neutrons. This detector will be constructed with 10 doubly coated cascade layers.

FIGURES

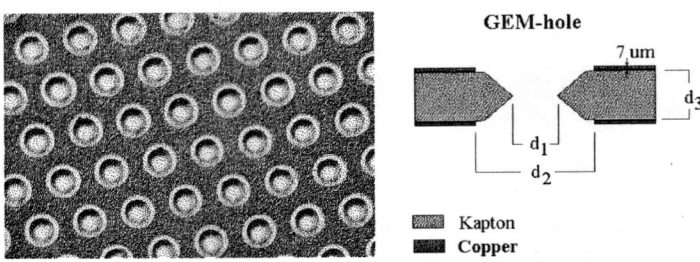

FIGURE 1. GEM from above (left) and GEM-hole cross-sectional view (right). The dimensions are $d_1 = 55$ µm, $d_2 = 95$ µm, $d_3 = 50$ µm.

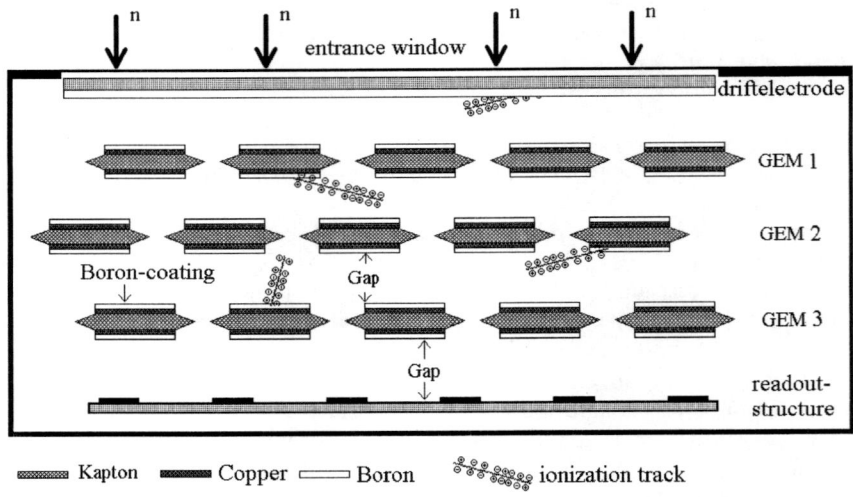

FIGURE 2. Sketch of the CASCADE detector concept: a cascade of three GEM-foils coated on both sides with Boron and a Boron-coated driftelectrode. Primary electrons, produced by the escaping α-particle of the neutron conversion, are guided undistorted through all of the GEMs onto an arbitrary readout-structure.

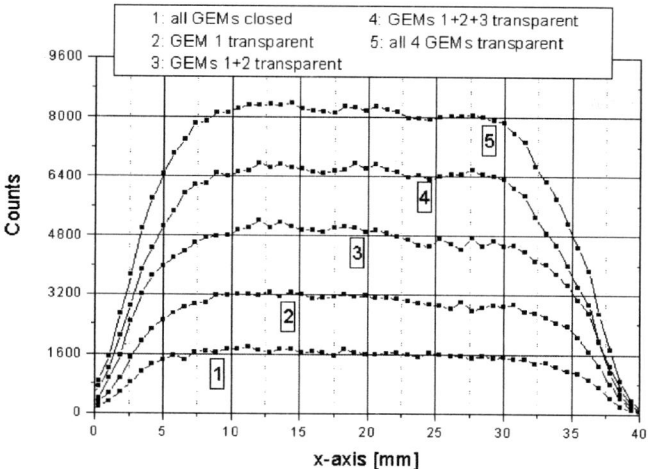

FIGURE 3. One-dimensional spatial distribution of the measured counts in a detector build up of 4 singly Boron-coated GEM-foils. Step by step, each of the 4 GEM-foils is switched transparent electrically, so the Boron coated driftelectrode gets also visible. Each coated GEM-foil increases dramatically the detection efficiency.

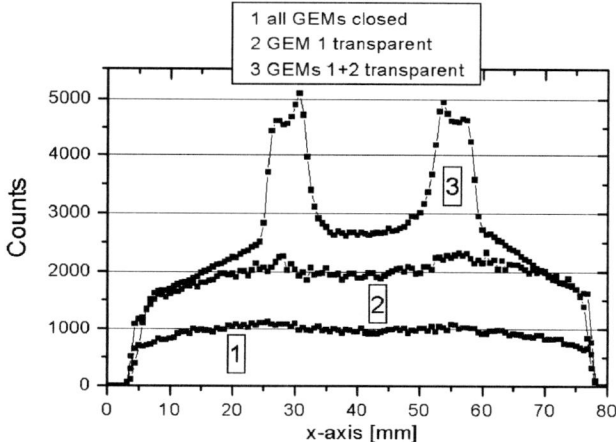

FIGURE 4. One-dimensional spatial distribution of the measured counts (projection onto the x-axis) in a detector (sensitive area: 80×80 mm^2) build up of 2 singly Boron-coated GEM-foils. Step by step, each of the 2 GEM-foils is switched transparent electrically, so the Boron coated driftelectrode (coating has the form of an "H") gets also visible.

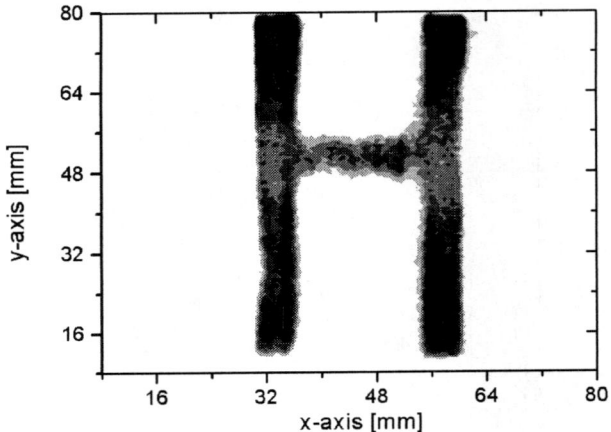

FIGURE 5. Measured Image of the Boron-coated driftelectrode (coating has the form of an "H"). The homogeneous image of the two coated GEM-foils is here substracted.

REFERENCES

1. Sauli, F., *Nucl. Instr. and Meth. A* **386**, 1997, pp 531.
2. Bressan, A., et. al., *Nucl. Instr. and Meth. A* **425**, 356.
3. Oed, A., *Nucl. Instr. and Meth. A* **263**, 1988, pp. 351.

THEORETICAL PERSPECTIVES

Symmetries and the Connection Between Spin and Statistics in Rigorous Quantum Field Theory[1]

E. H. Wichmann

Department of Physics, University of California at Berkeley, Berkeley, California 94720-7300

Abstract. Some basic features of relativistic quantum mechanics and of quantum field theory in the Wightman formulation are reviewed. The connection between spin and statistics, and the PCT-Theorem, is discussed within this framework.

INTRODUCTION

This article is an expansion of material presented in a brief talk at the Symposium in Honor of Professor Eugene Commins. The topics are symmetry operations and the connection between spin and statistics within the framework of rigorous quantum field theory. There is nothing new in this presentation, and detailed discussions can be found in standard texts on the subject[1-6]. Most of the results discussed emerged around the middle of the last century. The two most famous results, the connection between spin and statistics and the *PCT-Theorem*, have by now achieved almost a status of firmly established axioms. It is not my intention to raise any doubts about these predictions, which indeed seem to be strongly confirmed by experimental facts, but I felt it would be of interest to recall the underlying assumptions. With the passage of time, these are, perhaps, sometimes forgotten. I have in mind a particular audience: experimental physicists interested in these questions, but who might have felt frustrated in trying to read about them in one of the above-mentioned monographs. Understanding *all* the details of the proofs does call for a substantial preparation in operator theory, and some other areas in mathematics, but I think that the basic ideas and the logic of the reasoning can well be understood also by people who are not dedicated theoreticians.

This article is my attempt to explain things to my Experimental Friends. My arguments are in many places heuristic, and I do commit some sins, of the kind common in the physics literature. In each case the heuristic reasoning can be made perfectly rigorous, but to do this would involve mathematics which might be unfamiliar to my intended readers. I believe that I have not committed any *mortal sins*, of the kind which cannot be forgiven because there is no known way to make the argument rigorous. My original intention was to present more details of the reasoning, but I soon discovered that this was not possible, given the space limitations. I hope that the reader can nevertheless gain an understanding of what is really involved.

---

[1] Dedicated to my good friend, my respected colleague, and former fellow student Eugene D. Commins.

Among the texts mentioned above, the reader may find that the monograph *Local Quantum Physics* by Haag[4] is particularly accessible, in view of its strong emphasis on the physical ideas. This book also contains an extensive list of references.

SOME VERY ELEMENTARY CONSIDERATIONS

2.1. I want to begin with some very basic considerations about reflection, time reversal and charge conjugation. I here show, in Fig. 1, some rather naive pictures, which I have used in my lectures on special relativity. The purpose is to illustrate the actions of space reflection R, time reversal T, and charge conjugation C, on such physical quantities as angular momentum, electric and magnetic dipole moments, and the electric and magnetic fields. Here R refers to an inversion with respect to the origin in the spatial subspace of a particular inertial frame, and T refers to the time inversion in this frame, relative to the origin $x_o = (\mathbf{0}, 0)$. (I employ the symbol R instead of the customary P for space reflection, because P is needed for other purposes. However, since the *"PCT-Theorem"* is an established *name*, I will not write *RCT* for the *PCT* in this name). The point is that the notions of space reflection, time reversal, and charge conjugation, do derive from very elementary considerations within classical physics, and in particular within classical electrodynamics. The reasoning is the same as in the derivation of the transformation laws for the electromagnetic field under Poincaré transformations (inhomogeneous Lorentz transformations). When I speak of the Poincaré group P, I mean *proper* inhomogeneous Lorentz transformations, composed of spatial rotations, velocity transformations (also called "boosts"), and translations in spacetime. The *extended* Poincaré group P_e includes space reflection and time reversal. *Relativistic invariance* means invariance under the proper group P, but not necessarily under the extended group. However, classical electromagnetic theory is, in fact, also invariant under the extended group P_e. We recall the demonstration of this. We know how to perform any transformation in P_e on trajectories of a charged particle, assuming that the charge is an *invariant*. The electric and magnetic fields are operationally defined in terms of the Lorentz force equation. For a given "original" electromagnetic field we consider all possible trajectories of charged particles, and we then ask whether there exists, for any particular transformation in P_e, a transformed field for which the possible trajectories are precisely the ones obtained by applying the transformation on the "original" trajectories. The answer is *yes*, and we arrive at the obviously unique transformation laws for the electromagnetic field. For the case of proper transformations in P this means, as is well known, that the components of the electric and magnetic fields can be identified with the components of a second-rank skew-symmetric tensor field on spacetime. The transformation laws under space reflection and time reversal are shown in Table 1 below. We must, of course, verify that the transformed fields also satisfy Maxwell's equations, with the transformed sources. This involves a consistency check as follows. If we imagine that the sources are swarms of charged particles we know, from the transformation laws for the trajectories, how the charge density and current density must transform, and in this way we indeed arrive at a transformation law for the sources such that Maxwell's equations are satisfied by all the transformed quantities. For the case of proper transformations in P the charge density

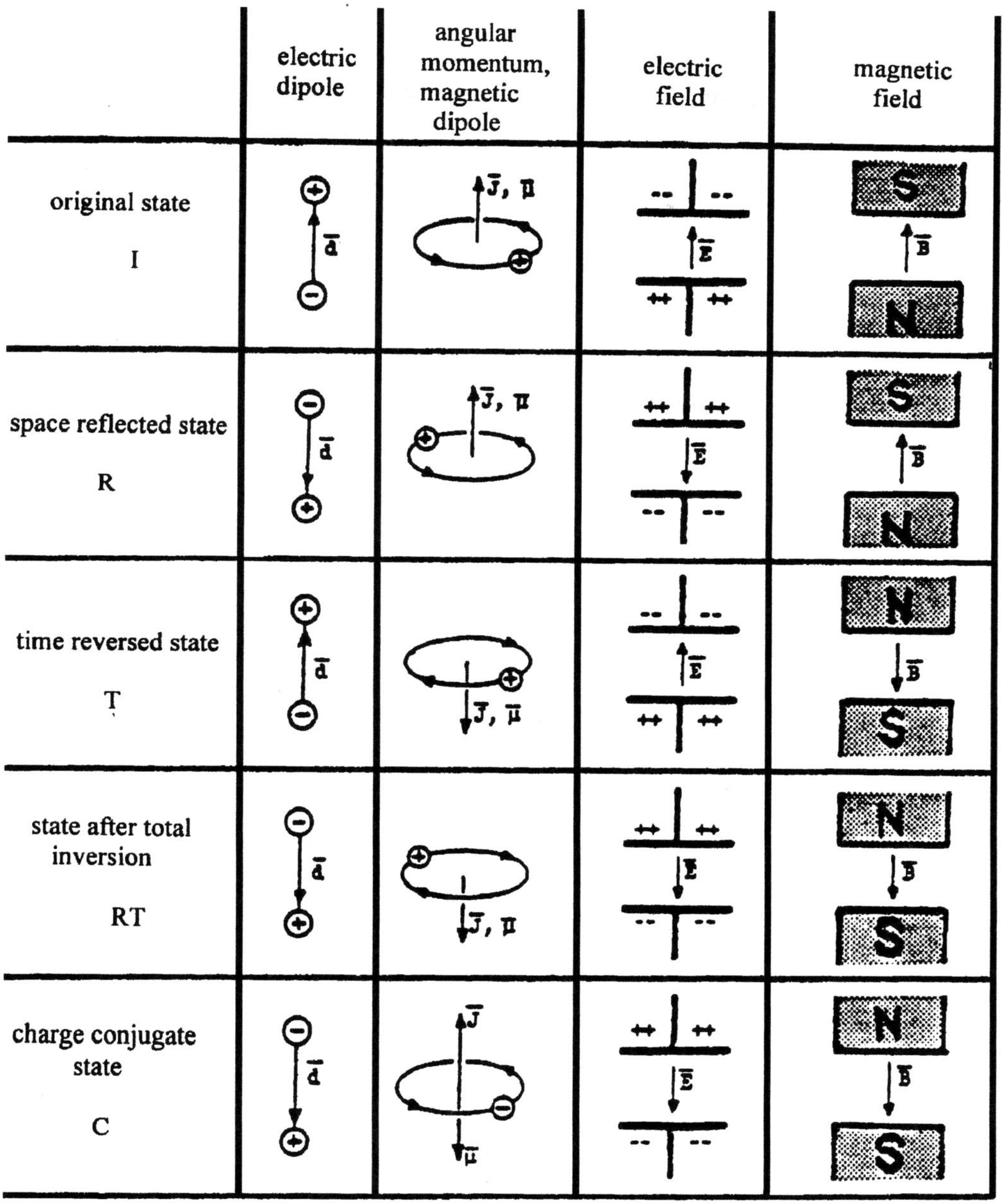

	electric dipole	angular momentum, magnetic dipole	electric field	magnetic field
original state I	$\bar{d}$ ↑ (+ above, − below)	↑ $\bar{J}, \bar{\mu}$	↑ $\bar{E}$	↑ $\bar{B}$ (S top, N bottom)
space reflected state R	$\bar{d}$ ↓ (− above, + below)	↑ $\bar{J}, \bar{\mu}$	↓ $\bar{E}$	↑ $\bar{B}$ (S top, N bottom)
time reversed state T	$\bar{d}$ ↑ (+ above, − below)	↓ $\bar{J}, \bar{\mu}$	↑ $\bar{E}$	↓ $\bar{B}$ (N top, S bottom)
state after total inversion RT	$\bar{d}$ ↓ (− above, + below)	↓ $\bar{J}, \bar{\mu}$	↓ $\bar{E}$	↓ $\bar{B}$ (N top, S bottom)
charge conjugate state C	$\bar{d}$ ↓ (− above, + below)	↑ $\bar{J}$, ↓ $\bar{\mu}$	↓ $\bar{E}$	↓ $\bar{B}$ (N top, S bottom)

FIGURE 1. Pictorial representation of the actions of space reflection, time reversal, total inversion, and charge conjugation on some physical variables. Presented as an aid to intuition.

and the components of the current density transform as the components of a four-vector field, in a well known fashion. The point of view in the above is that the transformations act on the states of the physical systems to produce new states, with reference to a *fixed* inertial frame, with a *fixed* standard coordinate system.

Table 1. Transformation laws under R, T, C, and RCT.

physical variable	space reflect. $X = R$	time reversal $X = T$	charge conj. $X = C$	RCT $X = \theta$
$\mathbf{E}_X(\mathbf{x},t) =$	$-\mathbf{E}(-\mathbf{x},t)$	$\mathbf{E}(\mathbf{x},-t)$	$-\mathbf{E}(\mathbf{x},t)$	$\mathbf{E}(-\mathbf{x},-t)$
$\mathbf{B}_X(\mathbf{x},t) =$	$\mathbf{B}(-\mathbf{x},t)$	$-\mathbf{B}(\mathbf{x},-t)$	$-\mathbf{B}(\mathbf{x},t)$	$\mathbf{B}(-\mathbf{x},-t)$
$\rho_X(\mathbf{x},t) =$	$\rho(-\mathbf{x},t)$	$\rho(\mathbf{x},-t)$	$-\rho(\mathbf{x},t)$	$-\rho(-\mathbf{x},-t)$
$\mathbf{j}_X(\mathbf{x},t) =$	$-\mathbf{j}(-\mathbf{x},t)$	$-\mathbf{j}(\mathbf{x},-t)$	$-\mathbf{j}(\mathbf{x},t)$	$-\mathbf{j}(-\mathbf{x},-t)$

Classical electrodynamics is also trivially invariant under charge conjugation C, under which the fields and the charge- and current densities merely change sign. The Table 1 shows the actions of the transformations R, T, and C, in a very pedantic manner. To avoid confusion it is necessary to exhibit the independent variables $\mathbf{x}$ and t explicitly. As far as we know particles with magnetic charge do not exist. If they did exist, they could be incorporated in a modified theory. A magnetic monopole changes sign under space reflection R, time reversal T, and charge conjugation C.

Note that C is *frame-independent* and commutes with all extended Poincaré transformations. The transformations R and T in the extended Poincaré group are frame-dependent, and the combined transformation $I = RT$, *total inversion*, depends on the choice of the origin. The space reflection and time reversal transformations relative to any other inertial frame are conjugate (under P) with the transformations defined above.

2.2. When we go to the quantum mechanical description we have to adhere to a *group theoretical correspondence principle*. The expectation values of physical variables which have classical counterparts must transform in the same manner as the corresponding classical variables, otherwise there will be conceptual chaos. The classical transformations R, T, and C generate an abelian group $\mathcal{G}_C$ of order eight. The corresponding quantum mechanical group $\mathcal{G}_Q$ is not necessarily isomorphic with the classical group $\mathcal{G}_C$, and not necessarily abelian. As an example we recall here that the square of time reversal acting on a state of half-integral angular momentum is not the identity I, but $-I$. The group $\mathcal{G}_Q$ will, in general, have an invariant subgroup $\mathcal{G}_G$ of transformations on the Hilbert space of states which have no observable consequences. These are *gauge transformations*, connected with *superselection principles*, first discussed in a paper by Wick, Wigner, and Wightman[7]. The existence of superselection principles contradicts the idea in what we might call "orthodox quantum mechanics" that *every* self-adjoint operator is an observable (in principle). Instead we have the principle that only those self-adjoint operators which commute with all elements of the gauge group can be observables. As discussed in the quoted paper this has important consequences for the notion of an intrinsic parity of a particle.

The relationship between the quantum mechanical group $\mathcal{G}_Q$ and the corresponding classical group $\mathcal{G}_C$ is that $\mathcal{G}_C$ is isomorphic with the quotient group $\mathcal{G}_Q/\mathcal{G}_G$. Since an observable must commute with the gauge transformations, it follows that the observables do transform under conjugation with the elements in G_Q like their classical counterparts. We can then draw firm conclusions from experimental results with reference to the picture in Fig. I, or by similar elementary considerations.

a) That parity is not conserved, i.e., that space reflection R is not a valid symmetry, was shown conclusively in an experiment by Wu, Ambler, Hayward, Hoppes, and Hudson[8], and in another experiment by Garwin, Lederman, and Weinrich[9]. The facts are simply that a reflection transformation performed on the experimental data produces "data" which are in contradiction to what is actually observed.

b) That charge conjugation C is not a valid symmetry was shown equally conclusively by Macq, Crowe, and Haddock[10]. In this case too one finds that the transformed data do not correspond to what is observed. It is of interest to note that the conclusion in this case (as well as in the case of parity violation) is *independent* of any detailed theory. It only involves the definitions of C and R, as explained above.

c) For a time it was believed, and hoped, that perhaps CR is a valid symmetry. This hope was shattered by some results by Christenson, Cronin, Fitch, and Turlay[11], and by Bennett, Nygren, Saal, Steinberger, and Sunderland[12]. The analysis of these experiments is not as simple as in the case of nonconservation of parity or violation of charge conjugation invariance, but the conclusion that the CR symmetry is violated seems inescapable.

d) In a hypothetical world in which the weak interactions could be switched off we believe that R, T, and C would be valid symmetries, i.e., for the strong and electromagnetic interactions. In the real world the weak interactions are present, and this implies (weak) violations of R symmetry also in atomic and nuclear physics. The object of our celebration, and his students and coworkers, have found examples of this, and so have other groups[13]. Violations of charge conjugation invariance cannot be tested in atomic physics, and the same is true for the combined symmetry CR. I have suggested to Professor Commins that he should repeat his thallium experiments[14] with antithallium, but he has not followed up on this suggestion. This is a typical example of how experimentalists fail to do experiments desired by theoreticians.

e) Tests of time reversal invariance in scattering experiments, or in decay processes, are tricky, because in principle one has to observe the reversed processes. This is in general impossible, except for some strictly elastic scattering processes. At present there is no unambiguous case of violation of time reversal invariance. The demonstration that the electron has an electric dipole moment would be clear evidence of a violation of *both* time reversal invariance and reflection invariance. There is no intrinsic prohibition against a particle having *both* an electric dipole moment and a magnetic dipole moment, but under T or R the relative direction of the electric and magnetic dipole moments is reversed, as shown in Fig. 1, corresponding to two *different* states of the particle. Since there is good evidence for only *one* kind of electron (and only *one* kind of neutron, or proton) we would conclude that time reversal invariance is not valid if these particles are found to have electric dipole moments. No electric dipole moment of any particle has been detected to date, and we have good upper limits for the neutron and the electron. For the electron a new stringent upper limit has very recently be obtained by Regan, Commins, DeMille, and Schmidt[15]. For a review of the status of electric dipole moments for leptons I here refer to an article by Commins[16], who has been interested in this very fundamental issue for many years.

2.3. For a review of the present status of non-invariance under CR I refer to an review article by Wolfenstein[17]. It appears that the interaction responsible for the observed violation is very weak even compared with the weak interaction responsible for beta decay, and hence it manifests itself only under very special conditions, as in the decays of the neutral K-mesons. All particles in nature are "coupled" in a sense, and a violation of CR invariance in one domain "infects" all of physics. If we believe in RCT-invariance, noninvariance under CR implies noninvariance under T, but in view of the extreme weakness of the interaction leading to a CR-violation, we expect that the violation of T invariance is also extremely small, and hence difficult to detect. The theoretical predictions of the magnitude of the electric dipole moment of the electron are not reliable, but we do expect a dipole moment, although it might be so extremely small that it will never be observed.

Invariance under RCT implies a number of things. To every particle corresponds an antiparticle, of the same mass and spin, and opposite charge. These are predictions for which very accurate experimental tests are possible. There is no evidence to date of any violation of RCT invariance. The extremely small upper limit on the mass difference of K^o and $\overline{K}^o$ is particularly noteworthy. From the theoretical side RCT invariance is particularly sacred, and a detected violation would be Bad News for local quantum field theory. I will discuss the basis of our belief in RCT invariance later in this article.

2.4. As is well known, symmetry operations are described in quantum mechanics by unitary or antiunitary operators, in accordance with a fundamental result of Wigner. It is easily shown that a *continuous* group of symmetries must always be described by *unitary* operators. For the symmetries R, T, and C, the *group theoretical correspondence principle* dictates that R and C must be *unitary*, but time reversal T *antiunitary*. In the case of T we can argue as follow. Under time reversal the angular momentum **J** changes sign, and we thus have $TJ_\alpha T^{-1} = -J_\alpha$. However, time reversal must commute with all rotations, and we thus have $\exp(-i\theta J_3) = T\exp(-i\theta J_3)T^{-1} = \exp(-T(iJ_3)T^{-1}) = \exp((TiT^{-1})\theta J_3)$, in the case of a rotation about the 3-axis. Hence we must have $TiT^{-1} = -i$, and T is thus antiunitary. By similar reasoning we conclude that R and C must be unitary, since the operations R and C leave the angular momentum unchanged, and commute with all rotations. The combined transformation $RCT = \theta$ must then be antiunitary.

ABOUT THE LORENTZ- AND POINCARÉ GROUPS

3.1. A short review of the relevant group theory is in order. First I want to state here that we are very fortunate to live in a *four*-dimensional Minkowski spacetime $\mathcal{M}$. This space has many special features, manifested in a certain mathematical simplicity which we do not find in spaces of other dimensions.

The *quantum mechanical* rotation group is the group $SU(2)$, consisting of all unitary 2×2 matrices of unit determinant. We can say that this is our first Lucky Break, because this group is the least complicated of all (compact) nonabelian Lie groups. All irreducible (unitary) representations $u \to D^j(u)$ of this group can be found very explicitly. Here the labelling index j is the angular momentum quantum number, and the dimensionality of the representations is $2j+1$. The matrix elements of $D^j(u)$ in the *standard form* are homogeneous polynomials of degree $2j$, with *real* coefficients, of the matrix elements of $u \in SU(2)$. The complex conjugate representation $D^j(u)^* = D^j(u^*)$ is similar to the representation $D^j(u)$, and explicitly we have $D^j(u)^* = \sigma_2 D^j(u)\sigma_2$. The matrix $u(\mathbf{e},\theta) = \exp(-i\theta\mathbf{e}\cdot\sigma/2) \in SU(2)$ describes a rotation by angle θ in the clockwise sense about the unit vector $\mathbf{e}$, and the matrix $-u(\mathbf{e},\theta)$ describes the *same* rotation physically. There is a two-to-one homomorphism from $SU(2)$ onto the *classical* rotation group $O^+(3)$, i.e., to every physical rotation correspond precisely two elements u and $-u$ of $SU(2)$. The element $-I \in SU(2)$ is a rotation by angle 2π about *any* axis, with no physically observable effect. We note here that $D^j(-I) = (-1)^{2j} \cdot I$. I assume that all of this is well known.

3.2. The second Lucky Break is that the *quantum mechanical* Lorentz group $\widehat{L}$ is the group $SL(2;C)$ of all complex 2×2 matrices g of unit determinant. It is "twice as large" as the *classical* Lorentz group L. The group L is the group of all *proper* Lorentz matrices M. Such an M is a real 4×4 matrix which satisfies the conditions $\widetilde{M}I_L M = I_L$, $M_{44} \geq 1$, and $\det(M) = 1$. Here I_L is the diagonal Lorentz metric matrix with diagonal matrix elements $(I_L)_{44} = 1$ and $(I_L)_{\alpha\alpha} = -1$ for $\alpha = 1,2,3$. There is a two-to-one homomorphism $g \to M(g)$ from $\widehat{L}$ onto L, given explicitly by

$$M(g)_{mn} = \frac{1}{2}Tr(\sigma_m g \sigma_n g^\dagger), \tag{1}$$

where $\sigma_4 = I$, and σ_m are the Pauli matrices for $m = 1,2,3$. To every $g \in \widehat{L}$ thus corresponds a unique $M(g) \in L$, and to every $M \in L$ correspond *two* matrices g and $-g$ in $\widehat{L}$, such that $M(g) = M(-g) = M$, and we have $M(g')M(g'') = M(g'g'')$. The group $SU(2)$ is a subgroup of the group $SL(2;C) = \widehat{L}$, and if $u \in SU(2)$, then the matrix $M(u)$, as given in (1), describes a rotation in the three-dimensional spatial subspace of $\mathcal{M}$ (relative to a particular inertial frame).

We regard the fact that $\widehat{L} = SL(2;C)$ as a lucky break because the group $SL(2;C)$ is relatively uncomplicated. Many facts about the classical Lorentz group L can actually be established most easily if we first consider the corresponding issue for the 2×2 matrices in $SL(2;C)$, and then exploit the homomorphism in (1).

3.3. A more fundamental Lucky Break is that the Lorentz group L is, as an abstract group, a *complex extension* of the rotation group $O^+(3)$. This is not "visible" if we look at the 4×4 Lorentz matrices, but it is immediately obvious that if we let the parameter θ in $\exp(-i\theta\mathbf{e}\cdot\sigma/2)$ become complex, then these "complex rotations" generate the group $SL(2;C)$. In particular, the "imaginary rotation" $\exp(\lambda\mathbf{e}\cdot\sigma/2)$ (with λ real) describes a velocity transformation (also called a "boost") by velocity $\mathbf{v} = \mathbf{e}\tanh(\lambda)$.

A consequence of this is that all the *finite*-dimensional irreducible representations of $SL(2;C)$ can be easily constructed from the representations $D^j(u)$ of $SU(2)$. As I said above, the matrix elements of $D^j(u)$ are homogeneous polynomials (of degree $2j$) of the matrix elements of u. The replacement of u by $g \in SL(2;C)$ in these polynomials yields an irreducible representation $g \to D^j(g)$ of $SL(2;C)$. The complex conjugate representation $D^j(g)^* = D^j(g^*)$ is *not* similar to the representation $D^j(g)$ if $j > 0$, and we thus obtain a new family of irreducible representations. One finds that the complete family of all *finite*-dimensional mutually inequivalent irreducible representations of

$SL(2;C)$ consists of all representations of the form $g \to D^{(s',s'')}(g) = D^{s'}(g) \otimes D^{s''}(g^*)$, indexed by the pairs (s',s''), where $2s'$ and $2s''$ are non-negative integers. Furthermore one finds that every finite-dimensional representation is completely reducible, and hence a direct sum of the above irreducible representations. The representation $D^{(0,0)}(g)$ is the trivial identity representation. None of the other representations is similar to a unitary representation. The representation $g \to M(g)$ in (1) is the representation $D^{(1/2,1/2)}$, and the familiar Dirac representations is the reducible representations $D^{(0,1/2)} \oplus D^{(1/2,0)}$. Note that $D^{(s',s'')}(-I) = (-1)^{2s'+2s''} \cdot I$. The representations for which $s'+s''$ is an *integer* are called *tensorial* representations, and the representations for which $s'+s''$ is a *half-integer* are called *spinorial* representations. They are distinguished from each other by the sign of $D^{(s',s'')}(-I) = \pm I$.

Note the important fact, which follows trivially from the nature of the representation $D^{(s',s'')}(g)$, that the representation $D^{(s',s'')}(g)^*$ is equivalent to the representation $D^{(s'',s')}(g)$.

3.4. For later reference we consider here a particular two-parameter family of elements in $SL(2;C)$, of the form $g_3(\lambda,\theta) = \exp(\lambda\sigma_3/2)\exp(-i\theta\sigma_3/2)$, where λ and θ are arbitrary *real* numbers. This family is obviously a two-dimensional commutative subgroup. The element $g_3(\lambda,\theta)$ describes a rotation by angle θ about the 3-axis, followed by a velocity transformation of velocity $\tanh(\lambda)$ in the 3-direction. The two factors commute, and we can write $g_3(\lambda,\theta) = \exp((\lambda - i\theta)\sigma_3/2)$. The corresponding 4×4 matrix $M(g_3(\lambda,\theta))$ is given by

$$M(g_3(\lambda,\theta)) = M_3(\lambda,\theta) = \begin{bmatrix} \cos\theta, & -\sin\theta, & 0, & 0 \\ \sin\theta, & \cos\theta, & 0, & 0 \\ 0, & 0, & \cosh\lambda, & \sinh\lambda \\ 0, & 0, & \sinh\lambda, & \cos\lambda \end{bmatrix}, \quad (2)$$

and for any (s',s'') the corresponding matrix $D^{(s',s'')}(g_3(\lambda,\theta))$ is given by

$$D^{(s',s'')}(g_3(\lambda,\theta)) = D^{s'}(\exp((\lambda-i\theta)\sigma_3/2)) \otimes D^{s''}(\exp((\lambda+i\theta)\sigma_3/2)) = D_3^{(s',s'')}(\lambda,\theta) \quad (3)$$

We first consider the 4×4 matrix $M_3(\lambda,\theta)$, and we note that it is well defined also for *complex* values of λ, and that it is then an *entire analytic function* of λ. For complex values of λ (and complex values of θ, if we like), the matrix is an element of the *connected complex Lorentz group* L_C, consisting of all complex 4×4 matrices M which satisfy the conditions $\widetilde{M}I_L M = I_L$ and $\det(M) = 1$. If we set $\lambda = i\pi$ and $\theta = \pi$ we obtain $M_3(i\pi,\pi) = -I$, and if θ varies continuously from 0 to π, and λ varies continuously from 0 to $i\pi$, then $M_3(\lambda,\theta)$ varies continuously from I to $-I$. The matrix $-I$ describes the total inversion transformation RT on Minkowski spacetime. It is *not* an element of the *proper* (real) Lorentz group L, but it *is* an element of the proper complex Lorentz group L_C. This fact is the key ingredient in the beautiful proof by Jost[18] of the *PCT*-Theorem.

We next note that as a function of λ the expression in the second member in (3) is also an entire analytic function of λ (since the matrix elements of $D^s(g)$ are polynomials of the matrix elements of the argument g), and we denote this function by $D_3^{(s',s'')}(\lambda,\theta)$. We then have the relation

$$D_3^{(s',s'')}(i\pi,\pi) = (-1)^{2s''} \cdot I = D_3^{(s',s'')}(-i\pi,-\pi) \tag{4}$$

which plays an important role in the proofs of both the spin-statistics theorem and the PCT-theorem. We will return to this later.

3.5. The elements of the *classical* Poincaré group P are of the form $\Lambda = \Lambda(x,M)$. Here $M \in \mathsf{L}$ is a 4×4 Lorentz matrix describing a homogeneous *proper* Lorentz transformation (which leaves the origin fixed), and x is any vector in Minkowski spacetime $\mathcal{M}$. The element $\Lambda(x,M)$ is interpreted as the homogeneous transformation M followed by a parallel translation by x. The law of composition is thus $\Lambda(x',M')\Lambda(x'',M'') = \Lambda(x'+M'x'',M'M'')$.

The *quantum mechanical* Poincaré group, denoted $\widehat{\mathsf{P}}$, is "twice as large." Its elements are of the form $\lambda = \lambda(x,g)$, where x is any vector in $\mathcal{M}$ as before, and where g is any element in the group $\widehat{\mathsf{L}} = SL(2,C)$. To the relationship between $\widehat{\mathsf{L}}$ and L corresponds a two-to-one homomorphism from $\widehat{\mathsf{P}}$ onto P, given explicitly by $\lambda(x,g) \to \Lambda(x,M(g))$. The law of composition for the group $\widehat{\mathsf{P}}$ is: $\lambda(x',g')\lambda(x'',g'') = \lambda(x'+M(g')x'',g'g'')$.

The relationships between the classical and quantum mechanical Lorentz groups and Poincaré groups are thus extensions of the very familiar relationship between the classical rotation group, i.e., the orthogonal group $O^+(3)$, and the group $SU(2)$ of all unitary unimodular 2×2 matrices. As we said in Sect. 2.2, a group which in the quantum mechanical framework corresponds to some classical group is not necessarily isomorphic with the classical group. The relationship is instead that the classical group is isomorphic with the quotient group of the quantum mechanical group with respect to some invariant subgroup of gauge transformations. In the case of $SU(2)$ and $SL(2;C) = \widehat{\mathsf{L}}$ the invariant subgroup in question is the central subgroup with the two elements $\{I,-I\}$. In the case of $\widehat{\mathsf{P}}$ the invariant subgroup is the central subgroup with the two elements $\{I,\lambda(0,-I)\}$. The element $\lambda(0,-I)$ can be described as a rotation by angle 2π about *any* (spacelike) axis, in *any* inertial frame.

3.6. Finally some remarks about the geometry of Minkowski spacetime $\mathcal{M}$. A characteristic feature of this geometry, which has no counterpart in Euclidean geometry, is the existence of *light cones*. In particular we consider the *open forward lightcone* V_+ with the origin as apex. It consists of all *timelike* four-vectors, i.e., all four-vectors $x = (\mathbf{x},t)$ such that $t > |\mathbf{x}|$. The *closed forward lightcone* $\overline{V}_+$ consists of V_+ and its boundary points. Similarly we also consider the light cones in four-momentum space, denoted by the same symbols. The action of the Lorentz group L on Minkowski space and on four-momentum space is described by the same Lorentz matrices M discussed above. The lightcones V_+ and $\overline{V}_+$ are mapped onto themselves by any $M \in \mathsf{L}$.

SOME GENERAL FEATURES OF RELATIVISTIC QUANTUM MECHANICS

4.1. What I discuss next is not something specific for quantum field theory. Any relativistically invariant quantum theory would be based on the assumptions which I state below, and the consequences of these assumptions have a general validity. Understanding this structure is crucial for an understanding of quantum field theory, which is why I felt that it is necessary to discuss it here. I will now list and discuss the basic assumptions.

Assumption I. The space $\mathcal{H}$ of physical state vectors is an infinite-dimensional separable Hilbert space. The term "separable" means that $\mathcal{H}$ has a countable orthonormal basis. Such a space is "large enough" to describe all multi-particle states of *massive* particles.

Assumption II. The Hilbert space $\mathcal{H}$ carries a *continuous unitary representation* $\lambda \to \Gamma(\lambda)$ of the group $\widehat{\mathsf{P}}$, i.e., to every $\lambda(x,g)$ corresponds a unique unitary operator $\Gamma(\lambda(x,g)) \equiv \Gamma(x,g)$ such that $\Gamma(x',g')\Gamma(x'',g'') = \Gamma(x' + M(g')x'', g'g'')$. "Continuous" means that the matrix element $\langle \Phi | \Gamma(x,g) \Psi \rangle$ is a continuous function of x and of g for all vectors Φ and Ψ, which is obviously a reasonable physical requirement. The existence of the representation of $\widehat{\mathsf{P}}$, with the correct action on all physical variables (by conjugation), is, of course, the essence of "relativistic invariance."

By the law of composition for $\widehat{\mathsf{P}}$ we have $\lambda(x,g) = \lambda(x,I)\lambda(0,g)$, and hence $\Gamma(x,g) = \Gamma(x,I)\Gamma(0,g)$. It is convenient to introduce special symbols for the operators which represent translations and homogeneous transformations, since we will consider these separately. We thus write $\Gamma(x,I) = T(x)$ for the translation by x, and $\Gamma(0,g) = U(g)$ for a homogeneous transformation, and hence we have $\Gamma(x,g) = T(x)U(g)$. From the law of composition it then follows that

$$U(g)T(x)U(g)^{-1} = T(M(g)x) \tag{5}$$

Assumption III. There exists a unit vector Ω, called the *vacuum vector*, uniquely determined up to a factor of modulus unity by the condition that $\Gamma(\lambda)\Omega = \Omega$ for all $\lambda \in \widehat{\mathsf{P}}$.

The translations have a common spectral resolution, and we have $T(x) = \exp(ix \cdot P)$, where $P = (\mathbf{P}, H)$ is the four-momentum operator. The scalar product $x \cdot P$ is the Lorentz scalar product $x \cdot P = tH - \mathbf{x} \cdot \mathbf{P}$, where $\mathbf{x}$ the spatial part of the translation x, t the purely temporal part, H the energy operator, and $\mathbf{P}$ the three-momentum operator. The assumption which follows is usually referred to as the *spectrum condition*.

Assumption IV. The spectrum of P is confined to the (closed) forward lightcone $\overline{V}_+$. Heuristically speaking, every common eigenvector of the components of P corresponds to eigenvalues which are the components of a vector in the forward lightcone. (Actually the only such eigenvector is the vector Ω, for which we have $P\Omega = 0$, but the reader can consider "plane waves," although these are not really vectors in $\mathcal{H}$). We thus have $\langle \Phi | H \Phi \rangle > 0$ for any vector Φ not proportional to Ω. There are no states of negative energy, and only one state of zero energy, namely the vacuum state Ω. Conversely this condition together with Assumptions II and III imply Assumption IV.

4.2. As I said in Sect. 3.5, the classical Poincaré group P is the quotient group of the quantum mechanical Poincaré group $\widehat{\mathsf{P}}$ with respect to the central subgroup with the two elements $\{I, \lambda(0, -I)\}$. For a representation of $\widehat{\mathsf{P}}$ as described above we write $U_o = U(-I) = \Gamma(0, -I)$, because this important operator deserves a special symbol. We have $U_o^2 = I$, and U_o commutes with all Poincaré transformations. It describes as a rotation by angle 2π about *any* (spacelike) axis, in *any* inertial frame. It is a gauge transformation, corresponding to the *angular momentum superselection principle*, according which the relative phase of a state with integer angular momentum and a state with half-integer angular momentum has no physical meaning, i.e., it cannot be measured. If Φ_b is any state of integer angular momentum we have $U_o\Phi_b = \Phi_b$, whereas we have $U_o\Phi_f = -\Phi_f$ for any state Φ_f of half-integer angular momentum. In case the reader has doubts about *this* superselection principle (or other superselection principles), and if these doubts persist even after a reading of the paper by Wick, Wigner and Wightman[7] mentioned before, I can only say that the reader should then build a device for the measurement of something prohibited by the superselection principle. Ultimately this is a question of *experimental* physics. That relativistic invariance should be described in terms of the group $\widehat{P}$, as above, rather than P, is dictated by the fact that particles of half-integer spin do exist.

4.3. We must now discuss a consequence of the spectrum condition (nonnegative energy condition) which is of crucial importance in quantum field theory. The issue is the extension of the translation operator $T(x)$ to an operator-valued function $T(z)$ on a certain domain $\overline{V}_{+i}$ in *complex* Minkowski spacetime $\mathcal{M}_C$. The domain $\overline{V}_{+i}$ is called the *closed forward imaginary tube*, and it consists of all complex four-vectors $z = x + iy$ such that $x \in \mathcal{M}$ and $y \in \overline{V}_+$. The interior of $\overline{V}_{+i}$ is the *open forward imaginary tube* V_{+i}, consisting of all complex four-vectors $z = x + iy$ such that the imaginary part y is forward timelike and the real part x is arbitrary. The closed tube $\overline{V}_{+i}$ thus consists of V_{+i} and its boundary points. Note that real Minkowski space $\mathcal{M}$ is contained in $\overline{V}_{+i}$, as a *part* of its boundary. We also note the following simple facts. Since V_+ is mapped onto itself by any proper homogeneous Lorentz transformation, it follows that V_{+i} is also mapped onto itself by such a transformation. Since the sum of two forward timelike vectors is forward timelike, it also follows that $z' + z'' \in V_{+i}$ whenever $z', z'' \in V_{+i}$, and that $z' + z'' \in \overline{V}_{+i}$ whenever $z', z'' \in \overline{V}_{+i}$.

Now the reader may rightly wonder why we should consider something so clearly "unphysical" as the complex spacetime. The reason is that the excursion into this unphysical domain will permit us to use the theory of analytic functions as an *extremely* powerful mathematical tool. I therefore ask the reader to be patient. A key result can be stated as follows.

Theorem 4.3: We assume the general framework in Sect. 4.1. There then exists a unique operator-valued function $T(z)$ on $\overline{V}_{+i}$ with the following properties: I) For any real x, $T(x)$ is the unitary operator describing a translation by x. II) For every $z \in \overline{V}_{+i}$, the operator $T(z)$ is a *bounded* operator (with norm $\|T(z)\| = 1$), and $T(z)$ is a *continuous* function of z on $\overline{V}_{+i}$ (in the strong operator topology). III) The operator-valued function $T(z)$ is a *jointly analytic function* of the components of z for all $z \in V_{+i}$ (relative to the norm topology). IV) For any two $z', z'' \in \overline{V}_{+i}$ we have $z' + z'' \in \overline{V}_{+i}$ and $T(z')T(z'') = T(z' + z'')$. V) For any $z \in \overline{V}_{+i}$ and any $g \in \widehat{L} = SL(2,C)$ we have $M(g)z \in \overline{V}_{+i}$ and $U(g)T(z)U(g)^{-1} = T(M(g)z)$.

About the Proof: 1) A Mathematical Friend seeing this would say that the theorem is a *triviality*, easily proved by an application of the Functional Calculus. For some readers this reasoning is likely to involve unfamiliar mathematics, and I will therefore outline a heuristic proof, in a style which is common in the physics community. We introduce a basis for $\mathcal{H}$ consisting of "eigenvectors" $\phi(p;q)$ of the four-momentum P. Here q is an additional label corresponding to the "other quantum numbers" needed to describe the states. We thus have $P_k \phi(p;q) = p_k \phi(p;q)$ for any component P_k of P. The vectors $\phi(p;q)$ can be thought of as "many-particle plane-wave states," normalized such that $\langle \phi(p';q')|\phi(p'';q'')\rangle = \delta_4(p'-p'')\delta_{q',q''}$. The identity operator I is thus of the form

$$I = \int_{\overline{V}_+} d^4(p) \sum_q |\phi(p;q)\rangle\langle\phi(p;q)| \qquad (6)$$

which is called the *completeness relation* among physicists. Since the possible momenta are in the closed forward lightcone, the integration is over $\overline{V}_+$, as shown. Since $\phi(p;q)$ is an eigenvector of P it is also an eigenvector of the translation operator $T(x) = \exp(ix \cdot P)$, and we have $T(x)\phi(p;q) = e^{ix \cdot p}\phi(p;q)$. Multiplying both members in (6) by $T(x)$ we then obtain the relation

$$T(x) = \int_{\overline{V}_+} d^4(p) e^{ix \cdot p} \sum_q |\phi(p;q)\rangle\langle\phi(p;q)| \qquad (7)$$

which is the heuristic version of the common spectral resolutions of the elements of the translation group.

2) We want to extend $T(x)$ to the domain $\overline{V}_{+i}$ in complex Minkowski space. If $h(p)$ is any (complex-valued) function of p, we can *define* an operator $h(P)$ by $h(P)\phi(p;q) = h(p)\phi(p;q)$. If $h(p)$ is not bounded on $\overline{V}_+$ this may lead to an unbounded operator $h(P)$, and domain problems arise, but if $h(p)$ is bounded, then $h(P)$ is also bounded. Applying this to the function $\exp(iz \cdot p)$, with $z \in \overline{V}_{+i}$, we obtain an operator, denoted $T(z)$, by

$$T(z) = \exp(iz \cdot P) = \int_{\overline{V}_+} d^4(p) e^{iz \cdot p} \sum_q |\phi(p;q)\rangle\langle\phi(p;q)| \qquad (8)$$

The salient point is this. It is well-known, and trivial, that the Minkowski scalar product of two forward timelike vectors is *positive*, and more generally we have $y \cdot p \geq 0$ if $y, p \in \overline{V}_+$. From this it follows that $|e^{iz \cdot p}| = e^{-y \cdot p} \leq 1$ for all $z = x+iy \in \overline{V}_{+i}$ and all $p \in \overline{V}_+$, and the function $e^{iz \cdot p}$ in the integral in (8) is therefore *bounded*.

By a simple computation we find that $T(z')T(z'') = T(z'+z'')$. With $z = x+iy \in \overline{V}_{+i}$ we have $T(z) = T(x)T(iy)$. The operator $T(iy) = \exp(-y \cdot P)$ is Hermitian, and $\phi(p;q)$ is an eigenvector of $T(iy)$ corresponding to the eigenvalue $e^{-y \cdot p} \leq 1$ All the eigenvalues of $T(iy)$ are thus positive and less than or equal to 1. The vacuum vector Ω is an eigenvector of $T(iy)$ of eigenvalue 1, and we conclude that $T(iy)$, and hence also $T(z) = T(x)T(iy)$, is a bounded operator of unit norm.

The relation $U(g)T(x)U(g)^{-1} = T(M(g)x)$ implies the relation $U(g)PU(g)^{-1} = M(g)^{-1}P$ (in accordance with the group theoretical correspondence principle), and from this it follows readily that we also have $U(g)T(z)U(g)^{-1} = T(M(g)z)$ for any $z \in \overline{V}_{+i}$. This is an extension of the relation (5) to the domain $\overline{V}_{+i}$.

3) The analyticity of the operator $T(z)$ is, perhaps, the trickiest item in the proof, and the reader may not be very familiar with the theory of functions which are jointly analytic functions of *several* complex variables. The notion of such an ("ordinary") function $f(\zeta) = f(\zeta_1, \zeta_2, \ldots, \zeta_n)$ is not complicated. The function f is jointly analytic at a point

$(\zeta_1', \zeta_2', \ldots, \zeta_n')$ if and only if it can be represented by a convergent multiple Taylor series in powers of the variables $(\zeta_k - \zeta_k')$, $k = 1, \ldots, n$, on some neighborhood of the point $(\zeta_1', \zeta_2', \ldots, \zeta_n')$. If the function $f(\zeta)$ is jointly analytic, it is naturally analytic in each variable separately, with the other variables held fixed.

The assertion in the theorem is that the *operator* $T(z)$ can be represented by a power series on a neighborhood of any point in the open forward imaginary tube, and it is asserted that such a series, the terms of which are bounded operators, converges in *norm*. This involves an examination of the Taylor expansion of $\exp(iz \cdot p)$ in powers of $i(z - z') \cdot p$ when z is in some neighborhood of a point $z' \in V_{+i}$ and $p \in \overline{V}_+$. This expansion always converges, but we have to show that the convergence is *uniform* in p. This is mildly tricky, although only elementary calculus is involved. I will not present the details here, and we will now regard the theorem as having been convincingly proved.

4.4. The theorem states more than we actually need. For our purposes it would suffice to know that every *matrix element* $\langle \Phi' | T(z) \Phi'' \rangle$ is a jointly analytic function of the components of z. This follows from our theorem, but an alternative approach would be to deal only with the matrix elements, which has the advantage that these are "ordinary" complex-valued functions, which are more familiar objects than operator-valued functions. We consider the matrix element $\langle \Phi' | T(x) \Phi'' \rangle$ of the translation operator, which we express in the form

$$F(x) = \langle \Phi' | T(x) \Phi'' \rangle = \int_{(\infty)} d^4(p) f(p) e^{ix \cdot p}, \quad f(p) = \sum_q \langle \Phi' | \phi(p;q) \rangle \langle \phi(p;q) | \Phi'' \rangle \quad (9)$$

The matrix element $F(x)$ is thus the Fourier transform of a "function" $f(p)$ which vanishes for all p outside the forward lightcone $\overline{V}_+$. Note that there must be something "singular" about $f(p)$ since there is a contribution to $F(x)$ from the point $p = 0$, corresponding to the vacuum state Ω. The "function" $f(p)$ is a function in the same sense as the Dirac delta function is a "function," and our Mathematical Friend would tell us that the integral in (9) is the Fourier transform of a *measure*. We can then show that $F(x)$ can be extended to a function $F(z)$ on the complex domain $\overline{V}_{+i}$, analytic on the interior V_{+i}, because $F(x)$ is the Fourier transform of a function $f(p)$ which vanishes outside the (closed) forward lightcone. Stated in this form the result is analogous to a result, which I assume is well-known, about the Fourier transform of a function $h(p)$ of a *single* variable p, with the property that $h(p) = 0$ when $p < 0$. Let $H(x) = \int dp \, e^{ixp} h(p)$ be the Fourier transform of such a function. Let $z = x + iy$ be a complex variable, with real part x and imaginary part y. Let C_{+i} be the open half-plane consisting of all points z *above* the real axis in the complex z-plane, and let $\overline{C}_{+i}$ be its closure, consisting of all $z = x + iy$ such that $y \geq 0$. If $p \geq 0$ and $z \in \overline{C}_{+i}$ we trivially have $|e^{izp}| = e^{-yp} \leq 1$, and this means that we can define a function $H(z)$ on $\overline{C}_{+i}$ by $H(z) = \int dp \, e^{izp} h(p)$ which is analytic on the *open* half-plane C_{+i}. In general the function $H(z)$ cannot be continued analytically into the *lower* half complex plane, unless $h(p)$ satisfies some additional special conditions. Note also here that $H(iy)$ is the *Laplace transform* of $h(p)$.

The condition that $h(p) = 0$ when $p < 0$ corresponds to the condition that $f(p)$ in (9) vanishes *outside* $\overline{V}_+$. The closed positive half of the real line in the one-dimensional case thus corresponds to the closed forward lightcone in the four-dimensional case. Similarly the half-plane $\overline{C}_{+i}$ corresponds to $\overline{V}_{+i}$, which is *precisely* the set of all complex four-vectors z such that $\text{Im}(z \cdot p) \geq 0$ for all $p \in \overline{V}_+$. I hope that these analogies are enlightening.

4.5. We now recall another result in complex variable theory which is undoubtedly also very familiar.

Lemma 4.5.A. Let $A(z)$ be a function, defined and continuous on the closed upper-half complex plane $\overline{C}_{+i}$, and *analytic* on the open upper-half complex plane C_{+i}. If $A(x)$ vanishes identically on any nonempty open interval (x', x'') on the real axis, then $A(z)$ vanishes for *all* $z \in \overline{C}_{+i}$, and in particular for all real z.

The proof is very simple, but I will not present it here. Proofs of theorems of this nature can be found in any textbook on complex variable theory. The version stated above is a particularly simple version.

This lemma has a rather straightforward generalization to functions on four-dimensional space.

Lemma 4.5.B. Let $B(z)$ be a function, defined and continuous on the closed forward imaginary tube $\overline{V}_{+i}$ in complex Minkowski space, and *analytic* on the open tube V_{+i}. If $B(x)$ vanishes identically on some nonempty open subset X_0 of $\mathcal{M}$, then $B(z)$ vanishes for *all* $z \in \overline{V}_{+i}$, and in particular for all real z.

Proof: The strategy of proof is to consider one (suitably chosen) component of z at a time, and then apply Lemma 4.5.A. Let $x_0 \in X_0$ and let v be a *forward timelike* four-vector. Let $\zeta = \xi + i\eta$ be a complex variable. We have $x(\zeta) = x_0 + \zeta v \in \overline{V}_{+i}$ if $\zeta \in \overline{C}_{+i}$, i.e., if $\eta \geq 0$, and $x(\zeta) \in V_{+i}$ if $\zeta \in C_{+i}$. The function $A(\zeta) = B(x_0 + \zeta v)$ then satisfies the premises of the function A in Lemma 4.5.A, and we have $A(\xi) = 0$ when $\xi \in (-\delta, \delta)$ for some $\delta > 0$, since $x_0 + \xi v \in X_0$ for a sufficiently small ξ. By Lemma 4.5.A we conclude that $A(\xi) = B(x_0 + \xi v) = 0$ for *all* ξ. Since v was an arbitrary forward timelike vector, it follows that $B(x_0 + x') = 0$ for *every* timelike x'. Repeating this argument we conclude that $B(x_0 + x' + x'') = 0$ for every timelike x' and every timelike x''. Since an arbitrary four-vector can be written as a sum of two timelike four-vectors, it follows that $B(x) = 0$ for *all* $x \in \mathcal{M}$.

Let $z = x + iy \in V_{+i}$ and let $\zeta = \xi + i\eta$ be a complex variable. Then the function $A_1(\zeta) = B(x + \zeta y)$ satisfies the premises of the function A in Lemma 4.5.A, and $A_1(\xi) = 0$ for all real ξ. We conclude that $A_1(i) = B(x + iy) = B(z) = 0$, and hence $B(z) = 0$ for all $z \in V_{+i}$. By continuity it then follows that $B(z) = 0$ for all $z \in \overline{V}_{+i}$. This completes the proof.

4.6. There is an obvious generalization of Lemma 4.5.B to functions of several complex four-vectors.

Theorem 4.6. Let $F(z^{(1)}, z^{(2)}, \ldots, z^{(n)})$ be a function of the n complex four-vectors $\{z^{(1)}, \ldots, z^{(n)}\}$, defined and (jointly) continuous on the domain for which $z^{(k)} \in \overline{V}_{+i}$ for each k, and such that F is an *analytic* function of $z^{(k)}$ if $z^{(k)} \in V_{+i}$. Let $X^{(k)}$ be a nonempty open subset of the real space $\mathcal{M}$ for each k. If $F(x^{(1)}, x^{(2)}, \ldots, x^{(n)}) = 0$ when $x^{(k)} \in X^{(k)}$ for each k, then $F(z^{(1)}, z^{(2)}, \ldots, z^{(n)}) = 0$ on its entire domain of definition, and in particular $F(x^{(1)}, x^{(2)}, \ldots, x^{(n)}) = 0$ for all real $\{x^{(k)}\}$.

About the Proof: We consider each variable in turn, keeping the other variables fixed, and then apply Lemma 4.5.B. In the first step we thus consider the function $B(z^{(1)}) = F(z^{(1)}, x^{(2)}, \ldots, x^{(n)}) = 0$, with $x^{(k)} \in X^{(k)}$ for $k = 2, 3, \ldots, n$. By Lemma 4.5.B we conclude that $B(z^{(1)}) = 0$ for all $z^{(1)} \in \overline{V}_{+i}$. I think that the continuation is obvious, and I therefore omit further details.

4.7. We next state a most important consequence of the spectrum condition.

Theorem 4.7. We assume the general framework in Sect. 4.1. Let $\{B_1, B_2, \ldots, B_n\}$ be an n-tuplet of *bounded* operators on $\mathcal{H}$, and let $\Psi \in \mathcal{H}$. With $z^{(k)} \in \overline{V}_{+i}$ for each $k = 1, \ldots, n$, the expression at right in

$$F(z^{(1)}, z^{(2)}, \ldots, z^{(n)}) = \langle \Psi | T(z^{(1)}) B_1 T(z^{(2)}) B_2 \ldots T(z^{(n)}) B_n \Omega \rangle \tag{10}$$

is well defined. It defines a function F which satisfies the premises of the function F in Theorem 4.6, and hence the conclusions in that theorem apply.

Proof: Follows readily from Theorem 4.3.

It is of interest to consider a "real variable formulation" of the implications of this theorem. Let $\{B_1, B_2, \ldots, B_n\}$ be an n-tuplet of *bounded* operators, and let $F(x^{(1)}, x^{(2)}, \ldots, x^{(n)}) = \langle \Psi | T(x^{(1)}) B_1 T(x^{(2)}) B_2 \ldots T(x^{(n)}) B_n \Omega \rangle$ for all real $x^{(k)}$. Let $X^{(k)}$ be a nonempty open subset of the space $\mathcal{M}$ for each k. If $F(x^{(1)}, x^{(2)}, \ldots, x^{(n)}) = 0$ when $x^{(k)} \in X^{(k)}$ for each k, then $F(x^{(1)}, x^{(2)}, \ldots, x^{(n)}) = 0$ for *all* real $x^{(k)}$. This consequence of the spectrum condition would certainly seem very surprising if we knew nothing about complex variable theory, and in my opinion it is *not* "physically obvious." I think that this illustrates the remarkable power of the theory of analytic functions. Theorems 4.6 and 4.7 are the simplest cases of a more general scheme of things. The assumption in Theorem 4.6 that the function F assumes its boundary values for real $\{x^{(k)}\}$ in a *continuous* fashion can be relaxed very substantially without invalidating the conclusion. Under certain conditions (having to do with domain issues) Theorem 4.7 remains valid also when the operators B_k are *unbounded* operators. In quantum field theory we really need stronger versions of the above theorems, but in our heuristic discussion we will ignore such complications. The main message in this discussion is that the spectrum condition leads to conclusions of the kind presented in Theorem 4.7.

4.8. We will need one more general result.

Theorem 4.8. We assume the general framework in Sect. 4.1. Let v be any (fixed) *spacelike* four-vector, and let $\Psi_1, \Psi_2 \in \mathcal{H}$. Then,

$$\lim_{s \to \infty} \langle \Psi_1 | T(sv) \Psi_2 \rangle = \langle \Psi_1 | \Omega \rangle \langle \Omega | \Psi_2 \rangle \tag{11}$$

About the proof: Our Mathematical Friend would say that this follows from the Riemann-Lebesgue Lemma, because the spectral measure in the joint spectral resolution of the one-parameter group $\{T(sv) \mid s \in \mathbb{R}\}$ of translations is *absolutely continuous* when restricted to the orthogonal complement of Ω. This, in turn, follows from the fact that this translation group is a subgroup of a continuous unitary representation of the Poincaré group. I will not explain this here, but instead argue that the result is "obvious physically," at least in a theory of particles. Let Ψ'_k be the component of Ψ_k orthogonal to Ω, for $k = 1, 2$. These vectors are represented by some many-particle wave functions which describe "clusters" of particles approximately localized in some region in physical 3-space at a particular time. Without any loss of generality we can assume that v is orthogonal to the time-axis. The vector $T(sv)\Psi'_2$ describes the second cluster displaced by sv, and as s tends to infinity the displaced cluster "overlaps" less and less with the first cluster, and hence $\langle \Psi'_1 | T(sv) \Psi'_2 \rangle$ must tend to zero, which is what the theorem asserts.

In the context of quantum field theory this result is referred to as the *cluster decomposition property*.

ABOUT QUANTUM FIELD THEORY

5.1. Formulations of quantum field theory emerged very soon after the birth of quantum mechanics. In particular, so-called *free-field theories*, which describe *non*-interacting particles, were discussed in the spirit of a straightforward "quantization" of classical field theories. In such a theory the field is a linear expression in creation and destruction operators which satisfy certain commutation or anticommutation relations. I assume that this is quite familiar. Although the original formulations were rather heuristic, it turns out that a free-field theory is mathematically unproblematic, and perfect rigor is easily established. However, in the attempts to formulate field theories describing *interacting* particles serious mathematical problems became apparent, and these problems have not been completely overcome to date. It became clear that the mathematical tools required for quantum field theory are much more sophisticated than those required for nonrelativistic quantum mechanics. A paper by Wightman[19] in 1956 can be regarded as a seminal paper in the rigorous formulation of quantum field theory. Many people have contributed to this theory, often called *axiomatic field theory*. The kind of quantum fields in this theory are commonly referred to as *Wightman fields*, in view of the important contributions by Wightman.

An alternative theory of local quantum physics (not orthogonal to the Wightman theory) was formulated as a theory of local algebras of *bounded* operators by Haag and Kastler[20]. This *algebraic quantum field theory* is discussed in the monographs by Haag[4], by Araki[5], and by Horuzhy[21]. The accomplishments in this theory is a long story, which we cannot discuss here.

I want to state immediately that both the Wightman theory and the algebraic theory are *general frameworks*, rather then specific theories. They have not produced any cross sections or other Numbers to date, and the frameworks are furthermore wide enough to include special cases of a manifestly unphysical nature, such as field theories with no particle interpretation. In contrast with this, *Lagrangian field theories* have produced Numbers, although the mathematical sense of such theories is problematic. In particular, products of field operators at the same point occur in the theories, and this is illegal *if* the fields are distributions. Quantum electrodynamics is a Lagrangian field theory, and in view of the spectacular successes of this theory one cannot dismiss Lagrangian theories as nonsense. The question is rather how the objects in the theories should be interpreted mathematically. A specific Lagrangian theory is characterized by a specific *dynamical principle*, which seems to me to be totally absent in the above-mentioned rigorous theories.

Within a general framework one can only hope to derive (equally general) features of structure. Hopefully these results will still be valid in a more specific future theory, and they should be applicable in some sense to the Lagrangian theories. The most spectacular results of this kind are the *spin-statistics connection* and the *PCT-Theorem*. Similar results also emerge in the algebraic quantum field theory, under suitable assumptions. All *free field theories* fall within the Wightman framework, which means that the Wightman assumptions are consistent, but these theories are trivial. It is worth mentioning that no explicit *nontrivial* example has ever been constructed.

I will now review the assumptions on which the theory is based. For a more detailed (and more rigorous) discussion I refer in particular to Chapters 3 and 4 in the monograph *PCT, Spin and Statistics, and All That*, by Streater and Wightman[1].

5.2. The basic postulates include all the Assumptions I-IV made in Sects. 4.1. The first additional assumption introduces the quantum fields.

Assumption V. There exist a finite number of *finite*-component spinor or tensor fields $\{\psi^{(1)}(x), \psi^{(2)}(x), \ldots, \psi^{(n)}(x)\}$. The components of the field $\psi^{(k)}(x)$ are denoted $\psi_\alpha^{(k)}(x)$, with $\alpha = 1, 2, \ldots, d_k$, for some positive integer d_k. Every field component $\psi_\alpha^{(k)}(x)$ is an *operator-valued tempered distribution*. The action of the Poincaré group $\widehat{P}$ on the fields (by conjugation) is described by

$$T(x')\psi^{(k)}(x)T(x')^{-1} = \psi^{(k)}(x+x') \tag{12}$$

and

$$U(g)\psi_\alpha^{(k)}(x)U(g)^{-1} = \sum_\beta S_{\alpha\beta}^{(k)}(g^{-1})\psi_\beta^{(k)}(M(g)x). \tag{13}$$

For each k, $S^{(k)}(g)$ is *one* of the irreducible matrix representations $D^{(s',s'')}(g)$ of $\widehat{L} = SL(2;C)$ discussed in Sect. 3.3, i.e., we have $S^{(k)}(g) = D^{(s'_k, s''_k)}(g)$ for some pair (s'_k, s''_k), and hence $d_k = (2s'_k+1)(2s''_k+1)$. We will say that the field is of *Type* (s'_k, s''_k). To each field $\psi^{(k)}(x)$ in the set of fields corresponds a *hermitian conjugate field* in the set, denoted $\psi^{(k)\dagger}(x)$, and hence $\psi^{(k)\dagger}(x) = \psi^{(l)}(x)$ for some l. We then have $S^{(k)}(g)^* = S^{(l)}(g)$ and $s'_l = s''_k$, $s''_l = s'_k$, in accordance with the discussion in Sect 3.3, and if the field $\psi^{(k)}$ is of type (s'_k, s_k''), then $\psi^{(k)\dagger}$ is of type (s''_k, s'_k). We say that a field of type (s', s'') is a *spinor field* if $s' + s''$ is a half-integer, and a *tensor field* if $s' + s''$ is an integer. This latter class includes a scalar field, for which $s' = s'' = 0$ and a vector field, for which $s' = s'' = 1/2$.

That the field components are operator-valued tempered distributions means the following. Let $\phi(x) = \psi_\alpha^{(k)}(x)$ be such a component. For every complex-valued function $f(x)$ on Minkowski spacetime, in a set $\mathcal{S}$ of *tempered test functions*, there exists a linear operator $\phi[f]$ on $\mathcal{H}$, which is written symbolically as an average of $\phi(x)$ weighted by $f(x)$,

$$\phi[f] = \int_{(\infty)} d^4(x) f(x) \phi(x) \tag{14}$$

Although one customarily speaks of "the field operator $\phi(x)$" there is *no* linear operator $\phi(x)$ at *a point* x. An operator-valued tempered distribution is analogous (in a sense) to an "ordinary" tempered distribution, of which the Dirac delta function $\delta(s)$ is the canonical example, well-known in the physics community. The "function notation," according to which a distribution is written as if it were an ordinary function, is convenient and widely employed. A number of operations valid for ordinary functions are also valid for these *generalized functions*, but some operations are definitely illegal. In particular it is not permissible to form the product of two distributions of the *same* argument. Within the framework of distribution theory the statements of the transformation laws in (12) and (13) are perfectly legitimate, and much more transparent than a restatement of these laws in terms of the objects $\phi[f]$, often called the *smeared fields*. We will employ the "function notation," and the reader can heuristically think about "the field operator $\phi(x)$", as long as it is remembered that we really deal with distributions. Forming a product $\phi'(x')\phi''(x'')$ of two field operators with *independent* arguments is perfectly legitimate, but the product $\phi'(x)\phi''(x)$ is illegal and meaningless.

The operators $\phi[f]$ are, in general, *unbounded* operators, and some assumption about domains is necessary. It is assumed that there is a common domain $\mathcal{D}$ for all these operators, which includes the vacuum vector Ω, and which is mapped into itself by every $\phi[f]$, and every Poincaré transformation. This assumption is a "technical assumption," which makes the transformation laws in (12) and (13) meaningful, and which permits us to consider the (polynomial) *algebra* $\mathcal{P}$ generated by the operators $\phi[f]$ and the identity I. If X is any open nonempty subset of $\mathcal{M}$, we denote by $\mathcal{P}(X)$ the algebra generated by the identity and all field operators $\phi[f]$ for which the test functions $f(x)$ vanish identically outside X. Although it is not quite correct, the reader can think about $\mathcal{P}(X)$ as the (complex) linear span of I and all products of field operators $\phi(x)$ with *independent* arguments x in X. If $Q \in \mathcal{P}(X)$, then $Q^\dagger \in \mathcal{P}(X)$, and if $Q', Q'' \in \mathcal{P}(X)$, then $Q'Q'' \in \mathcal{P}(X)$ and $c'Q' + c''Q'' \in \mathcal{P}(X)$ for all complex numbers c', c''. We can think about $\mathcal{P}(X)$ as the "local field operators in X." The algebra $\mathcal{P} = \mathcal{P}(\mathcal{M})$ is thus the algebra of *all* field operators. In our discussion we will assume that the domain $\mathcal{D}$ consists of all vectors of the form $Q\Omega$ as Q ranges through $\mathcal{P}$.

The assumption about the *specific* test function space $\mathcal{S}$ is also a "technical assumption," which is convenient and reasonable. The functions in $\mathcal{S}$ are infinitely differentiable, and they, and all their partial derivatives (which are also in $\mathcal{S}$), decrease rapidly as the argument tends to infinity. I remark here that even in classical electromagnetic theory the field at *a point* makes no sense operationally. We measure the field with a test-charge of finite extent, and hence we measure a *smeared* field. The idea that the quantum fields are distributions is thus physically reasonable.

A field $\psi^{(k)}(x)$ as described above is commonly called an *irreducible field*, because the transformation law in (13) involves an *irreducible* representation $g \to S^{(k)}(g)$ of $\widehat{L}$. This notion of "irreducible" is something different from the notion of "irreducible" in Assumption VI below. In the discussion which follows the term "field" means such an object $\psi^{(k)}(x)$, which we can think about as the set of *all* linear combinations of the *particular* components $\psi^{(k)}_\alpha(x)$. Every such linear combination $\phi(x)$ is a *component* of $\psi^{(k)}(x)$. The label "α" of the component $\psi^{(k)}_\alpha(x)$ refers to a particular choice of the matrices $S^{(k)}(g)$ to realize the (abstract) representation $D^{(s'_k, s''_k)}$. Field theory is beset with notational problems, and we easily end up with cumbersome expressions with many subscripts and superscripts. We will reserve the symbols $\psi^{(k)}(x)$, or just $\psi^{(k)}$, for the irreducible fields. We will use the symbol $\phi(x)$ (embellished with primes, subscripts, or superscripts, when necessary) to denote *any* component of an irreducible field, and we will call $\phi(x)$ a *field component*, or *field operator*.

5.3. The purpose of the next assumption is to ensure that "there are sufficiently many field operators," i.e., that "everything can be described in terms of the fields."

Assumption VI. The set of all field operators is *irreducible* in the following sense. If B is a *bounded* operator which satisfies the condition that $\langle \phi(x)^\dagger \Phi | B\Psi \rangle = \langle \Phi | B \phi(x) \Psi \rangle$ for every field operator $\phi(x)$, and for all $\Phi, \Psi \in \mathcal{D}$, then B is a multiple of the identity.

The condition on B is a cautious way of saying that "B commutes with all field operators," and the assumption is thus that only multiples of the identity commute with all field operators. We note immediately a simple consequence of this assumption.

Lemma 5.3. The vacuum vector Ω is a *cyclic vector* of the algebra $\mathcal{P}$ of all field operators. This means that every vector in $\mathcal{H}$ can be approximated arbitrarily closely by a vector of the form $Q\Omega$, where Q is a polynomial expression in the field operators. Equivalently stated: If a vector Φ satisfies the condition that $\langle \Phi | Q\Omega \rangle = 0$ for all $Q \in \mathcal{P}$, then $\Phi = 0$.

Proof: The set of all vectors $Q\Omega$ is a linear space (since $\mathcal{P}$ is a linear manifold). If its closure would be a *proper* sub-Hilbert space, then the self-adjoint projection onto this subspace would satisfy the above condition on B, and hence be a multiple of the identity, which entails a contradiction. The closure is therefore the entire space $\mathcal{H}$, and the orthogonal complement is empty, which means that there is no nonzero vector orthogonal to every $Q\Omega$.

5.4. The final assumption deals with the question of *causality* and *locality*.

Assumption VII. For any two fields $\psi^{(k)}$ and $\psi^{(l)}$ (not necessarily different) one of the following two alternatives obtains: *Case I:* The fields $\psi^{(k)}(x)$ and $\psi^{(l)}(y)$ *commute* for all spacelike $x - y$, in the sense that $\phi'(x)\phi''(y) = \phi''(y)\phi'(x)$ for every component ϕ' of $\psi^{(k)}$ and every component ϕ'' of $\psi^{(l)}$. *Case II:* The fields $\psi^{(k)}(x)$ and $\psi^{(l)}(y)$ *anticommute* for all spacelike $x - y$, in the sense that $\phi'(x)\phi''(y) = -\phi''(y)\phi'(x)$ for every component ϕ' of $\psi^{(k)}$ and every component ϕ'' of $\psi^{(l)}$. It is understood that these commutation or anticommutation relations hold on the domain of the field operators. We will say more about the motivation for this assumption later.

5.5. Let $\{\phi_1, \phi_2, \ldots, \phi_r\}$ be an arbitrary r-tuplet of components of the field operators $\psi^{(k)}$, i.e., for each s the field operator $\phi_s(x)$ is a linear combination of the particular components $\psi_\alpha^{(k)}(x)$, for some k. In view of the transformation law in (12) we can write $\phi_s(x) = T(x)\phi_s(0)T(x)^{-1}$. With this mode of writing we consider the vector

$$\Phi(x^{(1)}, \ldots, x^{(r)}) \equiv \phi_1(x^1)\phi_2(x^{(1)} + x^{(2)}) \ldots \phi_r(x^{(1)} + \ldots + x^{(r)})\Omega = \qquad (15)$$
$$T(x^{(1)})\phi_1(0)T(x^{(2)})\phi_2(0)T(x^{(3)})\phi_3(0) \ldots T(x^{(r)})\phi_r(0)\Omega$$

We want to apply Theorem 4.7, heuristically, to the study of the scalar product of this vector with an arbitrary vector Ψ. Theorem 4.7 refers to bounded operators, but here we deal with distributions and unbounded field operators. A detailed examination shows, however, that Theorem 4.7 is nevertheless applicable, and it follows that the scalar product $\langle \Psi | \Phi(x^{(1)}, \ldots, x^{(r)})\Omega \rangle$ can be extended to a "function" $\langle \Psi | \Phi(z^{(1)}, \ldots, z^{(r)})\Omega \rangle$ when each $z^{(s)}$ is in the forward imaginary tube $\overline{V}_{+i}$, and if $z^{(s)} \in V_{+i}$ for each s, then the expression is actually an analytic *function* of the variables. In particular this is the case if $\Psi = \Omega$. The "functions" (distributions) $\langle \Omega | \Phi(x^{(1)}, \ldots, x^{(r)})\Omega \rangle$, which are thus vacuum expectation values of products of the field operators, are called the *Wightman functions* of the theory. It is obvious that the totality of all the Wightman functions in a field theory completely determine the theory, i.e., everything can be expressed in terms of the Wightman functions. We will now prove an important theorem due to Reeh and Schlieder[22].

Theorem 5.5. Let X be a nonempty open subset of Minkowski space, and let $\mathcal{P}(X)$ be the polynomial algebra generated by all the field operators $\psi_\alpha^{(k)}(x)$ with arguments x within X, as defined in Sect. 5.2. Then Ω is a *cyclic* vector of $\mathcal{P}(X)$, i.e., the set of all vectors $Q\Omega$ is dense in the Hilbert space. Equivalently stated: If Ψ is orthogonal to $Q\Omega$ for every $Q \in \mathcal{P}(X)$, then $\Psi = 0$.

Proof. Suppose that Ψ is orthogonal to $Q\Omega$ for every $Q \in \mathcal{P}(X)$. Let Φ be the vector in (15). We replace $x^{(1)}$ by $x^{(1)} + x$, with a fixed $x \in X$, and consider the vector
$$\Phi(x+x^{(1)},x^{(2)},x^{(3)},\ldots,x^{(r)}) = \phi_1(x+x^1)\phi_2(x+x^{(1)}+x^{(2)})\ldots\phi_r(x+x^{(1)}+\cdots+x^{(r)})\Omega.$$
If all the $x^{(s)}$ are sufficiently small, say, in some open neighborhood of the origin, then every argument of every field operator is in X, and hence we have $\langle\Psi|\Phi(x+x^{(1)},x^{(2)},\ldots,x^{(r)})\Omega\rangle = 0$. By Theorem 4.6 it then follows that this scalar product vanishes for *all* $x^{(1)},\ldots,x^{(r)}$. This means that $\langle\Psi|Q\Omega\rangle = 0$ for every Q in the algebra $\mathcal{P}$ of *all* field operators, and it then follows by Lemma 5.3 that $\Psi = 0$, as was to be proved.

5.6. We next show that a nonzero *local* field operator cannot annihilate the vacuum vector.

Lemma 5.6. Let X_b be a *bounded* nonempty open subset of $\mathcal{M}$, and let $f(x)$ be a test function which vanishes identically outside X_b. Let $\phi(x)$ be a field operator, and let $\phi[f]$ be the averaged field operator, as in (14). If $\phi[f]\Omega = 0$, then $\phi[f] = 0$.

Proof: Let X be a nonempty open subset of $\mathcal{M}$, spacelike separated from X_b, i.e., every point in X is spacelike separated from every point in X_b. If $\phi_1(x)$ is any field operator, and if $x \in X$, then we either have $\phi_1(x)\phi[f] = \phi[f]\phi_1(x)$, or else $\phi_1(x)\phi[f] = -\phi[f]\phi_1(x)$, in view of the Assumption VII. From this we easily conclude that if $Q \in \mathcal{P}(X)$, then $\phi[f]Q = Q'\phi[f]$ for some $Q' \in \mathcal{P}(X)$, and hence $\phi[f]Q\Omega = 0$. We regard this as proving that $\phi[f] = 0$, since the set of all vectors $Q\Omega$, with $Q \in \mathcal{P}(X)$, is dense in $\mathcal{H}$, by Theorem 5.5.

5.7. In preparation for the next theorem we have to deal with an issue in geometry. We define the open region W_R in Minkowski space as the *wedge-shaped* region consisting of all points $x = (x_1,x_2,x_3,x_4)$ such that $x_3 > |x_4|$. We note that every point (vector) in W_R is spacelike, and that $x' + x'' \in W_R$ if $x',x'' \in W_R$. We consider the Lorentz transformations $M_3(\lambda,\theta)$ in (2) in Sect. 3.4. We easily see that if λ and θ are *real*, then $M_3(\lambda,\theta)x \in W_R$ for all $x \in W_R$. We consider the case when $\lambda = i\eta$ is pure imaginary (but θ real). Let $x \in W_R$ and let $z' = x' + iy' = M_3(i\eta,\theta)x$. We have $\cosh(i\eta) = \cos\eta$ and $\sinh(i\eta) = i\sin\eta$, and the components of the imaginary part y' of the vector z' are thus $(0,0,x_4\sin\eta,x_3\sin\eta)$. Since $x_3 > |x_4|$ it follows that $y' \in V_+$, and hence $z' \in V_{+i}$, for all η such that $\pi > \eta > 0$. If $\lambda = \xi + i\eta$ is complex we have $M_3(\lambda,\theta) = M_3(i\eta,0)M_3(\xi,\theta)$, and it follows from the above that $M_3(\lambda,\theta)x \in V_{+i}$ for all $x \in W_R$ if θ is real and η is in the open interval $(0,\pi)$. From these considerations, and on the basis of Theorem 4.3, we draw the following conclusions.

Lemma 5.7. Let $x \in W_R$, and let θ be real, and $\lambda = \xi + i\eta$ be complex. The operator-valued function $T(M_3(\lambda,\theta)x)$ is well defined and a continuous function of θ and λ when λ is in the strip in the complex λ-plane defined by $\pi \geq \eta \geq 0$, and it is an *analytic* function of λ on the open strip for which $\pi > \eta > 0$. In particular we have $T(M_3(i\pi,\pi)x) = T(-x)$. If λ is *real*, then $M_3(\lambda,\theta)x$ and $M_3(\lambda+i\pi,\theta)x$ are both *real*.

5.8. We are now ready for a key theorem.

Theorem 5.8. Let $\{\phi_1, \phi_2, \ldots, \phi_r\}$ be an arbitrary r-tuplet of components of the fields $\psi^{(l)}$, and let the vector $\Phi(x^{(1)}, x^{(2)}, \ldots, x^{(r)})$ be defined as in (15). Let (s'_k, s''_k) be the type of the field operator ϕ_k. We assume that $x^{(k)} \in W_R$ for each k. Then,

$$\langle \Omega | \Phi(x^{(1)}, x^{(2)}, \ldots, x^{(r)}) \rangle = (-1)^{2s''} \langle \Omega | \Phi(-x^{(1)}, -x^{(2)}, \ldots, -x^{(r)}) \rangle, \quad (16)$$

where s'' is the sum of the numbers s''_k.

Proof: 1) The proof is basically very simple, but there are obvious notational problems. We first note that it suffices to prove the above identity for the case when each component $\phi_k(x)$ is one of the *special* components $\psi_\alpha^{(l)}(x)$, and we assume that this is the case in what follows. We will write $A_k = T(x^{(k)})\phi_k(0)$, for each $k = 1, \ldots, r$, and we thus have $\langle \Omega | \Phi(x^{(1)}, x^{(2)}, \ldots, x^{(r)}) \rangle = \langle \Omega | A_1 A_2 \cdots A_r \Omega \rangle$. Since $U(g)\Omega = \Omega$ for any $g \in SL(2, C)$ we have $\langle \Omega | A_1 A_2 \cdots A_r \Omega \rangle = \langle \Omega | U(g) A_1 A_2 \cdots A_r U(g)^{-1} \Omega \rangle = \langle \Omega | A_1(g) A_2(g) \cdots A_r(g) \Omega \rangle$, where $A_k(g) = U(g) A_k U(g)^{-1} = T(M(g)x^{(k)}) U(g) \phi_k(0) U(g)^{-1}$.

2) We want to consider the case when g is an element of the form $g_3(\lambda, \theta)$, discussed in Sect. 3.4. We write $A_k[\lambda, \theta] = A_k(g_3(\lambda, \theta))$, and with λ and θ *real* we thus have $A_k[\lambda, \theta] = T(M_3(\lambda, \theta)x^{(k)}) U(g_3(\lambda, \theta)) \phi_k(0) U(g_3(\lambda, \theta))^{-1}$. We will show that this function can be extended to certain complex values of λ. For a particular k we have $\phi_k(x) = \psi_\alpha^{(l)}(x)$ for some l and some α. By (13) we then have

$$U(g_3(\lambda, \theta))\phi_k(0)U(g_3(\lambda, \theta))^{-1} = \sum_\beta D_3^{(s'_k, s''_k)}(-\lambda, -\theta)_{\alpha\beta} \psi_\beta^{(l)}(0) \quad (17)$$

with the notation in (3). Here (s'_k, s''_k) is the type of the field ϕ_k. As we found in Sect. 3.4, the matrix-valued function $D_3^{(s'_k, s''_k)}(-\lambda, -\theta)$ is well defined for all complex values of λ, and it is an entire analytic function of this variable. By Lemma 5.7 the function $T(M_3(\lambda, \theta)x^{(k)})$ is well defined for all real θ and all complex $\lambda = \xi + i\eta$ in the strip defined by $\pi \geq \eta \geq 0$, and it is an analytic function of λ on the interior of the strip. The same is then true for the functions $A_k[\lambda, \theta]$. We thus define a function $F(\lambda, \theta) = \langle \Omega | A_1[\lambda, \theta] A_2[\lambda, \theta] \cdots A_r[\lambda, \theta] \Omega \rangle$ on this domain, and this function is an analytic function of λ on the interior of the strip. For *real* values of λ we have $F(\lambda, \theta) = \langle \Omega | A_1 A_2 \cdots A_r \Omega \rangle = F(0, 0)$, by the (trivial) conclusion in Step 1, and it follows that the function is *constant* and equal to $F(0, 0)$ on its entire domain. In particular we have $F(i\pi, \pi) = F(0, 0)$, and since $T(M_3(i\pi, \pi)x^{(k)}) = T(-x^{(k)})$ by Lemma 5.7, and since $D_3^{(s'_k, s''_k)}(-i\pi, -\pi) = (-1)^{2s''_k}$ by the discussion in Sect. 3.4, we have $A_k[i\pi, \pi] = (-1)^{2s''_k} T(-x^{(k)})\phi_k(0)$, and from this the relation in (16) follows.

5.9. The *Spin and Statistics Theorem* has several parts. The first concerns a field operator and its hermitian conjugate. It was proved by Burgoyne[23].

Theorem 5.9. a) Suppose that Assumptions I-VI obtain. Let $\phi(x)$ be a field component of type (s', s''), and let $\zeta = (-1)^{2(s'+s'')}$. If $\phi(x)\phi^\dagger(y) = -\zeta\phi^\dagger(y)\phi(x)$ for all spacelike $x - y$, then $\phi(x) = 0$ identically, i.e., there is *no* such field component.

b) If Assumption VII also obtains, then an irreducible tensor field *commutes* with its hermitian conjugate, and an irreducible spinor field *anticommutes* with its hermitian conjugate, for spacelike separated arguments.

Proof: 1) We apply Theorem 5.8 to the case of the two field operators ϕ and $\phi^\dagger$, and we thus have $\langle\Omega|\phi(x')\phi^\dagger(x'+x'')\Omega\rangle = \zeta\langle\Omega|\phi(-x')\phi^\dagger(-x'-x'')\Omega\rangle$, since the field $\phi^\dagger(x)$ is of type (s'', s'). This holds for all $x', x'' \in W_R$, and actually for *all* x', since the vacuum expectation value is invariant under any common translation of all the arguments. Making use of this fact we note that $\langle\Omega|\phi(-x')\phi^\dagger(-x'-x'')\Omega\rangle = \langle\Omega|\phi(x'+x'')\phi^\dagger(x')\Omega\rangle$.

2) We now *assume* that $\phi(x)\phi^\dagger(y) = -\zeta\phi^\dagger(y)\phi(x)$ for all spacelike $x-y$, and since x'' is spacelike we obtain the relation $\langle\Omega|\phi(x')\phi^\dagger(x'+x'')\Omega\rangle = -\langle\Omega|\phi^\dagger(x')\phi(x'+x'')\Omega\rangle$. Each vacuum expectation value can be extended so that x'' is replaced by a z'' in the forward imaginary tube, and the resulting functions are then analytic functions of z'' on the open tube. Since the two functions are equal on a nonempty open set of real values of z'' it follows that they are equal for all $z'' \in \overline{V}_{+i}$, and in particular for *all* real values of z''. We thus have

$$\langle\Omega|\phi(x)\phi^\dagger(y)\Omega\rangle + \langle\Omega|\phi^\dagger(x)\phi(y)\Omega\rangle = 0 \tag{18}$$

Let $f(x)$ be a test function which vanishes identically outside some bounded region. We multiply the expression at left in (18) by $f(x)^* f(y)$ and integrate over x and y, and we then obtain the relation $\|\phi^\dagger[f]\Omega\|^2 + \|\phi[f]\Omega\|^2 = 0$, and since both terms are non-negative we have $\phi[f]\Omega = 0$. By Lemma 5.6 this implies that $\phi[f] = 0$ for all test functions of compact support, from which it follows that $\phi[f] = 0$ for *all* test functions. This means that $\phi(x) = 0$ identically, as asserted in part a).

3) Let ϕ_1 and ϕ_2 be two components of the *same* (irreducible) field. We consider the component $\phi(x) = c_1\phi_1(x) + c_2\phi_2(x)$, where c_1, c_2 are complex numbers. Under Assumption VII it follows from part a) that we then have $\phi(x)\phi^\dagger(y) = \zeta\phi^\dagger(y)\phi(x)$ when $x-y$ is spacelike. Applying this to the cases $(c_1, c_2) = (1,0)$, $(0,1)$, $(1,1)$, and $(1,i)$, we conclude that $\phi_1(x)\phi_2^\dagger(y) = \zeta\phi_2^\dagger(y)\phi_1(x)$ when $x-y$ is spacelike. The field is a tensor field if $\zeta = 1$, and a spinor field in $\zeta = -1$, and hence the conclusion in part b) follows.

In part a) the "wrong" connection between spin and statistics is *prohibited*, and this conclusion does not depend on Assumption VII.

5.10. The second part of the spin and statistics theorem was proved by Dell'Antonio[24]. It settles the question whether a field operator commutes or anticommutes with *itself* for spacelike separated arguments. The theorem which follows actually deals with a more general question.

Theorem 5.10. a) Suppose that two field components $\phi_1(x)$ and $\phi_2(x)$ (of the same, or two different fields) are such that $\phi_1(x)\phi_2(y) = \phi_2(y)\phi_1(x)$ and $\phi_2^\dagger(x)\phi_1(y) = -\phi_1(y)\phi_2^\dagger(x)$ whenever $x-y$ is spacelike. Then at least one of the fields operators vanishes identically.

b) Two components of the same tensor field must commute with each other, and two components of the same spinor field must anticommute with each other, for spacelike separated arguments.

c) If $\phi_1(x)$ and ϕ_2 are components of the *same* field $\psi^{(j)}$, then $\phi_1^\dagger(x')\phi_2(x'')$ and $\phi_1(x')\phi_2(x'')$ commute with *every* field $\psi^{(k)}(y)$ whenever both $x'-y$ and $x''-y$ are spacelike.

Proof: 1) We consider part a). If the two fields satisfy the stated conditions we obviously have $\phi_2^\dagger(x)\phi_1^\dagger(y) = \phi_1^\dagger(y)\phi_2^\dagger(x)$ and $\phi_2(x)\phi_1^\dagger(y) = -\phi_1^\dagger(y)\phi_2(x)$ whenever $x - y$ is spacelike. Let X be a bounded, nonempty, open subset of $\mathcal{M}$, and let v be a *spacelike* four-vector. There then exists a constant s_o such that X displaced by sv is spacelike separated from X for all $s > s_o$. Let $f(x)$ and $h(x)$ be test functions which vanishes outside X, and let $\phi_1[f;s] = T(sv)\phi_1[f]T(sv)^{-1}$. Then $\phi_1[f;s]$ commutes with $\phi_2[h]$, but anticommutes with $\phi_2[h]^\dagger$, whenever $s > s_o$. We write $Q_1 = \phi_1[f;s]$ and $Q_2 = \phi_2[h]$. With $s > s_o$ we thus have $0 \leq \langle Q_2 Q_1 \Omega | Q_2 Q_1 \Omega \rangle = \langle \Omega | Q_1^\dagger Q_2^\dagger Q_2 Q_1 \Omega \rangle = -\langle \Omega | Q_2^\dagger Q_2 Q_1^\dagger Q_1 \Omega \rangle = -\langle Q_2^\dagger Q_2 \Omega | Q_1^\dagger Q_1 \Omega \rangle = -\langle \phi_2[h]^\dagger \phi_2[h]\Omega | T(sv)\phi_1[f]^\dagger \phi_1[f]\Omega \rangle$. By Theorem 4.8 the right member tends to $-\langle \phi_2[h]^\dagger \phi_2[h]\Omega | \Omega \rangle \langle \Omega | \phi_1[f]^\dagger \phi_1[f]\Omega \rangle = -\|\phi_2[h]\Omega\|^2 \cdot \|\phi_1[f]\Omega\|^2 \leq 0$ as s tends to infinity, and it follows that at least one of the vectors $\phi_2[h]\Omega, \phi_1[f]\Omega$ must equal zero. It is easily seen, on the basis of Lemma 5.6, that this implies that at least one of the field operators must vanish identically.

2) The conclusion in part b) follows immediately from Theorem 5.9, and the result in part a) if we take ϕ_1 and ϕ_2 (or ϕ_1 and $\phi_2^\dagger$) to be components of the same field. The conclusions in part c) are now trivial.

5.11. We say that the locality conditions of a field theory are *normal*, or that the fields satisfy *normal* commutation and anticommutation relations, if any two spinor fields *anticommute*, and if every tensor field *commutes* with any other field for spacelike separated arguments. So far we have proved that each field and its hermitian conjugate *by themselves* must satisfy normal locality conditions: the "wrong" connection between spin and statistics is ruled out for each field separately. This is as far as we can go in the proof of normality, because within the theory of free fields we can easily construct two different spinor fields which *commute* with each other, or two different scalar fields which *anticommute* with each other, for spacelike separated arguments, and which satisfy all our conditions I-VII. However, one can prove the following. If a field theory satisfies all the conditions I-VII, then it is possible to *redefine* the fields by a so-called *Klein transformation*, without changing the *physical* content of the theory, such that the new fields satisfy normal locality conditions. The most refined proof of this is due to Araki[25]. The reasoning is rather long and involved, and cannot be presented here. The two Theorems 5.9 and 5.10 together express the essence, the *physical* content, of the Spin and Statistics Theorem, and the commutation or anticommutation relations not covered by these theorems are irrelevant physically. However, from a mathematical point of view it is *very* convenient if the locality conditions are normal. In view of what we have found we could now replace Assumption VII by the assumption that all locality conditions are normal, and this is often done in papers on field theory. In what follows we will also assume this to be the case.

5.12. The proof of the *PCT*-Theorem is also based on Theorem 5.8. We first state an intermediate step as a lemma.

Lemma 5.12. We assume *normal* locality conditions. Let $\{\phi_1, \phi_2, \ldots, \phi_r\}$ be an arbitrary r-tuple of field components, and let (s'_k, s''_k) be the type of the field operator ϕ_k. We then have

$$\langle \Omega | \phi_1(x^{(1)}) \phi_2(x^{(2)}) \cdots \phi_r(x^{(r)}) \Omega \rangle = i^m (-1)^{2s''} \langle \Omega | \phi_r(-x^{(r)}) \phi_{r-1}(-x^{(r-1)}) \cdots \phi_1(-x^{(1)}) \Omega \rangle \tag{19}$$

for all (real) $\{x^{(k)}\}$, where m is the number of spinor field operators, and where s'' is the sum of the s_k''. The above vacuum expectation values vanish if m is odd.

Proof: 1) Let $\xi^{(k)} \in W_R$, and let $x^{(k)} = \xi^{(1)} + \xi^{(2)} + \ldots + \xi^{(k)}$, for $k = 1,\ldots,r$. By Theorem 5.8 we then have $\langle\Omega|\phi_1(x^{(1)})\phi_2(x^{(2)})\cdots\phi_r(x^{(r)})\Omega\rangle = (-1)^{2s''}\langle\Omega|\phi_1(-x^{(1)})\phi_2(-x^{(2)})\cdots\phi_r(-x^{(r)})\Omega\rangle$. For this special configuration of the arguments $\{x^{(k)}\}$ the difference $x^{(k)} - x^{(l)}$ is spacelike for every pair (k,l), $k \neq l$. We can therefore reorder the field operators in the right member, on the basis of the commutation and anticommutation relations corresponding to the assumed *normal* locality conditions, and we obtain the relation $\langle\Omega|\phi_1(-x^{(1)})\phi_2(-x^{(2)})\cdots\phi_r(-x^{(r)})\Omega\rangle = i^m\langle\Omega|\phi_r(-x^{(r)})\phi_{r-1}(-x^{(r-1)})\cdots\phi_1(-x^{(1)})\Omega\rangle$. The only issue here is the factor i^m in the right member. We first note that a vacuum expectation value of a product of field operators vanishes if the number of spinor field-factors is odd, since a spinor field changes sign under conjugation by the rotation U_o, whereas a tensor field commutes with U_o. We thus have $m = 2m'$ and $i^m = (-1)^{m'}$, where m' is an integer, and a moment's reflection shows that i^m is the correct sign factor for all (even) m. This proves the identity in (19) for the case when $x^{(1)} \in W_R$ and $x^{(k)} - x^{(k-1)} \in W_R$ for $k = 2,3,\ldots,n$.

2) The vacuum expectation value at left in (19) is equal to

$$\langle\Omega|T(\xi^{(1)})\phi_1(0)T(\xi^{(2)})\phi_2(0)T(\xi^{(3)})\phi_3(0)\cdots T(\xi^{(r)})\phi_r(0)\Omega\rangle = F_L(\xi^{(1)},\ldots,\xi^{(r)}). \tag{20}$$

The vacuum expectation value at right in (19) remains unchanged if we add $x^{(r)} + x^{(1)}$ to each argument of the field operators, and this vacuum expectation value is therefore of the form $\langle\Omega|T(\xi^{(1)})\phi_r(0)T(\xi^{(r)})\phi_{r-1}(0)T(\xi^{(r-1)})\phi_{r-2}(0)\cdots T(\xi^{(2)})\phi_1(0)\Omega\rangle = F_R(\xi^{(1)},\ldots,\xi^{(r)})$. By Theorem 4.7 both functions F_L and F_R can be extended to the complex domain for which $\xi^{(k)} \in \overline{V}_{+i}$. We have shown that $F_L = i^m(-1)^m F_R$ when $\xi^{(k)} \in W_R$ for every k, and by Theorem 4.6 it follows that this identity holds for *all* real $\{\xi^{(k)}\}$. Hence the identity in (19) holds for *all* $\{x^{(k)}\}$, as asserted.

5.13. We can now prove the *PCT-Theorem*, first proved generally and rigorously in a remarkable paper by Jost[18].

Theorem 5.13. We assume *normal* conditions of locality. There then exists a unique antiunitary operator θ such that

$$\theta\Omega = \Omega, \text{ and } \theta\phi(x)\theta^{-1} = \eta_\phi \phi^\dagger(-x) \tag{21}$$

for every field component $\phi(x)$. Here $\eta_\phi = (-1)^{2s''}$ if $\phi(x)$ is a component of a tensor field of type (s',s''), and $\eta_\phi = i(-1)^{2s''}$ if $\phi(x)$ is a component of a spinor field of type (s',s''). The operator θ also satisfies the relations

$$\theta U(g) = U(g)\theta, \quad \theta T(x)\theta^{-1} = T(-x), \text{ and } \theta^2 = U_o = U(-I) \tag{22}$$

and can thus be interpreted as a *PCT-operator*.

Proof: 1) To prove the above we try to *define* an antilinear operator θ, first on Ω by $\theta\Omega = \Omega$, and on vectors of the form $\phi_1(x^{(1)})\phi_2(x^{(2)})\cdots\phi_r(x^{(r)})\Omega$ by $\theta\phi_1(x^{(1)})\cdots\phi_n(x^{(n)})\Omega = i^m(-1)^{2s''}\phi_1^\dagger(-x^{(1)})\cdots\phi_n^\dagger(-x^{(n)})\Omega$, for any n field components ϕ_k. Here m is the number of spinor fields among these, and s'' is the sum of the s_k'', where (s_k', s_k'') is the type of ϕ_k. We then extend the definition of θ to all linear combinations of the above special vectors, taking into account the antilinear nature of θ, i.e., the coefficients go to their complex conjugates. For this attempt to be successful we have to show that $\langle \theta\Phi'|\theta\Phi''\rangle = \langle \Phi''|\Phi'\rangle$ for all vectors Φ' and Φ'' of the above special form. These vectors span the Hilbert space, by Lemma 5.3, and it then follows that the relation $\langle \theta\Phi'|\theta\Phi''\rangle = \langle \Phi''|\Phi'\rangle$ holds for *all* $\Phi', \Phi'' \in \mathcal{H}$. The proof, based on Lemma 5.12, is actually very straightforward and simple, apart from the characteristic notational problems.

2) We write $A = \phi_q(x^{(q)})\phi_{q-1}(x^{(q-1)})\ldots\phi_1(x^{(1)})$ for a product of q field operators, and $B = \phi_{q+1}(x^{(q+1)})\phi_{q+2}(x^{(q+2)})\cdots\phi_r(x^{(r)})$ for a product of $r - q$ field operators. Furthermore we write $A_R = \phi_1(-x^{(1)})\ldots\phi_q(-x^{(q)})$ and $B_R = \phi_r(-x^{(r)})\cdots\phi_{q+1}(-x^{(q+1)})$. We thus obtain A_R from A by reversing the order of the factors and changing the sign of the arguments, and B_R is similarly obtained from B. Let m_a be the number of spinor-factors in A, and let m_b be the number of spinor-factors in B. Let (s_k', s_k'') be the type of the field operator ϕ_k, in which case $\phi_k^\dagger$ is of type (s_k'', s_k'). We write $s_A' = \sum_{k=1}^q s_k'$, $s_A'' = \sum_{k=1}^q s_k''$, and $s_B'' = \sum_{k=q+1}^r s_k''$, and we then have $\langle A\Omega|B\Omega\rangle = \langle \Omega|A^\dagger B\Omega\rangle = i^{m_a+m_b}(-1)^{2(s_A'+s_B'')} \langle \Omega|B_R A_R^\dagger \Omega\rangle = i^{m_a+m_b}(-1)^{2(s_A'+s_B'')} \langle B_R^\dagger \Omega|A_R^\dagger \Omega\rangle$, where the equality of the second and third members follows from Lemma 5.12. A moment's reflection shows that $m_a + 2(s_A' - s_A'')$ is an *even* integer. The matrix element $\langle A\Omega|B\Omega\rangle$ vanishes unless $m_a - m_b$ is an *even* integer, and hence we have $\langle A\Omega|B\Omega\rangle = i^{m_b-m_a}(-1)^{2(s_A''+s_B'')} \langle B_R^\dagger \Omega|A_R^\dagger \Omega\rangle$. According to our attempted definition of θ the right member is equal to $\langle \theta B\Omega|\theta A\Omega\rangle$. We trivially have $\langle \Omega|B\Omega\rangle = i^{m_b}(-1)^{2s_B''} \langle B_R^\dagger \Omega|\Omega\rangle$, with the right member equal to $\langle \theta B\Omega|\theta\Omega\rangle$ according to our definition of θ. Our definition of the antilinear operator θ is therefore consistent.

3) We next note that $\theta^2 B\Omega = (-1)^{2(s_B'+s_B'')} B\Omega = U_o B\Omega$. We trivially have $\theta^2\Omega = \Omega = U_o\Omega$, and we conclude that $\theta^2 = U_o$, and hence θ has the inverse $\theta^{-1} = U_o\theta = \theta U_o$, and is *antiunitary* since it is an isometry. It follows that $\theta^{-1}\Omega = \Omega$, and the relation at right in (21) then follows from our definition of θ. I omit the easy proofs of the uniqueness of θ, and of the two relations at left in (22).

The electromagnetic field tensor is a sum of two fields of type $(0,1)$ and $(1,0)$, and the factor η_ϕ in (21) is thus equal to $+1$. The four-current density is a vector field of type $(1/2, 1/2)$, and for this field the factor η_ϕ in (21) is thus equal to -1. These results agree with the transformation laws for RCT in Table 1, and hence θ is interpreted as RTC.

5.14. Both the spin and statistics theorem and the PCT-theorem are honest results in the sense that the conclusions were not somehow sneaked into the assumptions from the beginning in some barely disguised form. The two theorems are by *no means* obvious. Both theorems depend on the result in Theorem 5.8, which is also not at all obvious. An essential step in the proof of Theorem 5.8 involved a complex Lorentz transformation which was possible because of the spectrum condition, *and* the assumption that the fields are *finite*-component fields. It is of interest that examples of *infinite*-component fields, which otherwise satisfy the Wightman conditions, can be constructed such that the spin and statistics theorem and/or the PCT-theorem are violated. Examples have been given by Streater[26] (the "wrong" spin-statistics connection), and by Oksak and Todorov[27] (invalidity of the PCT-theorem). These particular examples are manifestly unphysical in that they involve multiplets with an infinite number of particles of the same mass and different spins, but as discussed by Grodsky and Streater[28], there is good reason to believe that infinite-component fields *always* suffer from such defects, and hence are not relevant for physics.

We may have some doubts about Assumption VII, according to which two field operators either commute or else anticommute for spacelike separated arguments. In a theory of local observables two observables associated with spacelike separated regions ought to be independent, and a *minimum* condition for this is that they commute. A spinor field changes sign under conjugation with $U_o = U(-I)$, and hence cannot be an observable. It is then not so clear what the locality condition should be for a spinor field. We have seen that a spinor field *cannot* commute with itself, or with its hermitian conjugate, for spacelike separations, but this does not violate (physical) locality, since it is not an observable. *Bilinear* expressions formed from a spinor field commute with U_o, and *might* thus be observables (unless this is ruled out by some other superselection principle). Anticommutation relations for spinor fields imply commutation relations for bilinears of spinor fields, and we have the result in part c) of Theorem 5.10. The two alternatives in Assumption VII are therefore consistent with the possibility that local operators which commute with U_o *might* be local observables. There is also an historical explanation for these two alternatives. When Assumption VII was formulated it had been known for some time that free spinor fields (of which the Dirac field is the prime example) are characterized by *anticommutation* relations, and proofs of various forms of the spin-statistics theorem had been given for the case of free fields, first by Pauli.

In the algebraic quantum field theory one deals (at first) only with *observable* operators, and the locality condition is then that observables associated with spacelike separated region commute. The theory can be "enlarged" to include operators which do not represent observables, such as operators of a "spinor character," and the spin-and-statistics issue then concerns the commutation or anticommutation relations of these. It turns out that there are other possibilities than Bose statistics or Fermi statistics for the description of identical particles, namely the so-called *parastatistics*, but at the price of superselection principles with nonabelian (global) gauge groups. For a discussion of parastatistics I refer to papers by Green[29], and by Greenberg and Messiah[30,31]. The conclusion of the last-mentioned authors is that there is *no* evidence that any stable or semi-stable particles obeys parastatistics.

I finally note here that there is absolutely *no* spin-and-statistics theorem (and no *PCT*-theorem!) in nonrelativistic quantum mechanics. It is *trivial* that one can construct a many-particle Schrödinger theory of spin zero particles which obey Fermi statistics, or a theory of spin 1/2-particles which obey Bose statistics.

PARTICLES IN QUANTUM FIELD THEORY

6.1. We now have to relate our result about *fields* to properties of *particles*. In a truly fundamental paper Wigner[32] finds and discusses all the *irreducible unitary* representations of the group $\hat{P}$ associated with particles. This epoch-making paper is not easy reading, but the results are, however, easily described. To every $m > 0$, and every s such that $2s$ is a non-negative integer, corresponds a unique irreducible unitary representation $\lambda \to \Gamma_{m,s}(\lambda)$ on a Hilbert space $\mathcal{H}_{m,s}$, interpreted as the space of all state vectors of a (single) particle of mass m and spin s. Furthermore, to every μ such that 2μ is an integer (positive, negative or zero), corresponds a unique irreducible unitary representation $\lambda \to \Gamma_\mu(\lambda)$ on a Hilbert space $\mathcal{H}_\mu$, interpreted as the space of all state vectors of a *massless* particle of helicity μ. (The helicity of a particle is the projection of its angular momentum on the direction of its 3-momentum vector **p**). We can say that this theory concerns the *kinematics* of a *stable* particle: it says nothing about a possible internal structure. It is of interest to note here that the representation $\Gamma_{m,s}$ can be extended to include a space reflection R_o and a time reversal T_o on the *same* Hilbert space $\mathcal{H}_{m,s}$, and the same is true for the representation Γ_0 corresponding to a massless particle of helicity zero. For $\mu \neq 0$ the representation Γ_μ can be extended to include a time reversal T_o, on the *same* Hilbert space $\mathcal{H}_\mu$, but *not* space reflection R, since the helicity changes sign under R. The *reducible* representation $\Gamma_\mu \oplus \Gamma_{-\mu}$, can, however, be extended to include space reflection on the direct sum $\mathcal{H}_\mu \oplus \mathcal{H}_{-\mu}$. We recall here that the space of all one-photon states carries the representation $\Gamma_1 \oplus \Gamma_{-1}$. The two subspaces $\mathcal{H}_1$ and $\mathcal{H}_{-1}$ correspond to the two states of opposite circular polarization.

6.2. For each space $\mathcal{H}_{m,s}$ (or $\mathcal{H}_\mu$) one can then construct a *Fock space*, of all many-particle states of particles of the *same* species, so that $\mathcal{H}_{m,s}$ (or $\mathcal{H}_\mu$) appears as the one-particle subspace. The vacuum vector Ω is also included in the Fock space. The representation of $\widehat{P}$ on the one-particle subspace can be promoted in a specific manner to a (highly reducible) representation of $\widehat{P}$ on the Fock space, and the same holds for time reversal and space reflection. A Hilbert space $\mathcal{H}_{GF}$ to describe all multiparticle states of several particles of different species is constructed as (essentially) the tensor product of the Fock spaces for each species. We can call $\mathcal{H}_{GF}$ a *general Fock space*, to have a name for it. The formalism of creation and destruction operators for the different particles is convenient in the description of $\mathcal{H}_{GF}$, and of operators on $\mathcal{H}_{GF}$ with a physical interpretation. I assume that all of this is quite familiar. The resulting theory is a theory of *non-interacting* particles, for which the assumptions I-IV in Sect. 4.1 are satisfied. It is important to realize that at *this stage* there is *no* connection between spin and statistics. We are free to assume that the creation operators for particles of a particular species either commute (in which case the particles are bosons), or else anticommute (in which case the particles are fermions), irrespective of the spin of the particle. The situation is different if we also want to introduce local fields. We first consider the case when there are no massless particles. One can then introduce a free field for each particle species such that the Wightman assumptions V-VII are satisfied, and the Theorems 5.9 and 5.10 concerning spin and statistics then apply. Particles associated with tensor fields have integer spins, and they must be bosons, whereas particles associated with spinor fields have half-integer spins, and must be fermions. This is, perhaps, obvious, but to show this pedantically we can consider the operator $U_o = U(-I)$. This operator commutes with creation operators for particles with integer spin and with tensor fields, whereas it anticommutes with creation operators for particles with half-integer spins and with spinor fields. Since the free fields are linear in the creation and destruction operators the conclusion follows. However, we can very well arrange it so that the creation operators for two *different* fermions commute, and hence the corresponding spinor fields commute for spacelike separated arguments. The *third part* of the spin-

and-statistics theorem, discussed in Sect. 5.11, says that if *abnormal* locality conditions obtain, then it is always possible to redefine the fields in such a way that the locality conditions are *normal*. For free fields this amounts to a redefinition of the creation and destruction operators by a Klein transformation, such that all fermion creation operators anticommute with each other, and such that all boson creation operators commute with *all* creation operators. Assuming normal locality conditions the *PCT*-Theorem then applies as stated in Theorem 5.13.

In the case when some of the particles have zero mass it is not possible to construct local free fields such that the Wightman conditions hold, *unless* every massless particle of helicity μ has a counterpart of helicity $-\mu$, but if this is the case, then the general conclusions above apply. Note here that the relations in (22) imply that $\theta J \theta^{-1} = -J$ and $\theta P \theta^{-1} = P$, which means that $\theta = RCT$ reverses the helicity of a state.

6.3. The above theory describes non-interacting particles, but in the real world particles do interact, and we are thus interested in a field theory describing interacting particles. It is important to understand that in such a (realistic) theory the very notion of a *particle* is an *asymptotic notion*. In nonrelativistic quantum mechanics we describe a state (at an instant of time) in terms of a (many-particle) wave function, which is a function of the position variables of the *fundamental* particles present. The number and kind of the fundamental particles do not change, although bound states may be formed. In a relativistic theory, in which creation and destruction phenomena occur, this kind of description makes no sense. In a collision process we only "see" the (stable) particles in the distant future, or in the distant past, when the state looks like a state of freely moving particles, no longer interacting with each other because they are well separated or "spread out." The theory of a general Fock space applies to these asymptotic states. The Hilbert space $\mathcal{H}$ can be identified with two different Fock spaces, an *in-Fock space* $\mathcal{H}_{in}$ and an *out-Fock space* $\mathcal{H}_{out}$, such that the asymptotic initial state of a vector $\Phi \in \mathcal{H}$ refers to $\mathcal{H}_{in}$, and the asymptotic final state of Φ refers to $\mathcal{H}_{out}$. One can say that a state $\Phi \in \mathcal{H}$ is described from two different points of view: as an initial *particle* state, or as a final *particle* state. The *S-matrix*, or *scattering operator S*, relates these two points of view. A *one*-particle state in $\mathcal{H}_{in}$ is also a *one*-particle state in $\mathcal{H}_{out}$, for the obvious reason that nothing happens to a particle if there is no other particle present with which it can interact. To each stable particle, of mass m, and spin s, corresponds a one-particle subspace $\mathcal{H}_{m,s}$ of the Hilbert space $\mathcal{H}$ of all states, distinguished by the feature that it carries the irreducible representation $\Gamma_{m,s}$ of the Poincaré group $\widehat{P}$. No state orthogonal to all the one-particle states and the vacuum state Ω has a definite mass.

The assumptions I-VII alone in the Wightman theory do not guarantee that there are any particles. With suitable *additional* assumptions the field theory has a particle interpretation, as outlined above. The theory of this is known as the *Haag-Ruelle Theory*. It is a long story which we cannot discuss here. I refer to the monograph by Jost[2] for a good and clear presentation. In this theory the asymptotic particle states are constructed as states obtained when suitable field operators act on the vacuum vector, which is somewhat analogous to the situation in a free field theory. In this manner one demonstrates the Fock space structures of the sets of initial and final asymptotic states. The *fourth* part of the spin-statistics theorem concerns the demonstration that the "right" spin and statistics connection for the *fields* induces the "right" spin and statistics connection for the *particles*. This is discussed in the monograph by Jost, and I feel that it is a very plausible result, but we have to leave it at that. For free-field theories the result is trivial, as we discussed above. In a Lagrangian field theory there is also a field associated with every particle, and although the field is not a free field, it is plausible that the "right" locality condition for the field implies the "right" statistics for the particle. In the Wightman theory the relationship between the particles and fields is more complicated, and it is *not* assumed that there is a field associated with every particle.

By Theorem 5.13 the antiunitary *PCT*-operator θ exists. By the relations in (22) it commutes with the four-momentum operator P, and therefore with the *mass operator* $P \cdot P$. This means that θ must map a one-particle state onto a one-particle state of the *same* mass. It also follows from the relations in (22) that the magnitude of the spin is preserved. To every one-particle subspace $\mathcal{H}_{m,s}$ of $\mathcal{H}$ there thus corresponds a one-particle subspace $\theta \mathcal{H}_{m,s}$, which also carries the representation $\Gamma_{m,s}$. This means that every particle has an antiparticle of the same mass and spin, which may, or may not be equal to the particle. Likewise every massless particle of helicity μ has an antiparticle of helicity $-\mu$.

Within the Haag-Ruelle theory we have the result that $\theta S \theta^{-1} = S^{-1}$, which expresses the implication of the *PCT*-symmetry for scattering processes.

6.4. In a theory of *free* fields it is always possible to define a space reflection operator R and a time reversal operator T, but this is not necessarily the case for Wightman fields in general. It may be illuminating to see what happens if we try to construct a space reflection operator R in the same manner as we constructed the *PCT*-operator θ in Sect. 5.13. For simplicity we consider the case when there is only one field $\psi(x)$, which is a hermitian scalar field. Let I_L be the 4×4 matrix in Sect. 3.2 which describes space reflection on Minkowski spacetime, i.e. $I_L(\mathbf{x},t) = (-\mathbf{x},t)$. The operator R ought to be a unitary operator which satisfies the conditions $R\Omega = \Omega$, and $R\psi(x)R^{-1} = \xi\psi(I_L x)$, where ξ is a sign factor. We try to define R by $R\Omega = \Omega$, and $R\psi(x^{(1)})\cdots\psi(x^{(n)})\Omega = \xi^n\psi(I_L x^{(1)})\cdots\psi(I_L x^{(n)})\Omega$, in analogy with the construction in Sect. 5.13. We then have to show that $\langle R\Phi'|R\Phi''\rangle = \langle \Phi'|\Phi''\rangle$ for all $\Phi',\Phi'' \in \mathcal{H}$, and in particular we have to show that $\langle \Omega|\psi(x^{(1)})\cdots\psi(x^{(n)})\Omega\rangle = \xi^n\langle \Omega|\psi(I_L x^{(1)})\cdots\psi(I_L x^{(n)})\Omega\rangle$ for any n-tuplet of field operators. There is, however, no known reason why the vacuum expectation values (Wightman functions) should satisfy *these* conditions, and that is the end of the attempted construction. In contrast with this, the success in Sect. 5.13 followed from the entirely nontrivial relation in (19).

Violation of reflection invariance means that *no* reflection transformation, with the *right* properties dictated by the group theoretical correspondence principle, can be defined. Similarly for time reversal. Within Lagrangian field theory it is easy to construct examples which violate reflection and/or time reversal invariance. The violations are thus a consequence of the nature of the interaction. We have shown that there is always an antiuniatry symmetry θ, which we can write as RCT if R, C, and T are valid symmetries. If these are not valid symmetries, we cannot "factor" θ.

REFERENCES

[1] R. F. Streater and A. S. Wightman: *PCT, Spin and Statistics, and All That.* (Addison-Wesley Pub., Reading, MA, 1989).

[2] R. Jost: *The General Theory of Quantized Fields.* (American Mathematical Society, Providence, RI, 1965).

[3] N. N. Bogolubov, A. A. Logunov and I. T. Todorov: *Introduction to Axiomatic Quantum Field Theory.* (Benjamin Cummings Publishing Co., Inc., Reading, MA 1975).

[4] R. Haag: *Local Quantum Physics.* (Springer-Verlag, Berlin-Heidelberg, 1992).

[5] H. Araki: *Mathematical Theory of Quantum Fields*, (Oxford Science Publications, 1999).

[6] C. Itzykson and J-B. Zuber: *Quantum Field Theory.* (McGraw-Hill Inc, 1980).

[7] G.C.Wick, E.P.Wigner, and A.S.Wightman: "Intrinsic parity of elementary particles," *Phys.Rev.* **88**, 101 (1952).

[8] C.S.Wu, E.Ambler, R.W.Hayward, D.D.Hoppes, and R.P.Hudson: "Experimental test of parity conservation in beta-decay," *Phys. Rev.* **105**, 1413 (1957)

[9] R.L.Garwin, L.M.Lederman, and M.Weinrich: "Observation of the failure of conservation of parity and charge conjugation in meson decays; the magnetic moment of free muons," *Phys.Rev.* **105**, 1415 (1957).

[10] P.C.Macq, K.M.Crowe, and R.P.Haddock: "Helicity of the electron and positron in muon decay," *Phys.Rev.Lett.* **112**, 2061 (1958)

[11] J.H.Christenson, J.W.Cronin, V.L.Fitch, and R.Turlay: "Evidence for the 2π decay of the K_2^o meson," *Phys.Rev.Lett.* **13**, 138 (1964)

[12] S.Bennett, D.Nygren, H.Saal, J.Steinberger, and J.Sunderland: "Measurement of the charge asymmetry in the decays $K_L^o \to \pi^\pm + e^\mp + \nu$," *Phys.Rev.Lett.* **19**, 993 (1976).

[13] E. D. Commins: "Parity nonconservation in atoms and searches for permanent electric dipole moments," in *Amazing Light: A volume dedicated to C. H. Townes on his 80th Birthday,* p. 125. Edited by R. Y. Chiao (Springer-Verlag, 1996).

[14] P. S. Drell and E. D. Commins: "Parity nonconservation in atomic thallium," *Phys. Rev.* **A32**, 2196 (1985).

[15] B. C. Regan, E. D. Commins, D. P. DeMille, and C. J. Schmidt: "A new limit on the electron electric dipole moment." (To be published).

[16] E. D. Commins: "Electric dipole moments of leptons," *Adv. Atom. Mol. and Opt. Phys.* **40**, 1 (1999).

[17] L.Wolfenstein: "Present status of CP violation," *Ann.Rev.Nucl.Part.Sci.* **36**, 137 (1986).

[18] R. Jost: "Eine Bemerkung zum CTP-Theorem," *Helv.Physica Acta* **30**, 409 (1957)

[19] A. S. Wightman: "Quantum field theory in terms of vacuum expectation values," *Phys. Rev.* **101**, 860 (1956).

[20] R. Haag and D. Kastler: "An algebraic approach to quantum field theory," *J. Math. Phys.* **5**, 848-861 (1964).

[21] S. S. Horuzhy: *Introduction to Algebraic Quantum Field Theory.* (Kluwer Academic Publishers, Dordrecht, Netherlands, 1990).

[22] H. Reeh and S. Schlieder: "Bemerkungen zur Unitäräquivalenz von Lorentzinvarianten Feldern," *Il Nuovo Cimento* **22**, 1051 (1961)

[23] N. Burgoyne: "On the connection of spin with statistics," *Il Nuovo Cimento* **8**, 807 (1958).

[24] G. F. Dell'Antonio: "On the connection between spin and statistics," *Annals of Physics* **16**, 153 (1961).

[25] H. Araki: "Connection of spin with commutation relations," *J.Math.Phys.* **2**, 267 (1961).

[26] R.F.Streater: "Local fields with the wrong connection between spin and statistics," *Commun.Math.Phys.* **5**, 88 (1967)

[27] A. I. Oksak and I. T. Todorov: "Invalidity of the TCP-theorem for infinite-component fields," *Commun. Math. Phys.* **11**, 125 (1968)

[28] I. T.Grodsky and R.F.Streater: "No-Go Theorem," *Phys.Rev.Lett.* **20**, 695 (1968).

[29] H. S. Green: "A generalized method of field quantization," *Phys.Rev.* **90**, 270 (1953).

[30] O. W. Greenberg and A. M. L. Messiah: "Selection rules for parafields and the absence of para particles in nature," *Phys.Rev.* **136B**, 248 (1964).

[31] O. W. Greenberg and A. M. L. Messiah: "Symmetrization postulate and its experimental foundation," *Phys.Rev.* **138B**, 1155 (1965).

[32] E. Wigner: "On unitary representations of the inhomogeneous Lorentz group," *Annals of Math.* **40**, 39 (1939).

Electric field distribution in nuclei produced by the P,T-odd nuclear Schiff moment [1]

V.V. Flambaum and J.S.M. Ginges

School of Physics, University of New South Wales, Sydney 2052, Australia

Abstract. Parity and time invariance violating (P,T-odd) nuclear forces produce P,T-odd nuclear moments. In turn, these moments can induce electric dipole moments (EDMs) in atoms through the mixing of electron wavefunctions of opposite parity. The nuclear EDM is screened by atomic electrons. The EDM of an atom with closed electron subshells is induced by the nuclear Schiff moment. Previously the interaction with the Schiff moment has been defined for a point-like nucleus. No problems arise with the calculation of the electron matrix element of this interaction as long as the electrons are considered to be non-relativistic. However, a more realistic model obviously involves a nucleus of finite-size and relativistic electrons. In this work we have calculated the finite nuclear-size and relativistic corrections to the Schiff moment. The relativistic corrections originate from the electron wavefunctions and are incorporated into a new "nuclear" moment, which we term the local dipole moment. For ^{199}Hg these corrections amount to $\sim 20\%$. We have found that the natural generalization of the electrostatic potential of the Schiff moment for a finite-size nucleus corresponds to an electric field distribution which, inside the nucleus, is well approximated as constant and directed along the nuclear spin, and outside the nucleus is zero.

I INTRODUCTION

The best limit on parity and time invariance violating (P,T-odd) nucleon-nucleon interactions (as well as quark-quark P,T-odd interactions) has been obtained from the measurement of the ^{199}Hg electric dipole moment (EDM) [1]. The mechanism of this EDM generation is the following. P,T-odd nuclear forces create P,T-odd nuclear moments, e.g. the EDM and Schiff moment (SM). According to the Schiff theorem [2, 3, 4], the EDM of a point-like nucleus is completely screened by atomic electrons, so it cannot be measured. However, the electrostatic interaction between atomic electrons and the nuclear Schiff moment induces an atomic EDM.

The electrostatic potential produced by the Schiff moment is usually presented in the form [5]

$$\varphi(\mathbf{R}) = 4\pi \mathbf{S} \cdot \nabla \delta(\mathbf{R}), \qquad (1)$$

where **S** is the Schiff moment (vector), $\delta(\mathbf{R})$ is a delta-function. The contact interaction $-e\varphi$ mixes s- and p-wave electron orbitals and produces EDMs in atoms; for example

---

[1] Dedicated to Eugene Commins.

the atomic EDM induced in an atom with a single electron in state ns has the form

$$d_{atom} = 2 \sum_m \frac{\langle ns| - e\varphi(\mathbf{R})|mp\rangle\langle mp| - e\mathbf{R}|ns\rangle}{E_{ns} - E_{mp}} . \quad (2)$$

The expression (1) is consistently defined for non-relativistic electrons. Using integration by parts, it is seen that the matrix element $\langle s| - e\varphi|p\rangle$ is finite,

$$\langle s| - e\varphi|p\rangle = 4\pi e \mathbf{S} \cdot (\nabla \psi_s^\dagger \psi_p)_{R=0} = \text{constant} . \quad (3)$$

However, atomic electrons near the nucleus are ultra-relativistic, the ratio of the kinetic or potential energy to mc^2 in heavy atoms is about 100. For the solution of the Dirac equation, $(\nabla \psi_s^\dagger \psi_p)_{R\to 0} \to \infty$ for a point-like nucleus. Usually this problem is solved by a cut-off of the electron wavefunctions at the nuclear surface. However, even inside the nucleus $\nabla \psi_s^\dagger \psi_p$ varies significantly, $\approx Z^2\alpha^2$, where α is the fine-structure constant, Z is the nuclear charge. In Hg ($Z = 80$), $Z^2\alpha^2 = 0.34$. Recently, proposals have been made to measure EDMs of very heavy atoms like Ra ($Z = 88$) [6, 7, 8, 9] and Pu ($Z = 94$) [10] where P,T-violating nuclear moments and the resulting atomic EDMs are very strongly enhanced.

A consistent treatment of the Schiff moment is needed especially because the Schiff moment itself is defined as the difference of two approximately equal terms (see Eq. (15)). The aim of the present work is to develop a consistent theory of the nuclear Schiff moment, and the atomic EDM it induces, which properly takes into account the relativistic character of the electron wavefunctions inside the nucleus.

For relativistic electrons we should introduce a finite-size Schiff moment potential. In this paper we show that the natural generalization of the Schiff moment potential for a finite-size nucleus is

$$\varphi(\mathbf{R}) = -\frac{15\mathbf{S} \cdot \mathbf{R}}{R_N^5} n(R - R_N) , \quad (4)$$

where R_N is the nuclear radius and $n(R - R_N)$ is a smooth function which is 1 for $R < R_N - \delta$ and 0 for $R > R_N + \delta$; $n(R - R_N)$ can be taken as proportional to the nuclear density. This potential (4) corresponds to a constant electric field inside the nucleus (see Fig. 1) which can be produced by P,T-odd nuclear forces or by an intrinsic EDM of an external nucleon. This expression has no singularities and may be used in relativistic atomic calculations.

A more accurate treatment requires the calculation of a new nuclear characteristic which we call the *local dipole moment* (LDM). This moment takes into account relativistic corrections to the nuclear Schiff moment which originate from the electron wavefunctions. So in the non-relativistic limit, $Z\alpha \to 0$, the LDM $L = S$. For ^{199}Hg, $L \approx S(1 - 0.6Z^2\alpha^2) \approx 0.8S$. When considering the interaction of atomic electrons with the LDM we define it as placed at the center of the nucleus, that is the electrostatic potential is

$$\varphi(\mathbf{R}) = 4\pi \mathbf{L} \cdot \nabla \delta(\mathbf{R}) . \quad (5)$$

This paper is organized in the following manner. In Section II we derive a general expression for the dipole component of the P,T-odd electrostatic potential inside the

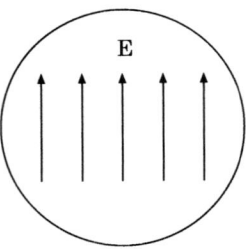

FIGURE 1. Constant electric field **E** inside the nucleus produced by the Schiff moment. **E** is directed along the nuclear spin **I**.

nucleus. In Section III we take the electronic matrix element of this potential and show that it is related to the nuclear Schiff moment. In this section the electronic and nuclear problems are separated. It is shown that it is convenient to define a "nuclear" moment (the local dipole moment) which is the nuclear Schiff moment with higher-order (relativistic) corrections which originate from the electron wavefunctions. In Section IV we calculate various nuclear LDMs which arise due to P,T-odd nuclear forces; we calculate the contribution of an external proton and that of core protons to the LDM of a spherical nucleus, and we calculate the collective LDM of an octupole-deformed nucleus. Then in Section V we calculate the electric field distribution associated with the nuclear Schiff moment.

II THE SCHIFF CONTRIBUTION TO THE NUCLEAR ELECTROSTATIC POTENTIAL

The nuclear electrostatic potential with electron screening taken into account can be presented in the following form (see, e.g., Appendix of Ref. [7] for derivation):

$$\varphi(\mathbf{R}) = \int \frac{e\rho(\mathbf{r})}{|\mathbf{R}-\mathbf{r}|} d^3r + \frac{1}{Z}(\mathbf{d}\cdot\nabla)\int \frac{\rho(\mathbf{r})}{|\mathbf{R}-\mathbf{r}|} d^3r, \qquad (6)$$

where $e\rho$ is the nuclear charge density, $\int \rho d^3r = Z$, and $\mathbf{d} = e\int \rho \mathbf{r} d^3r \equiv e\langle\mathbf{r}\rangle$ is the nuclear EDM. The second term cancels the dipole long-range electric field in the multipole expansion of $\varphi(\mathbf{R})$. The Coulomb potential $\frac{1}{|\mathbf{R}-\mathbf{r}|}$ can be expanded in terms of Legendre polynomials

$$\frac{1}{|\mathbf{R}-\mathbf{r}|} = \sum_l \frac{r_<^l}{r_>^{l+1}} P_l(\cos\theta), \qquad (7)$$

where $r_<$ ($r_>$) is $\min[r,R]$ ($\max[r,R]$). The P,T-odd part of the potential (7) originates from the odd harmonics l. The third harmonic $l=3$ corresponds to the octupole field which has been considered in [11]. The contribution of the $l=3$ term is usually small (in ^{199}Hg, which has nuclear spin $I=1/2$, it vanishes). Higher l always gives negligible contributions. Therefore, we concentrate on $l=1$ (it may be presented as $\frac{\mathbf{r}\cdot\mathbf{R}}{R^3}\Theta(R-r) +$

$\frac{\mathbf{r} \cdot \mathbf{R}}{r^3}\Theta(r-R)$, where $\Theta(r-R) = 1$ for $r > R$ and $\Theta(r-R) = 0$ for $r < R$:

$$\varphi^{(1)}(\mathbf{R}) = \frac{e\mathbf{R}}{R^3} \int_0^R \mathbf{r}\rho(\mathbf{r})d^3r + e\mathbf{R} \int_R^\infty \frac{\mathbf{r}}{r^3}\rho(\mathbf{r})d^3r - \frac{e\langle \mathbf{r}\rangle \mathbf{R}}{ZR^3} \int_0^R \rho(\mathbf{r})d^3r. \tag{8}$$

Note that in the second (screening) term in (6) we only keep the zero multipole $l = 0$. Also note that for $R \to \infty$ the first and third terms of Eq. (8) cancel each other. Therefore, we can use $\int_0^R = \int_0^\infty - \int_R^\infty = -\int_R^\infty$ and present $\varphi^{(1)}(\mathbf{R})$ as

$$\varphi^{(1)}(\mathbf{R}) = e\mathbf{R} \int_R^\infty \left(\frac{\langle \mathbf{r}\rangle}{ZR^3} - \frac{\mathbf{r}}{R^3} + \frac{\mathbf{r}}{r^3}\right)\rho(\mathbf{r})d^3r. \tag{9}$$

We see that $\varphi^{(1)}(\mathbf{R}) = 0$ for $R > R_N$ (nuclear radius) since $\rho(\mathbf{r}) = 0$ in this area. We will see in the next section that this potential (9) is related to the Schiff moment.

III ELECTRON MATRIX ELEMENTS OF THE P, T-ODD ELECTROSTATIC POTENTIAL

All the electron orbitals for $l > 1$ are extremely small inside the nucleus. Therefore, we can limit our consideration to the matrix elements between s and p Dirac orbitals. We will use the following notations for the electron wavefunctions:

$$\psi(\mathbf{R}) = \begin{pmatrix} f(R)\Omega_{jlm} \\ -i(\sigma \cdot \mathbf{n})g(R)\Omega_{jlm} \end{pmatrix}, \tag{10}$$

where Ω_{jlm} is a spherical spinor, $\mathbf{n} = \mathbf{R}/R$, and f(R) and g(R) are radial functions (see, e.g., [12]). Using $(\sigma \cdot \mathbf{n})^2 = 1$, then we can write the electron transition density as

$$\rho_{sp}(\mathbf{R}) \equiv \psi_s^\dagger \psi_p = \Omega_s^\dagger \Omega_p U_{sp}(R) \tag{11}$$

$$U_{sp}(R) = f_s(R)f_p(R) + g_s(R)g_p(R) = \sum_{k=1}^\infty b_k R^k. \tag{12}$$

The expansion coefficients b_k are calculated analytically and are presented in the Appendix; as is seen here, the summation is carried over the odd powers of k. Now we can find the matrix element of the electron-nucleus interaction,

$$\langle s| - e\varphi^{(1)}(\mathbf{R})|p\rangle = -e^2\langle s|\mathbf{n}|p\rangle \int_0^\infty \left[\left(\frac{1}{Z}\langle\mathbf{r}\rangle - \mathbf{r}\right)\int_0^r U_{sp}\,dR + \frac{\mathbf{r}}{r^3}\int_0^r U_{sp}R^3\,dR\right]\rho\,d^3r$$

$$= -e^2\langle s|\mathbf{n}|p\rangle \sum_{k=1}^\infty \frac{b_k}{k+1}\left[\frac{1}{Z}\langle\mathbf{r}\rangle\langle r^{k+1}\rangle - \frac{3}{k+4}\langle \mathbf{r}r^{k+1}\rangle\right], \tag{13}$$

where $\langle s|\mathbf{n}|p\rangle \equiv \int \Omega_s^\dagger \mathbf{n} \Omega_p d\phi \sin\theta d\theta$ and $\langle r^n\rangle \equiv \int \rho(\mathbf{r})r^n d^3r$. Note that all vector values $\langle \mathbf{r}r^n\rangle$ are due to the P, T-odd correction to the nuclear charge density ρ, while $\frac{1}{Z}\langle r^n\rangle$ are the usual P, T-even moments of the charge density starting from the mean-square radius $\frac{1}{Z}\langle r^2\rangle = r_q^2$ for $k = 1$.

In the non-relativistic case ($Z\alpha \to 0$) we have just $b_1 \neq 0$, and so

$$\langle s| - e\varphi^{(1)} |p\rangle = -e^2 \langle s|\mathbf{n}|p\rangle \frac{b_1}{2} \left[\frac{1}{Z} \langle \mathbf{r}\rangle \langle r^2\rangle - \frac{3}{5} \langle \mathbf{r}r^2\rangle \right]$$
$$= 4\pi e \mathbf{S} \cdot (\nabla \psi_s^\dagger \psi_p)_{R \to 0}, \qquad (14)$$

where the Schiff moment $\mathbf{S}$ is defined as

$$\mathbf{S} = \frac{e}{10} \left[\langle r^2 \mathbf{r}\rangle - \frac{5}{3Z} \langle r^2\rangle \langle \mathbf{r}\rangle \right] = S \mathbf{I}/I, \qquad (15)$$

$\mathbf{I}$ is the nuclear spin. The expressions (14), (15) agree with the results of Ref. [5].

Therefore, Eq. (13) gives us the possibility of a consistent relativistic treatment of the atomic effects produced by P,T-odd nuclear forces. The nuclear and electronic problems can be separated in the following way. The nuclear calculations can provide us with the value of the *local dipole moment* (LDM)

$$\mathbf{L} = e \sum_{k=1}^{\infty} \frac{b_k}{b_1} \frac{1}{(k+1)(k+4)} \left[\langle \mathbf{r}r^{k+1}\rangle - \frac{k+4}{3Z} \langle \mathbf{r}\rangle \langle r^{k+1}\rangle \right] = L \mathbf{I}/I \qquad (16)$$

which coincides with the Schiff moment $\mathbf{S}$ (15) in the non-relativistic limit ($Z\alpha \to 0$). Note that this "nuclear" moment contains relativistic corrections which arise from the *electron wavefunctions* which are calculated analytically. (It should further be noted that the corrections originating from the s-$p_{1/2}$ and s-$p_{3/2}$ matrix elements are different.) The electron matrix elements are then given by

$$\langle s| - e\varphi^{(1)}|p\rangle = 4\pi e \mathbf{L} \cdot (\nabla \psi_s^\dagger \psi_p)_{R\to 0} = 3e\mathbf{L} \cdot \langle s|\mathbf{n}|p\rangle \left(\frac{f_s f_p + g_s g_p}{R} \right)_{R\to 0}. \qquad (17)$$

These formulae (16,17), in principle, solve the problem of the consistent approach for the calculation of the interaction of the relativistic electrons with the Schiff moment. Note that to achieve $\sim 10\%$ accuracy it is enough to keep in $\mathbf{L}$ just the first correction, $b_3/b_1 = -3/5 \ Z^2\alpha^2/R_N^2$ for the s-$p_{1/2}$ matrix element and $b_3/b_1 = -9/20 \ Z^2\alpha^2/R_N^2$ for the s-$p_{3/2}$ matrix element (see Appendix); and at this level of accuracy ($\sim 10\%$) the values of the coefficients b_3/b_1 (for s-$p_{1/2}$ and s-$p_{3/2}$) can be taken to be the same.

IV LOCAL DIPOLE MOMENTS INDUCED BY P,T-ODD NUCLEAR FORCES

We can now calculate local dipole moments induced by P,T-odd nuclear forces. We will begin in Section IV A with the calculation of the contribution of an external proton to the local dipole moment of a spherical nucleus. Because the best limit on the P,T-odd nucleon-nucleon interaction has been extracted from ^{199}Hg (which has an external neutron, so only the core protons contribute to the LDM) the result of Section IV A is not so interesting by itself. However, as will be explained in Section IV B, it provides us with a check of the method for the calculation of the contribution of core protons to the LDM. Then in Section IV C we calculate the collective LDM of an octupole deformed nucleus.

The P,T-odd nucleon-nucleon interaction, to first-order in the velocities p/m, can be presented as [5]

$$\hat{W}_{ab} = \frac{G}{\sqrt{2}}\frac{1}{2m}\left((\eta_{ab}\sigma_a - \eta_{ba}\sigma_b)\cdot\nabla_a\delta(\mathbf{r}_a-\mathbf{r}_b) + \eta'_{ab}[\sigma_a\times\sigma_b]\{(\mathbf{p}_a-\mathbf{p}_b),\delta(\mathbf{r}_a-\mathbf{r}_b)\}\right), \quad (18)$$

where $\{\,,\,\}$ is an anticommutator, G is the Fermi constant of the weak interaction, m is the nucleon mass, and σ, $\mathbf{r}$, and $\mathbf{p}$ are the spins, coordinates, and momenta of the nucleons a and b. The dimensionless constants η_{ab} and η'_{ab} characterize the strength of the P,T-odd nuclear potential (experiments on EDMs are aimed to measure these constants).

A Nuclear LDM produced by external proton

In this section we calculate the LDM arising due to an external proton. We are therefore interested in the P,T-odd interaction of the external proton with the core nucleons. We can average the two-particle interaction (18) over the core nucleons to obtain the effective single-particle P,T-odd interaction between the proton and core [5],

$$\hat{W} = \frac{G}{\sqrt{2}}\frac{\eta}{2m}\sigma\cdot\nabla\rho_A(\mathbf{r})\,. \quad (19)$$

Here it has been assumed that the proton and neutron densities are proportional to the total nuclear density $\rho_A(\mathbf{r})$; the dimensionless constant $\eta = \frac{Z}{A}\eta_{pp} + \frac{N}{A}\eta_{pn}$. Notice that there is only one surviving term from the P,T-odd nucleon-nucleon interaction (18); this is because all other terms contain the spin of the internal nucleons for which $\langle\sigma\rangle = 0$. The shape of the nuclear density ρ_A and the strong potential U are known to be similar; we therefore take $\rho_A(\mathbf{r}) = \frac{\rho_A(0)}{U(0)}U(\mathbf{r})$. Then we can rewrite Eq. (19) in the following form:

$$\hat{W} = \xi\sigma\cdot\nabla U\,, \quad \xi = \eta\frac{G}{2\sqrt{2}m}\frac{\rho_A(0)}{U(0)} = -2\times 10^{-8}\eta \text{ fm}\,. \quad (20)$$

Now it is easy to find the solution of the Scrödinger equation including the interaction $\hat{W}$:

$$(\hat{H}+\hat{W})\tilde{\psi} = E\tilde{\psi}\,,$$
$$\tilde{\psi} = (1+\xi\sigma\cdot\nabla)\psi\,, \quad (21)$$

where ψ is the unperturbed solution ($\hat{H}\psi = E\psi$). The valence proton density is then equal to

$$\rho = \tilde{\psi}^\dagger\tilde{\psi} = \psi^\dagger\psi + \xi\nabla(\psi^\dagger\sigma\psi)\,. \quad (22)$$

The second term gives the P,T-odd part of the density which generates a Schiff moment S [5],

$$S = -\frac{e\xi}{10}\left[\left(t_I + \frac{1}{I+1}\right)r_{ex}^2 - \frac{5}{3}t_I r_q^2\right], \tag{23}$$

where we denote r_{ex}^2 as the mean-square radius of the external nucleon (in this case that of the proton), r_q^2 is the mean-square nuclear charge radius, $\langle\sigma\rangle = t_I \frac{\mathbf{I}}{I}$, and

$$t_I = \begin{cases} 1 & I = l + 1/2 \\ -\frac{I}{I+1} & I = l - 1/2 \end{cases} \tag{24}$$

where I and l are the total and orbital angular momentum of the proton, respectively. It should be noted that for the Schiff moment the recoil effect (the motion of the nuclear core around the center-of-mass) disappears due to the cancellation of its contributions to the first and second (screening) terms in Eq. (6) [5].

To calculate the local dipole moment, it is enough to substitute the external proton density (22) into the expression for the LDM (16) and perform integration using integration by parts. The result is

$$L = -\sum_{k=1}^{\infty} \frac{b_k}{b_1} \frac{e\xi}{(k+1)(k+4)}\left[\left(t_I + \frac{k+1}{2(I+1)}\right)r_{ex}^{k+1} - \frac{k+4}{3}t_I r_q^{k+1}\right]. \tag{25}$$

(Notice that for a proton in the state $s_{1/2}$, the LDM is reduced to the difference of two approximately equal terms $(r_{ex}^k - r_q^k)$ for all $k = 2, 4, \ldots$. This makes an analytical calculation hopeless when trying to estimate the LDM of a nucleus with an external proton in state $s_{1/2}$, as is the case for ^{203,205}Tl.)

While the result (25) on its own is not very interesting, it will be used in the next section as a check of the method for the calculation of the LDM of ^{199}Hg.

B Nuclear LDM produced by core protons. Mercury moments.

In nuclei like ^{199}Hg and ^{129}Xe the external nucleon is a neutron. It does not contribute to the Schiff moment directly. In Ref. [13] it was shown that virtual excitations of the core nucleons caused by a P,T-odd interaction with the external nucleon produce a Schiff moment which is comparable to that produced by an external proton. The actual calculation of the Schiff moments in [13] was carried out numerically. In this section we perform a simple analytical calculation which allows us to estimate the contribution of the relativistic corrections $\sim Z^2\alpha^2$ to the Schiff moment. Here we follow an approach which was used in [13] to estimate the contribution of the giant dipole resonance to the nuclear EDM.

The expression for the local dipole moment **L** induced in the nuclear state $|0\rangle$ by the P,T-odd interaction $\hat{W}_{ab}$ (18) between the nucleons a and b is

$$\mathbf{L} = \sum_n \frac{\langle 0|\hat{\mathbf{L}}|n\rangle\langle n|\hat{W}_{ab}|0\rangle + \langle 0|\hat{W}_{ab}|n\rangle\langle n|\hat{\mathbf{L}}|0\rangle}{E_0 - E_n} \tag{26}$$

$$= \sum_n \frac{\langle 0|[\hat{H},\hat{\mathbf{L}}]|n\rangle\langle n|\hat{W}_{ab}|0\rangle - \langle 0|\hat{W}_{ab}|n\rangle\langle n|[\hat{H},\hat{\mathbf{L}}]|0\rangle}{(E_0 - E_n)^2}. \tag{27}$$

Here $\hat{H}$ is the Hamiltonian, and $[\hat{H},\hat{\mathbf{L}}]$ is a commutator; the LDM operator $\hat{\mathbf{L}}$ is defined from Eq. (16) as $\mathbf{L} = \langle \hat{\mathbf{L}} \rangle$. Now we assume that the transition strength in the sum over intermediate states $|n\rangle$ is concentrated around the excitation energy ω_r, and replace $(E_0 - E_n)^2$ by ω_r^2. (Note that the replacement of $(E_0 - E_n)$ by ω_r in Eq. (26) gives an incorrect result since in single-particle language there are transitions with $E_0 - E_n = \omega_r$ and $E_0 - E_n = -\omega_r$.) Use of closure, $\sum_n |n\rangle\langle n| = 1$, gives

$$\mathbf{L} = \frac{1}{\omega_r^2} \langle 0|\,[[\hat{H},\hat{\mathbf{L}}],\hat{W}_{ab}]\,|0\rangle. \tag{28}$$

To calculate the commutator we assume that the motion of each nucleon in the nucleus can be described by the Hamiltonian $\hat{H} = \frac{p^2}{2m} + V(r)$. The contribution to the LDM from a single proton is then

$$\mathbf{L} = -\frac{1}{m\omega_r^2} \langle 0|(\nabla_\alpha \hat{\mathbf{L}})(\nabla_\alpha \hat{W}_{ab})|0\rangle. \tag{29}$$

As a check of the validity of this approach for the calculation of the core contribution, we use this formula (29) to calculate both the contribution of the external proton (which we can compare with Eq. (25)) and the contribution of the core protons.

1 External proton contribution

For the external proton we can just substitute the effective potential $\hat{W}$ (19) into expression (29),

$$\mathbf{L} = -\frac{G}{\sqrt{2}} \frac{1}{2m^2\omega_r^2} \eta \langle 0|(\nabla_\alpha \hat{\mathbf{L}})(\nabla_\alpha(\boldsymbol{\sigma}\cdot\nabla)\rho_A)|0\rangle, \tag{30}$$

where here $|0\rangle$ corresponds to the state of the external proton. If we consider the nuclear density ρ_A to be proportional to the nuclear potential U, and use the oscillator model to approximate the potential so that

$$\rho_A(r) = \frac{\rho_A(0)}{U(0)} \frac{1}{2} m\omega^2 r^2, \tag{31}$$

then the LDM is reduced to

$$\mathbf{L} = -\frac{\omega^2}{\omega_r^2} \sum_{k=1}^{\infty} \frac{b_k}{b_1} \frac{e\xi}{(k+1)(k+4)} \left[\left(t_I + \frac{k+1}{2(I+1)} \right) r_{\text{ex}}^{k+1} - \frac{k+4}{3} t_{II} r_q^{k+1} \right]. \tag{32}$$

Because here we are considering the LDM produced by a single proton, we can set $\omega_r = \omega$, so the factor $(\omega^2/\omega_r^2) = 1$. Therefore, using the resonance method we can reproduce Eq. (25).

2 Core contribution

In this section we use the "Schiff resonance" formalism to estimate the local dipole moment produced by the core protons. Again we start from Eq. (29), where in this case the derivatives are with respect to the internal proton coordinates. Assuming that the proton density is proportional to the total nuclear density ρ_A, we obtain

$$\mathbf{L} = \frac{G}{2\sqrt{2}m} \frac{1}{m\omega_r^2} \eta \langle 0|\sigma \cdot \nabla \left[\nabla_\alpha \left(\rho_A \nabla_\alpha \hat{L} \right) \right] |0\rangle, \tag{33}$$

where $|0\rangle$ is the state of the external nucleon. Because here we are considering the P,T-odd interaction $\hat{W}_{ap}$ (18) between the external nucleon (proton or neutron, a) and the core protons, the P,T-odd dimensionless constant $\eta = (Z/A)\eta_{ap}$. Using (31) to approximate the nuclear density, the LDM becomes

$$L = \frac{\omega^2}{\omega_r^2} \sum_{k=1}^{\infty} \frac{b_k}{b_1} \frac{e\xi}{(k+1)(k+4)} \left[\frac{k^2+7k+8}{2} \left(t_I + \frac{k+1}{2(I+1)} \right) r_{\text{ex}}^{k+1} - \frac{k+4}{3} t_I r_q^{k+1} \right]. \tag{34}$$

Before we consider the size of the relativistic corrections, let us check that this result gives us a reasonable value for the Schiff moment ($Z\alpha \to 0$). In the approximation of a uniform and spherical charge distribution, $r_q^2 = (3/5)R_N^2$, assuming that $r_{\text{ex}}^2 = r_q^2$, and setting the resonance frequency for core protons $\omega_r = 2\omega$ (frequency of giant resonance), we obtain for the Schiff moment of ^{199}Hg (which has an external neutron in the state $p_{1/2}$)

$$S_{\text{Hg}} \approx -1.6 \times 10^{-8} \, \eta_{np} e\text{fm}^3. \tag{35}$$

From a numerical calculation of the Schiff moment for ^{199}Hg performed in Ref. [13] the result $S = -1.4 \times 10^{-8} \, \eta_{np} e\text{fm}^3$ was obtained. This value agrees with our analytical estimate (35). Therefore it seems that the resonance method can be used for Hg to give a crude estimate of the size of the relativistic corrections to the Schiff moment.

To estimate the size of the corrections, we use the approximation of a uniform and spherical charge distribution, $r_q^n = [3/(n+3)]R_N^n$, and we assume that $r_{\text{ex}}^n = r_q^n$, for $n = 2, 4, \ldots$. Substituting the coefficients from the Appendix into the expression for the LDM (34), we see that the correction to the Schiff moment for Hg is about 20%, $L \approx 0.8S$. Therefore the relativistic corrections to the Schiff moment for Hg (and for other spherical nuclei) are not very large.

C Collective LDM of an octupole deformed nucleus

Nuclei with octupole deformation have enhanced collective Schiff moments which may be up to 1000 times larger than the Schiff moments of spherical nuclei [6, 7]. In Ref.

[10] it was pointed out that the soft octupole vibration mode produces an enhancement similar to that of the static octupole deformation. This makes heavy atoms containing nuclei with collective Schiff moments attractive for future experiments on the search for T-violation.

The mechanism generating the collective Schiff moment is the following [6, 7]. In the "frozen" body frame the collective Schiff moment S_{intr} can exist without any P,T-violation. However, the nucleus rotates, and this makes the expectation value of the Schiff moment vanish in the laboratory frame if there is no P,T-violation. (This is because the intrinsic Schiff moment is directed along the nuclear axis, $\mathbf{S}_{\text{intr}} = S_{\text{intr}}\mathbf{n}$, and in the laboratory frame the only possible correlation $\langle \mathbf{n} \rangle \propto \mathbf{I}$ violates parity and time reversal.) The P,T-odd nuclear forces mix rotational states of opposite parity and create an average orientation of the nuclear axis $\mathbf{n}$ along the nuclear spin $\mathbf{I}$,

$$\langle n_z \rangle = 2\alpha \frac{KM}{I(I+1)}, \tag{36}$$

where

$$\alpha = \frac{\langle \psi_- | \hat{W} | \psi_+ \rangle}{E_+ - E_-} \tag{37}$$

is the mixing coefficient of the opposite parity states, $K = |\mathbf{I} \cdot \mathbf{n}|$ is the absolute value of the projection of the nuclear spin $\mathbf{I}$ on the nuclear axis, $M = I_z$, and $\hat{W}$ is the effective single-particle potential (19). The Schiff moment in the laboratory frame is

$$S_z = S_{\text{intr}} \langle n_z \rangle = S_{\text{intr}} \frac{2\alpha KM}{I(I+1)}. \tag{38}$$

In the "frozen" body frame the surface of an axially symmetric deformed nucleus is described by the following expression

$$R(\theta) = R_N \left(1 + \sum_{l=1} \beta_l Y_{l0}(\theta)\right). \tag{39}$$

To keep the center-of-mass at $r = 0$ we have to fix β_1 [14]:

$$\beta_1 = -3\sqrt{\frac{3}{4\pi}} \sum_{l=2} \frac{(l+1)\beta_l \beta_{l+1}}{\sqrt{(2l+1)(2l+3)}}. \tag{40}$$

We assume that the distributions of the protons and neutrons are the same, so the electric dipole moment $e\langle \mathbf{r} \rangle = 0$ (since the center-of-mass of the charge distribution coincides with the center-of-mass) and hence there is no screening contribution to the Schiff moment. We also assume constant density for $R < R(\theta)$. The intrinsic Schiff moment S_{intr} is then [6, 7]

$$S_{\text{intr}} = eZR_N^3 \frac{3}{20\pi} \sum_{l=2} \frac{(l+1)\beta_l \beta_{l+1}}{\sqrt{(2l+1)(2l+3)}} \approx eZR_N^3 \frac{9\beta_2 \beta_3}{20\pi\sqrt{35}}, \tag{41}$$

where the major contribution comes from $\beta_2\beta_3$, the product of the quadrupole β_2 and octupole β_3 deformations. For $\beta_2 \sim \beta_3 \sim 0.1$ and $Z = 88$ (Ra) we obtain $S_{\text{intr}} \sim 10 \, efm^3$. The estimate of the Schiff moment in the laboratory frame gives [7]

$$S \sim \alpha S_{\text{intr}} \sim 0.05 \, e\beta_2\beta_3^2 ZA^{2/3}\eta r_0^3 \frac{eV}{E_+ - E_-} \sim 700 \times 10^{-8} \, \eta efm^3 , \qquad (42)$$

where $r_0 \approx 1.2$ fm is the internucleon distance, $E_+ - E_- \sim 50$ keV. This estimate (42) is about 500 times larger than the Schiff moment of a spherical nucleus like Hg. Note that S in Eq. (42) is proportional to the squared octupole deformation parameter β_3^2. According to [10], in nuclei with a soft octupole vibration mode $\langle \beta_3^2 \rangle \sim (0.1)^2$, i.e., this is the same as in nuclei with static octupole deformation. This means that a number of heavy nuclei can have large collective Schiff moments.

With no screening term, it is easy to calculate the collective LDM. Use of Eq. (16) gives

$$L = S\left(1 + \sum_{k=3} \frac{5b_k}{(k+4)b_1} R_N^{k-1}\right) \approx S(1 - 0.3Z^2\alpha^2) . \qquad (43)$$

As with spherical nuclei, we see that the correction to the Schiff moment for collective nuclei is not very large (for Ra, Pu this correction $\sim 15\%$).

V P,T-ODD PART OF THE NUCLEAR ELECTRIC FIELD (SCHIFF FIELD)

In this section we calculate the actual distribution of the P,T-odd component of the electrostatic potential $\varphi(\mathbf{R})$ inside the nucleus (arising from the P,T-odd nucleon-nucleon interaction) for two models: for an external proton in a spherical nucleus and for a collective Schiff moment which appears in a nucleus with octupole deformation. It is found that in the collective case the electric field is constant and directed along the nuclear spin. This field distribution is also approximately correct in the spherical case when the external proton is in state $s_{1/2}$; this is also true without the P,T-odd interaction but when the external nucleon (proton or neutron) possesses an intrinsic EDM and is in state $s_{1/2}$.

A P,T-odd electric field produced by a valence proton

To calculate the P,T-odd part of the electrostatic potential φ_T produced by the external proton, we substitute the P,T-odd perturbed external proton density (22) into (6) and integrate by parts,

$$\varphi_T(\mathbf{R}) = e\xi\nabla\int\frac{\rho_\sigma}{|\mathbf{R}-\mathbf{r}|}d^3r - \frac{1}{Z}e\xi\langle\sigma\rangle\nabla\int\frac{\rho}{|\mathbf{R}-\mathbf{r}|}d^3r , \qquad (44)$$

where $\rho_\sigma = \psi^\dagger\sigma\psi$ is the spin density, $\langle\sigma\rangle = t_l\frac{\mathbf{I}}{I}$. Note the similarity of this expression and that for a P,T-odd potential produced by an external proton electric dipole moment

d_p (see, e.g., [3, 12]). In the latter case one should only replace $e\xi$ by d_p (or by d_n in the case of Hg or Xe). Note, however, that generally speaking $\rho_\sigma = \psi^\dagger \sigma \psi \neq \langle \sigma \rangle \psi^\dagger \psi$ (this assumption was used in [3]), i.e. the direction of the external nucleon EDM depends on the coordinate $\mathbf{r}$. The separation of the spin and the coordinate variables is possible in the case of $I = l + 1/2$. Taking $I = I_z$ we obtain $\rho_\sigma = \frac{\mathbf{I}}{I}\rho_M$, where ρ_M is the density of the valence proton. If we now assume that this density is constant within the sphere of the radius R_M, $\rho_M(r) = \frac{3}{4\pi R_M^3}$, $r < R_M$, and $\rho_M(r) = 0$ for $r > R_M$, and, similarly, the nuclear charge density $\rho_q(r) = \frac{3Z}{4\pi R_N^3}$, $r < R_N$, and $\rho_q(r) = 0$ for $r > R_N$, then we obtain for the dipole term (and for $R_N < R_M$):

$$\varphi^{(1)}(\mathbf{R}) = -e\xi \mathbf{R} \cdot \frac{\mathbf{I}}{I} \begin{cases} (\frac{1}{R_M^3} - \frac{1}{R_N^3}) & R < R_N \\ (\frac{1}{R_M^3} - \frac{1}{R^3}) & R_N < R < R_M \\ 0 & R > R_M \end{cases} \quad (45)$$

Thus, the P,T-odd part of the electrostatic potential is $\varphi^{(1)}(\mathbf{R}) \propto R \cos\theta$ inside the nucleus. This gives us a very simple picture for the P,T-odd electric field (Schiff field): $E_z = -\frac{\partial \varphi}{\partial z} \propto I$ inside the nucleus. Thus, the Schiff moment gives a constant electric field along the nuclear spin, $\mathbf{E} \propto \mathbf{I}$, and this field vanishes within the nuclear "skin" (see Fig. 1).

We can easily establish a relation between the P,T-odd electrostatic potential inside the nucleus $\varphi^{(1)} \propto R \cos\theta$ and the Schiff moment $\mathbf{S}$. Comparing Eq. (45) with Eq. (23) (with $t_I = 1$, $I = 1/2$) we obtain

$$\varphi^{(1)}(\mathbf{R}) = -\frac{15 \mathbf{S} \cdot \mathbf{R}}{R_N^5} n(R - R_N), \quad (46)$$

where $n(R - R_N)$ is a smoothed step-function $\Theta(R_N - R)$, that is $n(R - R_N) \approx \Theta(R_N - R)$; $n(R - R_N) = 1$ for $R < R_N$ and $n(R - R_N) = 0$ for $R > R_N + \delta$, where $\delta = R_M - R_N \ll R_N$. It gives the natural generalization of the Schiff moment potential (1) for the case of a finite-size nucleus. Of course, in the general case, the radial function $n(R - R_N)$ in the first harmonic of the P,T-odd potential (46) is more complicated (this gives some "wiggling" of the electric field inside the nucleus).

B P,T-odd electric field produced by a collective Schiff moment

Now we wish to calculate the electrostatic potential $\varphi^{(1)}$ arising due to a collective Schiff moment in a nucleus with octupole deformation. We use Eqs. (39,40) and assume that the distributions of the protons and neutrons are the same (therefore $e\langle \mathbf{r} \rangle = 0$, and so there is no screening term) and that the density for $R < R(\theta)$ is constant. Calculating the integral in Eq. (9) for $R < \min R(\theta)$ gives

$$\varphi^{(1)}(\mathbf{R}) = -\frac{9Ze}{4\pi R_N^2} \mathbf{R} \cdot \frac{\mathbf{I}}{I} \sum_{l=2} \frac{(l+1)\beta_l \beta_{l+1}}{\sqrt{(2l+1)(2l+3)}}. \quad (47)$$

In the laboratory frame the result differs by an extra factor $\langle n_z \rangle$ (36). Using Eq. (41) we can present the final result for $\varphi^{(1)}$ as

$$\varphi^{(1)}(\mathbf{R}) = -\frac{15\mathbf{S}\cdot\mathbf{R}}{R_N^5} n(R - R_N), \qquad (48)$$

where $n(R - R_N) = 1$ for $R < \min R(\theta)$, and $n(R - R_N) = 0$ for $R > \max R(\theta)$. This result is similar to Eq. (46). Thus, the collective Schiff moment produces a constant electric field along the nuclear spin inside the nucleus and zero field outside (Fig. 1). In this case the width of the transition area of the nuclear surface $\sim \beta_l R_N \sim 0.2 R_N$.

APPENDIX: ELECTRON TRANSITION DENSITY FOR S-$P_{1/2}$ AND S-$P_{3/2}$

To calculate the electron wavefunctions inside the nucleus $R \leq R_N$ we assume that the nuclear charge is uniformly distributed about a sphere. This charge distribution corresponds to the harmonic-oscillator potential

$$V = -\frac{Z\alpha}{R_N}\left(\frac{3}{2} - \frac{1}{2}x^2\right), \qquad (49)$$

where we have set $x \equiv R/R_N$. Solving the Dirac equation for an electron in states $s_{1/2}$, $p_{1/2}$ and $p_{3/2}$ moving in this potential gives the s-$p_{1/2}$ and s-$p_{3/2}$ radial transition densities (see Eqs. (10,11,12) for definition)

$$U_{sp_{1/2}} = \frac{2}{3} f_s(0) g_{p_{1/2}}(0) m R_N x \left\{ 1 - \frac{3}{5} Z^2 \alpha^2 x^2 + \frac{9}{70} Z^2 \alpha^2 x^4 + \ldots \right\} \qquad (50)$$

$$U_{sp_{3/2}} = f_s(0) \left(f_{p_{3/2}}/x \right)_{R=0} x \left\{ 1 - \frac{9}{20} Z^2 \alpha^2 x^2 + \frac{69}{700} Z^2 \alpha^2 x^4 + \ldots \right\}, \qquad (51)$$

where $f_s(0)$, $g_{p_{1/2}}(0)$, and $(f_{p_{3/2}}/x)_{R=0}$ are the s, $p_{1/2}$, and $(p_{3/2}/x)$ radial wavefunctions at zero, and m is the electron mass. The omitted terms in (50,51) correct the Schiff moment (25,34,43) by only a few percent.

REFERENCES

1. M.V. Romalis, W.C. Griffith, J.P. Jacobs, and E.N. Fortson, Phys. Rev. Lett. **86**, 2505 (2001).
2. E.M. Purcell and N.F. Ramsey, Phys. Rev. **78**, 807 (1950).
3. L.I. Schiff, Phys. Rev. **132**, 2194 (1963).
4. P.G.H. Sandars, Phys. Rev. Lett. **19**, 1396 (1967).
5. V.V. Flambaum, I.B. Khriplovich, and O.P. Sushkov, ZhETF **87**, 1521 (1984) [Sov. Phys. JETP **60**, 873 (1984)].
6. N. Auerbach, V.V. Flambaum, and V. Spevak, Phys. Rev. Lett. **76**, 4316 (1996).
7. V. Spevak, N. Auerbach, and V.V. Flambaum, Phys. Rev. C **56**, 1357 (1997).
8. V.V. Flambaum, Phys. Rev. A **60**, R2611 (1999).
9. V.A. Dzuba, V.V. Flambaum, and J.S.M. Ginges, Phys. Rev. A **61**, 062509 (2000).

10. J. Engel, J.L. Friar, and A.C. Hayes, Phys. Rev. C **61**, 035502 (2000).
11. V.V. Flambaum, D.W. Murray, and S.R. Orton, Phys. Rev. C **56**, 2820 (1997).
12. I.B. Khriplovich, *Parity Nonconservation in Atomic Phenomena* (Gordon and Breach, Philadelphia, 1991).
13. V.V. Flambaum, I.B. Khriplovich, and O.P. Sushkov, Nucl. Phys. **A449**, 750 (1986).
14. A. Bohr and B. Mottelson, *Nuclear Structure* (Benjamin, New York, 1975), Vol. 2.

Night Thoughts on Consciousness and Time Reversal

A. Zee

Institute for Theoretical Physics
University of California
Santa Barbara, California 93106

Abstract. A theoretical physicist worries about his consciousness.

As a theorist, I am honored to be invited to speak at the 60th birthday celebration of a distinguished experimentalist. This must mean that some of my work, mirabile dictu, actually has something to do with reality. By the way, just as I was leaving Santa Barbara yesterday I learned that not only is Gene Commins a distinguished experimentalist, but he has had, indirectly, a major impact on my life.

Actually, I've just found out that the number 60 was a purely theoretical and erroneous assumption on my part, but since I had written 60 into my talk, as you will see, I can't change it now.

I know that the organizers wanted me to talk about the theoretical prediction of the electric dipole moment of the electron by Barr and Zee, but the calculation is really too straightforward to discuss here. Not only that, but I have always felt that our present theoretical understanding of time reversal invariance violation is very unsatisfying. In Schrödinger's equation the partial time derivative goes with the imaginary unit *i* and so time reversal is associated with complex conjugation. On the other hand, the Lagrangian of the world is required only to be Hermitian as a whole. So, in almost any moderately complicated Lagrangian there are plenty of opportunities to break time reversal invariance. But I find this colossally boring and totally unilluminating. There is got to be more to it.

In ancient civilizations, such as the Chinese and the Babylonian, time is measured in units of 60 (hence the 60 seconds and minutes we still have.) Thus, at 60 a man starts his life anew. One of the few things Confucius said that I strongly agree with is that at age 60 a man can do whatever his heart desires. (My scholarly friends always remind me that I conveniently forget the modifying clause "within the bounds of society" in the original.) I have not yet reached the age of 60 but for a theorist, first approximation is always good enough. So I shall do what my heart desires today, which is to talk nonsense.

My starting point is the last chapter of my 1986 book Fearful Symmetry, recently re-published by Princeton University Press. I quote from page 278:

I have saved the discussion of time reversal invariance for the closing pages of *Fearful Symmetry* because I do not understand it. Neither does anybody else. As a physicist, I know what I have told you about time reversal invariance: The fundamental laws of Nature do not pick out a direction of time except in the decay of a certain subnuclear particle, blah, blah, blah. But as a conscious being, I know darn well that there *is* a direction of time. I don't care what physicists say, I know that the flow of time is irrevocable. For lovers and nonlovers alike, time goes by.

In physics, time is simply treated as a mathematical parameter; as time changes, various physical quantities change in accordance with various physical laws. Einstein's work deepened the mystery by treating time and space on equal footing. Yet, again as a conscious being, I *know* that time is different from space: I can go east or west, as I please, but I can only go in one direction in time.

We are confronted here with an impasse enforced by a fundamental guiding tenet of science: the exclusion of consciousness. Physicists are careful to say that their knowledge is limited to the physical world. The realization that the world may be divided into the physical and, for lack of a better term, the nonphysical surely ranks as a major turning point in intellectual history, and one that has made possible the advent of Western science. But eventually we will have to cross the dividing line. I believe that a deep understanding of time reversal invariance will take us across that line.

I then went on to explain the different arrows of time known to physicists. I even talked about the tantalizing clue from quantum physics.

Ever since the early days of the quantum, when it was realized that the act of observation unavoidably disturbs the observed (as quantified by the uncertainty principle), physicists and philosophers have speculated about the possible link between consciousness and the probabilistic mystery of the quantum. There is no lack of speculation and musing on the subject, but it is fair to say that the overwhelming majority of working physicists find what has been written exceedingly difficult, if not impossible, to understand. The distinguished physicist Murph Goldberger was once asked by a television interviewer why he had never worked in this area. He answered that every time he decided to think about these questions, he would sit down, get out a clean piece of paper, sharpen his pencil, and then he just couldn't think of anything to put down. That is as good a summary as any of our present understanding of the role of consciousness in physics.

In the intervening 15 years, needless to say I have not had anything intelligent to add to this. Deep thinkers east and west have grappled with the mystery of consciousness for millennia. For example, Schrödinger in his influential book, <u>What is Life?</u>, found himself forced to the often expressed position that there exists only one universal consciousness, of which our individual consciousness is but one manifestation. In fact, the universal consciousness controls the laws of physics. Deus factum sum, "We have become God," he said.

Indeed, in looking through various books on consciousness written by various eminent biologists and philosophers I learned that it is even difficult to define consciousness. To paraphrase one biologist, perhaps we should define consciousness by what it is not and say for this audience that consciousness is whatever you have when you are doing physics. By definition, it is impossible to do physics while unconscious.

Incidentally, in my reading about consciousness I find one fact rather intriguing: routine tasks are increasingly assigned to the zombic system (and that is a technical term!) Probably we have all had the experience of arriving at work without being conscious of how we got there: the absence of what the Zen masters call "mindfulness." Much of bodily functions are governed by the zombic system, and we are aware of say our kidneys only when they are in trouble. As brain functions evolve, might it be possible that we are conscious of more and of less?

A physicist's self is largely vast empty space with a few particles whirling around here and there. Yet, many of us feel that we have another self beyond the physicist's self. I am curious how many people here feel that they have only the physicist's self. A theoretical physicist far more distinguished than I once told me I should be ashamed of myself to doubt the physicist's self. Inside your head electrons whirl around nuclei in eternal silence, not even knowing advanced physics, merely Coulomb's law and the like. And that is it!, he asserted.

To forestall obvious criticism, let me say that I am of course well aware of the orthodox view of brain function and consciousness, that it is all governed by a complex network of neurons sending signals to each other, and that the actual firing can all be understood in excruciating detail in terms of known chemical processes. Nothing beyond chemistry needs to be invoked. I am also familiar with the possibility of emergent and collective phenomenon: The solidity and transparency of ice are certainly not evident from the Schrödinger equation describing the motion of electrons in the hydrogen and oxygen atoms. Consciousness may be a supreme example of an emergent phenomenon.

While it is difficult to define consciousness, certainly one hallmark of consciousness is the awareness of the passage of time.

What is the smallest unit of time t_{min} whose passage we as humans are conscious of? Subjectively, I would imagine that it is in the range of tenths of a second. It would be interesting if a precise figure can be determined experimentally and the range of variation between humans established.

In physics, we discuss many arrows of time, among which the cosmological and the thermodynamic arrows of time. It is difficult for me to imagine that the recession of the galaxies out "there" has something to do with what goes on in my brain. Or perhaps the cosmological arrow of time really refers to a universal relentless flow of time, somehow synchronized across all consciousnesses. The relentless rise of entropy no doubt manifests itself in the aging of our brains. But again, I find it difficult to imagine that the thermodynamic arrow of time has something to do with our awareness of the passage of time over the scale of one second.

But I would still like to think that the time reversal violation in the fundamental laws of physics has something to do with consciousness.

Instead, I would like to entertain the highly speculative possibility that consciousness, at least the part that has to do with the awareness of time, may have something to do with the violation of time reversal invariance at the level of the laws of physics. I can't imagine that the K-meson could be involved; so that leaves the electron and its electric dipole moment. I am aware that the time and energy scales involved may be way off, so this may not make any sense at all. But I throw it out as a speculation for you to rule out.

To make the speculation quantitative, perhaps we can turn a value for the electric dipole moment of the electron into an estimate of t_{min}. I have in mind asking questions along the following line. In the weak electric field in our brain, how many times must the electron precess and how many electrons must precess together in some sort of coherent phenomenon before a definitive mental arrow of time can be established?

I hope that this wild speculation at least serves to underline the potential importance of measuring the electric dipole moment of the electron.

Note added in proof:

During a recent trip to China I learned that it is at 70 that one can do what one's heart desires, not at 60. As befitting a good theorist, I committed two errors, but in the same direction, so that indeed, to first approximation, Gene can follow his bliss as he is almost 70. But alas, it is still too early for me. I thank Professor H. T. Nieh for his hospitality at the Center for Advanced Study, Tsinghua University, Beijing.

ASTROPHYSICS AND COSMOLOGY

Supernovae, Dark Energy, And The Accelerating Universe: What Next?

Saul Perlmutter

Lawrence Berkeley National Laboratory
1 Cyclotron Rd, MS 50-232
Berkeley, CA 94720

Abstract. The recent supernova measurements of cosmological acceleration, together with the CMB and galaxy cluster measurements, raise new important questions: Do we have a reliable, consistent picture of cosmology? What is the identity of the previously unknown energy that is causing the acceleration? To address these questions, we are designing a new supernova experiment using a satellite with a telescope (called "SNAP," for SuperNova / Acceleration Probe). The goal of this project is to measure the expansion history of the universe at a 1%—2% level of precision. The various theories of the accelerating energy predict different histories of decelerations and accelerations in the universe's expansion.

INTRODUCTION

It is a genuine pleasure to be here to speak for Gene's celebration. I do want to comment that I feel a little bit awkward speaking at this Festschrift for a couple of reasons. First, I may be one of the few speakers here who has never had a chance either to take a course or work with Gene yet -- although he would have been my first choice, as he was for every other physics graduate student that I knew at Berkeley.

The only saving grace is that I may soon have the chance to correct this oversight, since Gene has said he may like to work with us on some of the projects I'll be describing today. So this is my excuse for presenting a somewhat unusual Festschrift contribution, in that it is not retrospective—it's not describing some project already completed. Instead I will present a future project that we hope will happen. Perhaps this is actually a fitting Festschrift contribution, since the best reason to retire from active teaching may be to pursue fun new projects.

The other reason I feel awkward presenting here is that unlike most people in the room who have been working in areas like Gene's, where what you're known for is the elegance and precision of the experiment, I'll be talking about a project and field which is known for its inaccuracy of measurement. Cosmology has been notorious, not just in this century, but for millennia, for how rough the measurements have been. But once again, the saving grace is that what I'll be describing is an attempt to move in a direction that looks a little bit more like precision measurement—and a way to address some really fun physics questions.

These questions involve the expansion of the universe, its possible fate, and also the spatial extent of the universe. Let me briefly remind you how it is that you go about obtaining answers to these questions. Figure 1 shows my one-plot explanation.

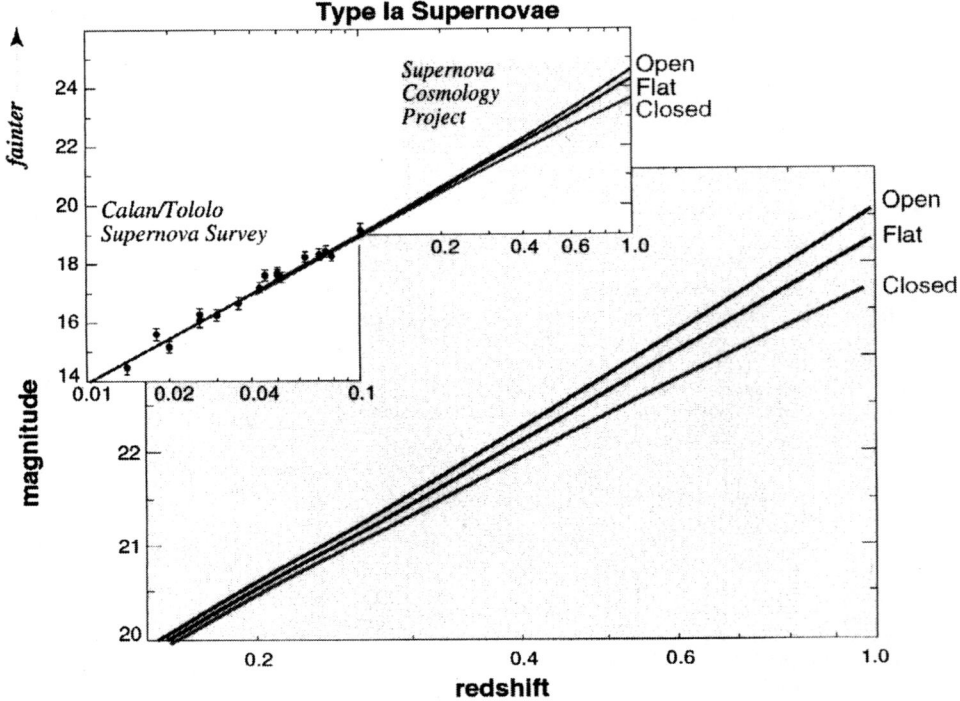

FIGURE 1. The astronomers' Hubble diagram for type Ia supernovae can be thought of as a very simple plot of the stretch of the universe as a function of time

The basic concept is that you draw the astronomers' Hubble diagram of brightness ("magnitude") against redshift. This is a plot on two axes that can be understood very easily. One is the apparent brightness, which tells you how far the object is—if you are using objects of known brightness. In this case, we are plotting type Ia supernovae, whose peak brightness can be calibrated to within about 12%. The fainter they are, the farther away they are and the farther back in time you are "seeing." Thus the magnitude axis is the time axis. The redshift is a very direct measure of the total expansion of the universe since the time that the light left the supernova. That's because, as the photons are traveling to us, their wavelength is being stretched in exact proportion to the stretch of universe. So you can think of this plot, then, as a very simple measure of the stretch of the universe as a function of time.

Now you can look for the deviations that you might expect at the faint/far side of Figure 1, due to, for example, the mass-density of the universe slowing its expansion. This is what we first set out to do some twelve years ago: we thought we were trying to find out how much the universe was slowing down. This measurement would tell us whether the universe was spatially "open" or "flat" (in both cases, a spatially infinite universe), or "closed" (a spatially finite universe). It seemed also that this measurement could predict the fate of the universe, whether it would last forever or someday come to an end. We were very excited to see what the results would be among these options.

And as you probably know, the result came out to be None of the Above.

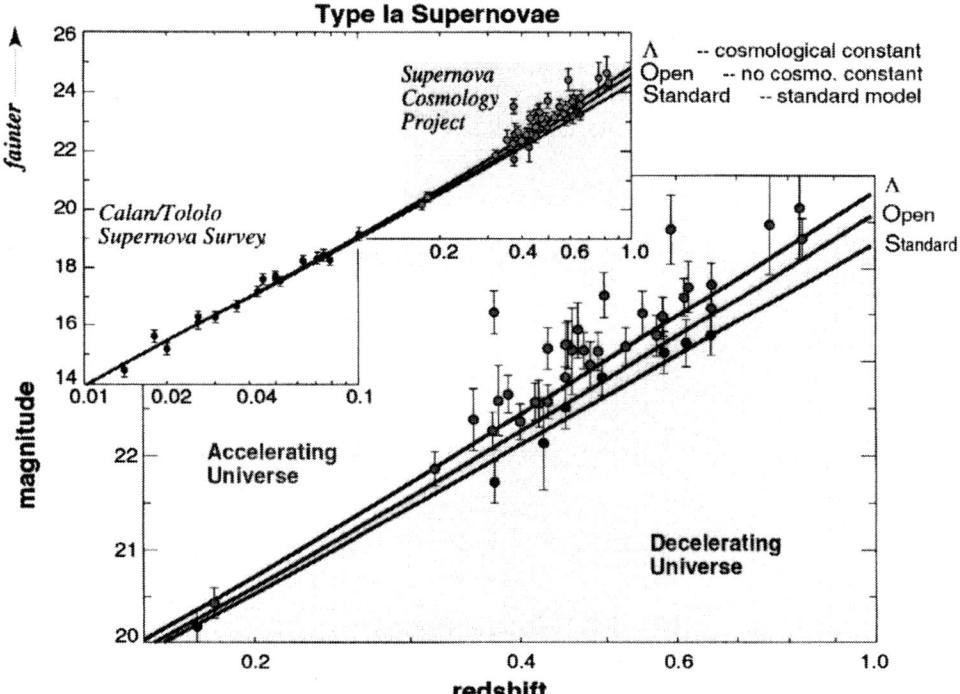

FIGURE 2. The type Ia supernova Hubble diagram after the 42 Supernova Cosmology Project high-redshift supernovae were added in 1998 [1, 2]. See also [3] for the high-redshift supernovae from the High-Z Supernova Search.

As shown in Figure 2, the high-redshift supernova data was consistent with neither open, flat, nor closed, but rather it lay in the yellow "accelerating" region of the Hubble diagram that we hadn't even considered as a possibility. It appears that we live in a universe that is actually speeding up in its expansion at this point. Figure 3 shows the corresponding confidence region on a plot of the mass density that could lead to deceleration versus the vacuum energy density that would potentially accelerate the expansion.

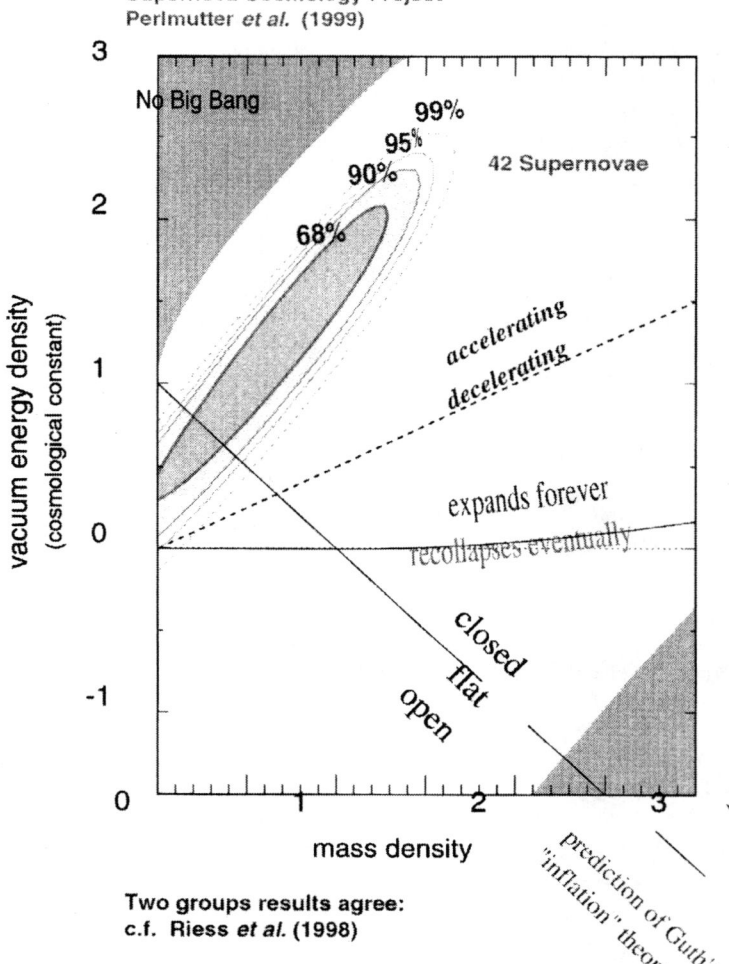

FIGURE 3. The confidence region in the mass density—vacuum energy density plane, for the Supernova Cosmology Project supernovae shown in Figure 2 [1, 2]. A very similar confidence region was also obtained by the High-Z Supernova Search team [3].

Figure 3 clearly shows that there is strong evidence for a vacuum energy of some sort that leads to acceleration dominating over the mass density's deceleration. However, you cannot tell from this supernova-based confidence region whether or not we live in a closed, flat, or open universe. Now that you have two parameters in the problem, mass density alone no longer determines what the spatial curvature of the universe is.

The spatial curvature is, however, very well measured by the cosmic microwave background (CMB). People here in Berkeley have also been very active in this measurement (from a year ago) shown as the green confidence region in Figure 4.

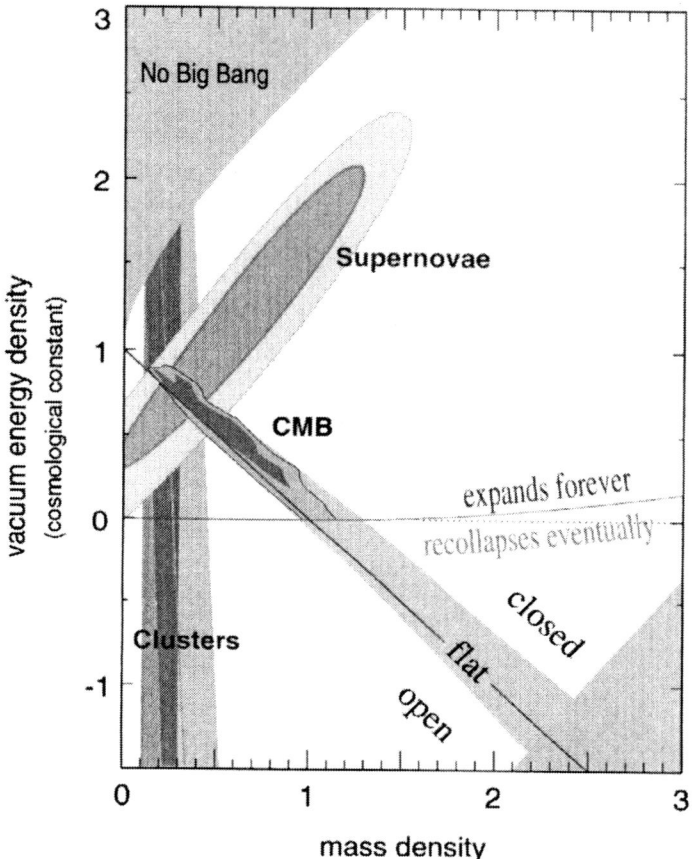

FIGURE 4. Complementary measurements of the mass density and vacuum energy density of the universe. The CMB confidence region is from [4], and the galaxy cluster confidence regions include the range discussed in [5].

The current CMB results are even a little bit tighter than this, but it is clear that the cosmic microwave background data indicates that we live in a universe that is very close to being spatially flat.

TWO QUESTIONS

Figure 4 highlights the two major questions for the future that I want to discuss. The first one is: how close are we to a consistent story for cosmology? Looking at Figure 4, it appears that the data from the clusters of galaxies, from the supernovae, and from the CMB all overlap. So one might think that we're done; We have a consistent picture of cosmology—we can now go home. But it is important to remember that these are some of the very first measurements we've made in cosmology and that it's a very young field. This is one of the simplest models that one can imagine for a cosmological picture and, in some sense, we may just be lucky that when we have these big sweeping error bars from these different measurements they happen to overlap. I believe that the first order of business is to ask: As we begin to tighten all these error bars and obtain more precise measurements, will we then find that we have a consistent picture of cosmology?

The supernova results are mostly dependent upon data from the most recent part of history, primarily from redshifts around 0.5, within the last 5 billion years or so. The CMB data is collected from right after the big bang. I think it would be fascinating to find out whether they agree with each other when you go back all the way from 5 to 15 billion years in time. If we improve these two measurements' uncertainties will their confidence regions on Figure 4 still be consistent with each other? That is Question Number One.

Question Number Two may be even more fascinating: What is the identity of the vacuum energy density that could be causing the acceleration of the universe? This is an entity that Einstein actually had in one of his equations, the cosmological constant, and you could interpret it as simply the vacuum energy density that you get from the zero points of all the particles and fields from quantum mechanics. But there are some serious problems with that identity. If it really is just a zero point of all the particles and fields in ordinary quantum mechanical analysis, then you could make some naive calculation of what you might expect the results to be. That has been done, and the answer that you come up with is off by a factor of 10^{120}. I think it was Mike Turner who said that this is probably the worst prediction in all of physics. So it's clear that this is not something we understand at this stage. It's a terrible hierarchy problem in our model.

Then you may ask: Why is this current vacuum energy density today just between a factor of 2 and 3 of the mass energy density? The mass density has been falling by the third power of the scale of the universe—so it has been dropping by orders of magnitude over the history of the universe—and the vacuum energy density has been staying constant, by definition. So you could ask: why would those be so close to each other today? It's a coincidence problem that is very troubling to the theorists, and these two problems have triggered a whole industry of new theories as people try to identify what alternatives there might be to a simple zero-point vacuum-energy explanation about the acceleration of the universe. You'll hear terms like "dark energy" describing these theories, generically; and "quintessence" is one of the terms that has most often been used to describe the dynamical scalar-field models in particular.

Often these dark energy theories are characterized by the relationship between the pressure and the density in the acceleration equation from Einstein's general relativity, and so one of the measurements we'll be making for the dark energy is of its equation-of-state variable w, which is just the ratio of its pressure to its density. This is something we can actually measure with the supernovae; Figure 5 shows the first measurement that we have made.

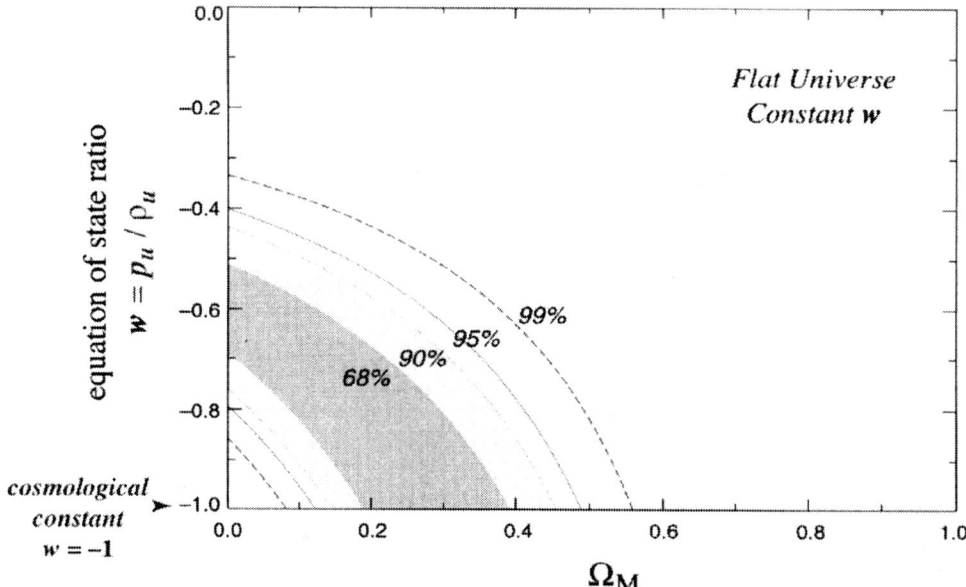

FIGURE 5. Confidence region in the plane of the dark energy's equation-of-state ratio, w, versus the mass density of the universe (in a flat cosmology). The equation-of-state variable, w, is the ratio of the dark energy's pressure to its density. This confidence region is from the Supernova Cosmology Project's results [2]; it is closely matched by the similar plot from the High-Z Supernova Search [6].

From Figure 5 it is clear that the confidence region still includes a fairly wide range of possible values of this equation-of-state ratio w. We cannot yet differentiate between a dark energy consistent with $w = -1$ (the cosmological constant's equation-of-state ratio) and a dark energy consistent with $w = -0.8$—as favored by one of the quintessence models today.

A NEW APPROACH TO THESE QUESTIONS

How do we want to address these two big questions? We're currently engaged in a large number of dark energy studies with supernovae using existing facilities, and we're constructing some new facilities that we expect to be using over the next few years. But the bottom line is that to obtain the level of precision that can really make a dent in this kind of problem we're going to have to push to a new sort of facility—and that is what I will describe to you today.

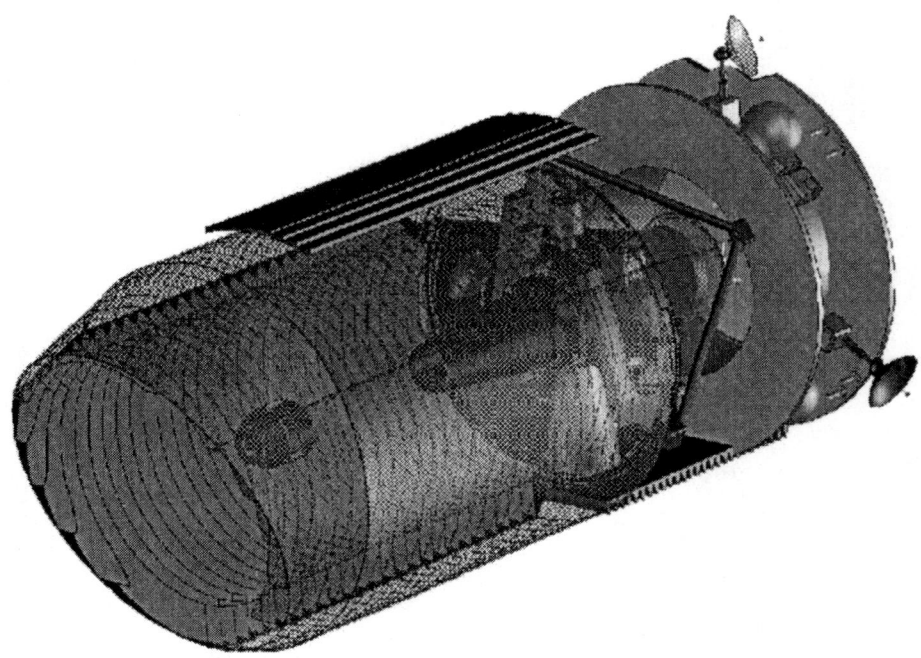

FIGURE 6. The SuperNova / Acceleration Probe (SNAP) concept, a satellite with a two-meter class telescope [7].

The concept we've been working with is a new satellite experiment, as shown in Figure 6. The aim is to make a very simple, elegant experiment that has just a telescope with a two-meter class mirror and a couple of instruments. One of instruments, however, is a fairly dramatic step forward. It's a one-square-degree mosaic camera that would have a billion pixels; it would be the largest mosaic camera being used in astronomy. Now it happens that, if you look at it from the point of view of a particle physicist, this is no more than the amount of silicon that you tile around detectors in a beam line. So the particle physicists are not daunted by an instrument of this scale, although it would be a big step forward for astronomy.

The spectrograph we want to use would be a very simple instrument by these standards. It would be very low-resolution in wavelength, but very high throughput from near UV all the way out to near IR. This would allow detailed analysis of very large numbers of supernovae (which have broad spectral features, so they can be studied with low-resolution spectra).

The satellite experiment can be kept simple by requiring as few moving parts as possible, and by designing a mission in which we observe one region of the sky over and over again. The consequence of building such a satellite-telescope would be that the Hubble diagram I showed you (in Figure 2) with some 42 supernovae in it would turn into something that looks more like Figure 7, with some thousands of supernovae populating this entire curve.

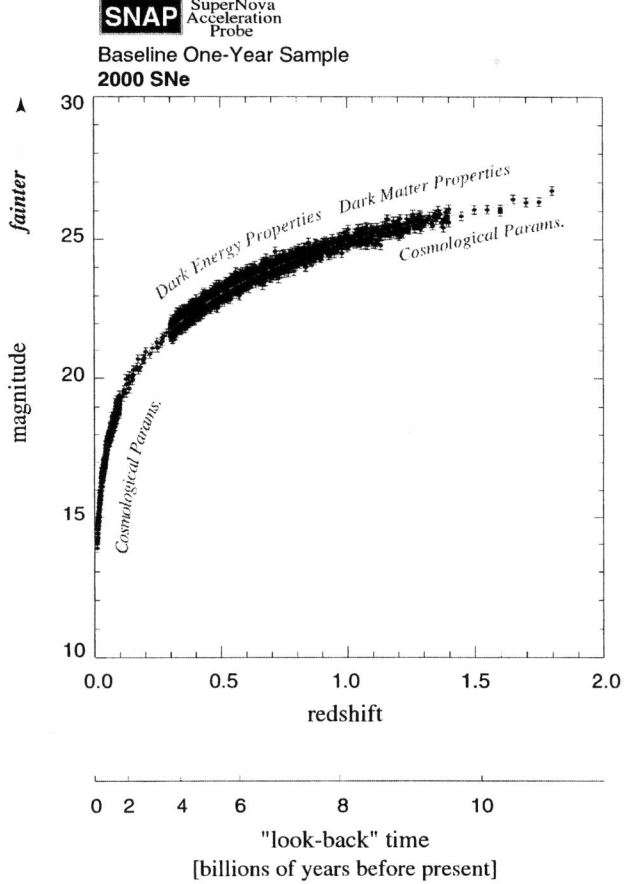

FIGURE 7. An example of a type Ia supernova Hubble diagram that could be obtained using the SNAP satellite [7]. This particular simulated data set emphasizes the redshift range from 0.3 to 1.4, but other mission designs would distribute the supernovae differently in redshift.

The range of redshifts would be greatly increased. Instead of studying supernovae out to about a redshift of 0.5 or 0.6 as in Figure 2, we would pursue supernovae to redshifts of about 1.7. This very large lever arm gives you a good handle on the cosmological parameters, including the spatial curvature of universe. The expansion of the universe between redshifts of about 0.3 and 1.3 is particularly sensitive to the dark energy properties. At still higher redshifts, when the universe's mass was more dense, the dark matter strongly predominated over the dark energy. This highest redshift range will allow studies of the properties of the dark matter, while its redshift leverage also greatly improves the measurements of the dark energy properties.

With this SNAP data, the current constraints on mass density vs. vacuum energy density that were shown in Figure 3 can be dramatically improved. Figure 8 shows this target uncertainty. It will be a very dramatic step forward in precision for this field if we can reach this goal.

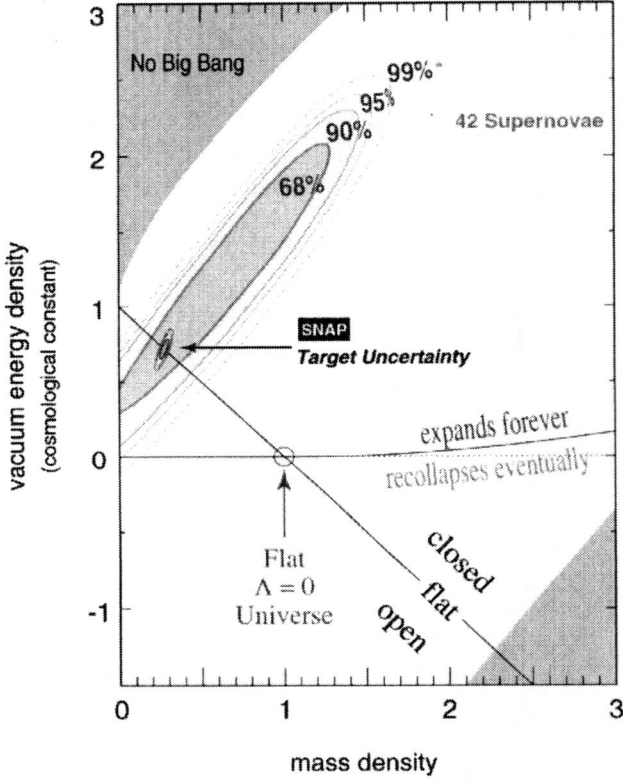

FIGURE 8. The targeted confidence region in the mass density—vacuum energy density plane that could be derived from the supernovae to be obtained by the SNAP satellite (see Figure 7 for a simulated example of such a dataset).

Of course, Figure 8 shows the result you might expect if our current cosmology theory turns out to be essentially right. If so, we can expect that the SNAP supernova data will show a spatially flat universe, consistent the current CMB data I showed you in Figure 4. However, it is conceivable that the SNAP supernova data will give a result that is just a little bit off from a flat universe, as in Figure 9. That, of course, would be a really fascinating answer to our Question Number One. In fact, it would be a huge surprise. Of course, the accelerating universe was a big surprise; clearly, our cosmological picture is not yet a finished, mature theory. It is crucial for us to test its consistency in all possible ways.

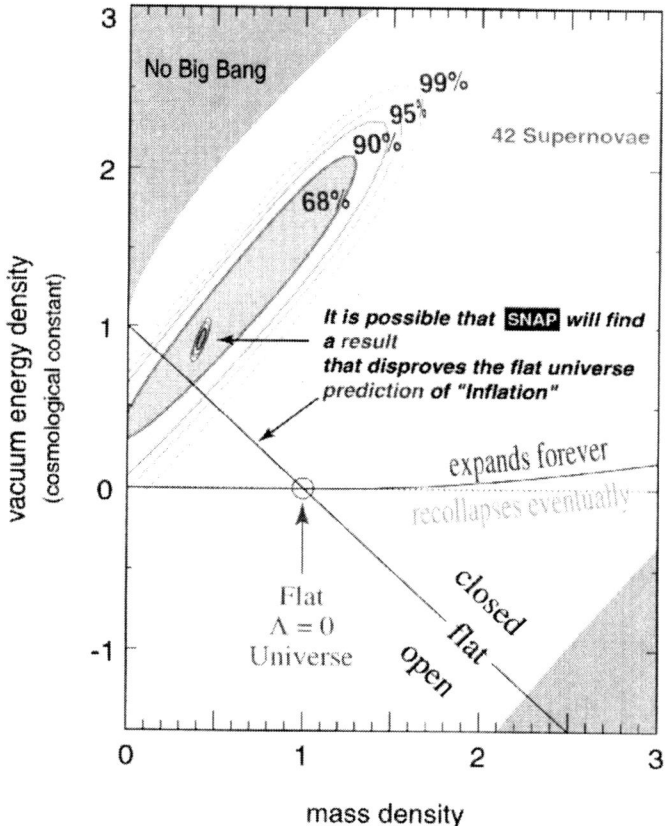

FIGURE 9. One conceivable—if very unlikely—scenario in which the SNAP supernova data shows a positive spatial curvature in the universe. This would indicate a significant problem in our cosmological theory, since it would disagree with the current CMB data (shown in Figure 4), which indicates that the universe is spatially flat.

STUDYING THE DARK ENERGY

The key goal, however, for SNAP is to begin to pin down the identity of that dark energy—what exactly is it? Figure 10 shows what we might expect from a SNAP measurement of the equation-of-state ratio w if it turns out to be consistent with the cosmological constant, allowing us to rule out a range of quintessence tracker models [8] that have been proposed.

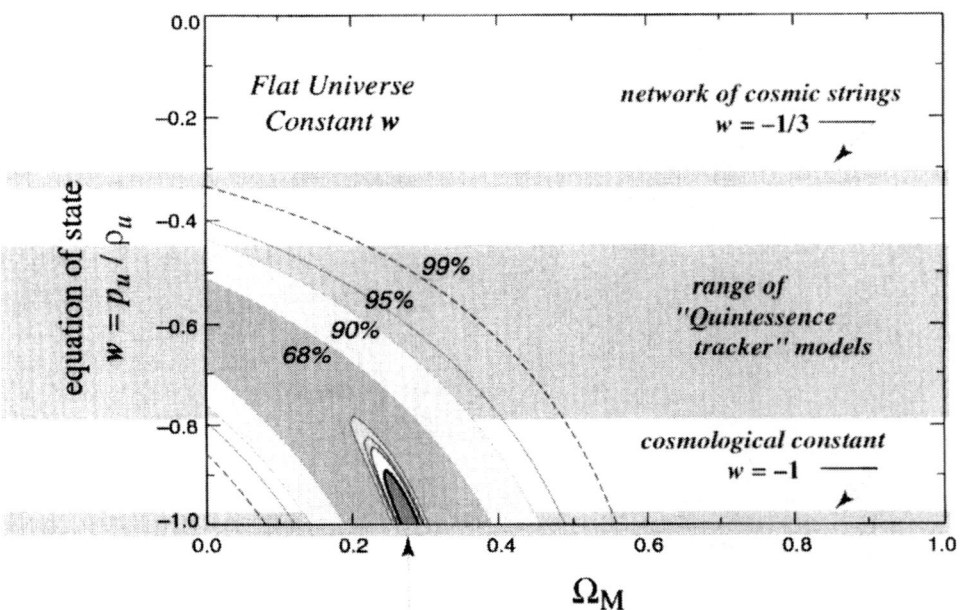

**SNAP Satellite
Target Uncertainty**

FIGURE 10. The targeted confidence region in the plane of equation-of-state ratio w versus mass density (in a flat universe) that could be derived from the supernovae to be obtained by the SNAP satellite (see Figure 7 for a simulated example of such a dataset).

It is perhaps even more interesting to look at this from the point of view of the wide range of dark energy models that have already appeared in the literature and see how they might be distinguished by the SNAP supernova data [9, 10]. Figure 11 shows a simulation of what the data might look like on a plot of the magnitude difference from our current favorite flat $\Omega_\Lambda = 0.7$ model, as a function of redshift. The data points labeled "Calan/Tololo" and "SCP" show you what the 1998 data looked like. We expect this data to improve, of course; more recent observations of supernovae have obtained somewhat smaller error bars. However, we will not be able to approach the statistical—and systematic—uncertainties that SNAP will obtain, as shown by the points labeled "SNAP". Each of these points represents some fifty to sixty supernovae binned together. At this level of binning these statistical error bars are at the systematics level that we are targeting with SNAP. With this quality of data we will begin to differentiate between the various solid lines and dashed lines of Figure 11, which show the range of dark energy models that were out in the literature just about nine months ago when Weller and Albrecht did this compilation.

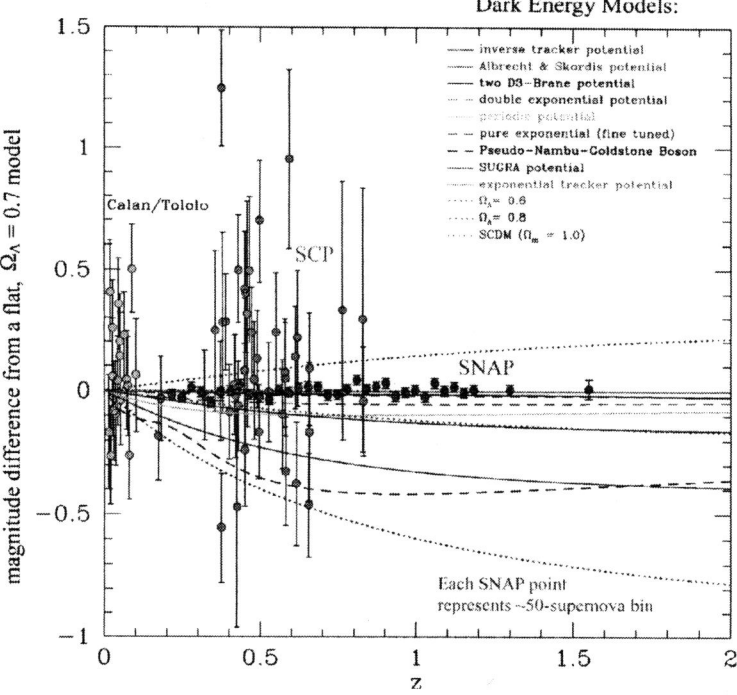

FIGURE 11. Examples of current ground-based data ("Calan/Tololo" and "SCP" points) compared with binned, simulated SNAP data on a plot of the magnitude difference from our current favorite flat $\Omega_\Lambda = 0.7$ model, as a function of redshift (based on Weller & Albrecht [9, 10]). The solid and dashed lines show a range of models of dark energy drawn from the literature.

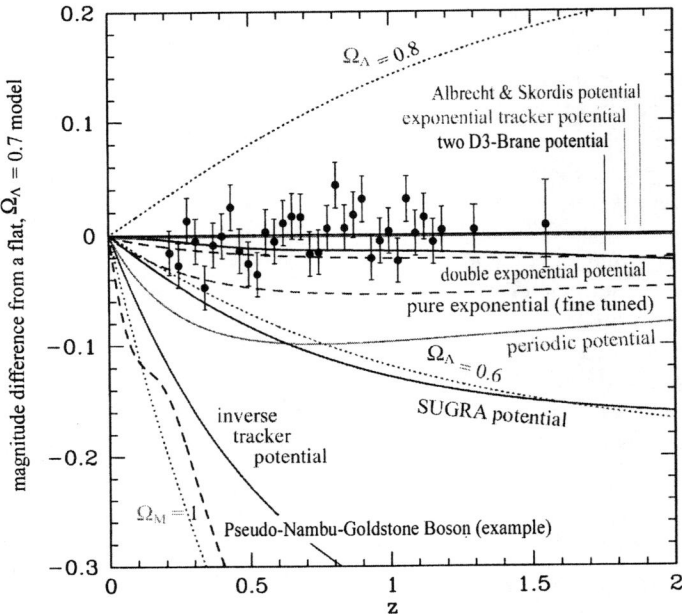

Figure 12. Binned, simulated SNAP data (red points) on a plot of the magnitude difference from our current favorite flat $\Omega_\Lambda = 0.7$ model, as a function of redshift (based on Weller & Albrecht [9, 10]). Each of the data points represents some fifty to sixty supernovae binned together, so that the statistical error bars are the same size as the targeted systematics error bars (not shown separately). The solid and dashed lines show a range of models of dark energy drawn from the literature.

The blow-up of this plot is shown in Figure 12. You can see here that some of the models cross each other as a function of redshift, for example, the supergravity model (labeled "SUGRA"), and the periodic potential dynamic scale field model. If you were to measure the equation of state only at the redshifts where we have most of our current data (i.e. $z < \sim 0.6$, as shown in Figure 11), you would not be able to distinguish these sorts of models. It is very important to have a large range of redshifts to take you out to the era in which the mass density dominates, in order to distinguish between dark energy models.

A SYSTEMATICS-LIMITED EXPERIMENT

I also want to emphasize that this is a systematics-limited experiment. The statistical error bars on the red SNAP points in Figures 11 and 12 are as small as would be useful if the systematic uncertainties are kept below an ambitious goal of 2%. This is a level of systematic uncertainties that has never before been achieved in this kind of work. This control of systematic uncertainties is the heart of the SNAP experiment design.

Score Card of Current Uncertainties
on $(\Omega_M^{flat}, \Omega_\Lambda^{flat}) = (0.28, 0.72)$

Statistical
- ☑ high-redshift SNe   0.05
- ☑ low-redshift SNe   0.065
- **Total**   **0.085**

Systematic
- ☑ dust that reddens   < 0.03
  $R_B(z=0.5) < 2 R_B(\text{today})$

- ⁇ evolving grey dust
  - ⁇ clumpy
  - ⁇ same for each SN

- ☑ Malmquist bias difference < 0.04

- ⁇ SN Ia evolution
  shifting distribution of
  prog mass/metallicity/C-O/..

- ☑ K-correction uncertainty   < 0.025
  including zero-points

- **Total**   **0.05**
  identified entities/processes

Cross-Checks of sensitivity to
- ☑ Width-Luminosity Relation < 0.03
- ☑ Non-SN Ia contamination   < 0.05
- ☑ Galactic Extinction Model < 0.04
- ☑ Gravitational Lensing   < 0.06
  by clumped mass

FIGURE 13. An accounting of the primary sources of uncertainties, both statistical and systematic, for the supernova measurement of the cosmological parameters Ω_M and Ω_Λ in a flat universe, based on Table 4 of Perlmutter et al. [2] and the detailed discussion of uncertainties in Sections 4 and 5 of that paper.

Figure 13 is a "score card" showing where the uncertainties are primarily coming from in today's supernova measurements of the cosmological parameters (in this case, the mass density and vacuum energy density in a flat universe). You can see that the statistical uncertainties are already comparable to the total systematic uncertainties that are listed. You cannot improve the statistical uncertainty by much more than an extra factor of two before hitting the systematics limit. To make any further progress, the systematic uncertainties have to be reduced.

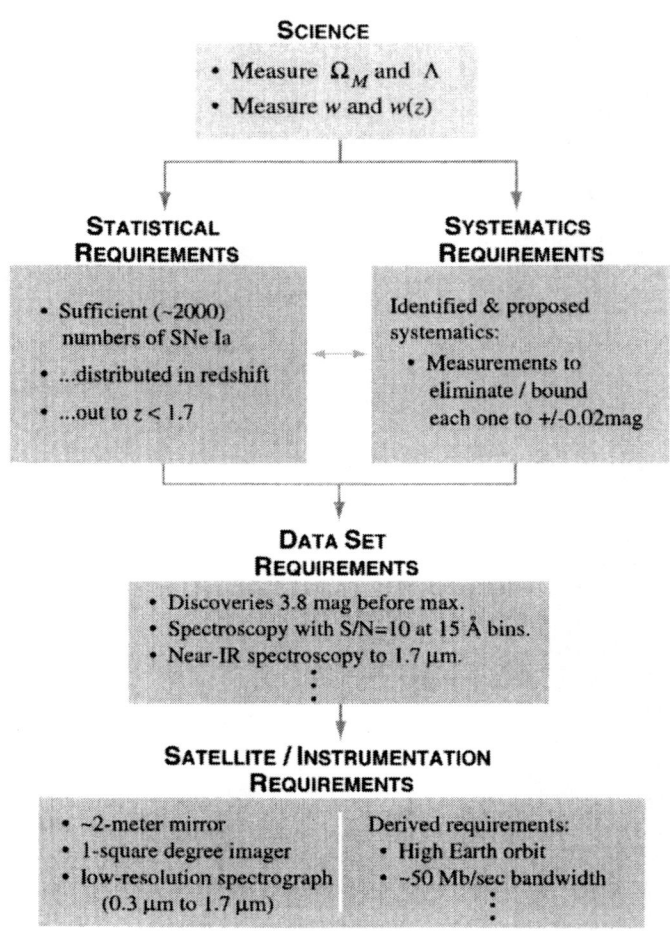

FIGURE 14. The design strategy for the SNAP experiment, a flow-down from the science goals to the specific instrumentation choices.

The set of systematic uncertainty requirements is therefore a key driver of the design of the whole SNAP experiment, in order to achieve our science goals. Figure 14 shows this "flow down" from the science goals to the SNAP design. We have science requirements and goals that we are targeting, measurements of curvature and the mass density and vacuum energy density of the universe. The most ambitious science goal is the measurement of the dark energy's equation of state as a function of redshift. The science goals can all be addressed by specifying the statistical requirements and the systematic requirements, which, in turn, drive the choice of data sets and then the satellite design.

I won't have time to go into the details of the baseline design—such as the interesting choice of orbit that allows the necessary high-bandwidth telemetry and an extremely stable thermal and optical system. However, I will briefly show a simulation of the kind of data we'd like to be able to get from the imaging camera and the spectrograph.

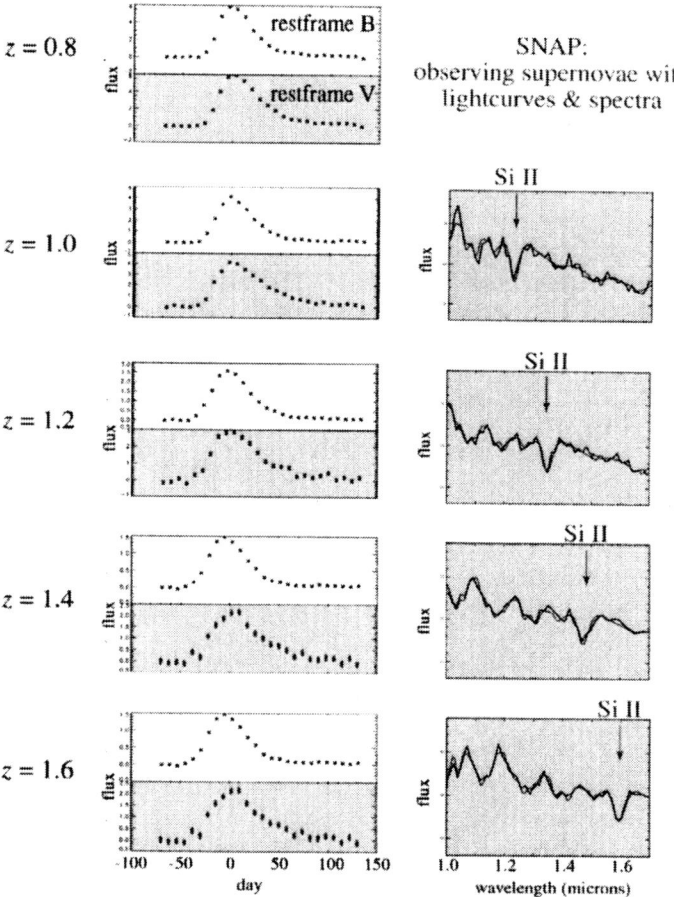

FIGURE 15. Simulated supernova photometry and spectroscopy data from the SNAP satellite, including effects of instrumental throughput and noise [7]. The red spectral curves show an idealized template type Ia supernova spectrum, redshifted to the appropriate wavelengths, while the black spectral curves show the simulated SNAP data for comparison. The Si II absorption is the characteristic identifying feature of the type Ia supernova spectrum.

Figure 15 shows that for the full range of redshifts, SNAP would be able to obtain detailed measurements of the supernova lightcurves in two filters. These wavelengths are where the supernovae emit most of their light, and you can see that the uncertainties here are at the 2 and 1% level, required to be able to recognize the extinction of the light by intervening dust. Every one of these supernovae also has a spectrum; Figure 15 shows the quality of the spectra (the black lines shows the simulation and the red lines show the template spectrum of a perfectly measured Type Ia supernova at that redshift). These simulated spectra are good enough that you can recognize all the primary features found in the template spectrum. The Si II feature that is labeled on the spectra is the hallmark of the Type Ia supernovae that we are using.

It is by matching the details of such spectral features, including their widths and amplitudes, and the details of the lightcurves, including their timescale, that we can recognize which subsets of type Ia supernovae are the same brightness as which others. We expect the population of Type Ia supernovae to shift somewhat among these Type Ia subsets as we look back in redshift, so that we find more or fewer matches to a given characteristic supernova lightcurve and spectrum. This will not effect our cosmological measurements, however, as long as we only compare "like with like" supernovae. We expect to be able to plot a series of Hubble diagrams like Figure 12, each with supernovae from just one Type-Ia-supernova subset. Each of these Hubble diagrams will yield independent cosmological measurements, which can then be compared and combined. This control over supernova "evolution" effects is one of those systematics drivers that is at the heart of this experiment as well.

OPTIMISM FOR PRECISION COSMOLOGY

I believe that the goal of this kind of precision cosmology is not outside our grasp. It is an exciting prospect, as we move through the different levels of precision outlined in Figure 16. We have already advanced from the earlier era of 20th century cosmology, when you might be happy if you ended up with a measurement uncertainty of 50%. The Hubble constant measurements, for example, have only within the last decade improved to point that their uncertainty is at the ~10% level. To begin to address Question Number Two, the identity of the dark energy, we must move down into the precision range that the CMB work is currently focusing on, the 1-2% level. Of course, we'd like to have a couple of these measurements that we can compare to each other.

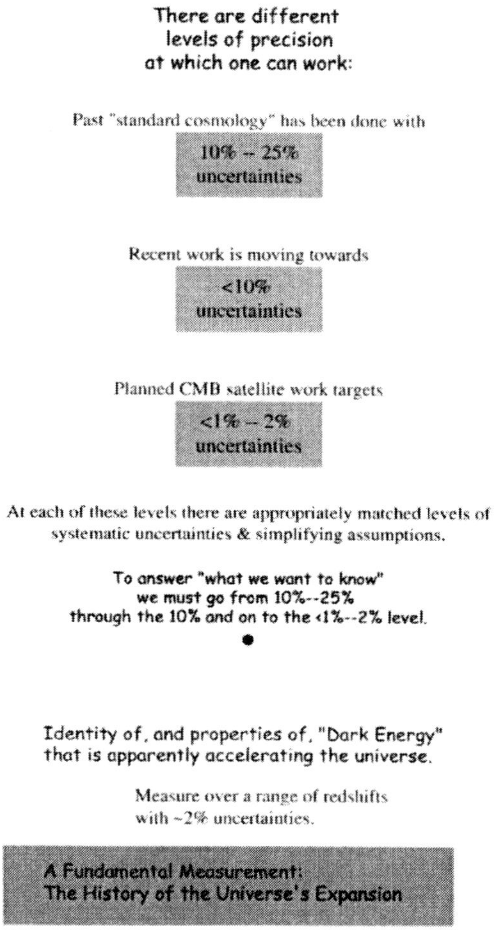

There are different
levels of precision
at which one can work:

Past "standard cosmology" has been done with

10% – 25%
uncertainties

Recent work is moving towards

<10%
uncertainties

Planned CMB satellite work targets

<1% – 2%
uncertainties

At each of these levels there are appropriately matched levels of
systematic uncertainties & simplifying assumptions.

To answer "what we want to know"
we must go from 10%--25%
through the 10% and on to the <1%--2% level.

•

Identity of, and properties of, "Dark Energy"
that is apparently accelerating the universe.

Measure over a range of redshifts
with ~2% uncertainties.

A Fundamental Measurement:
The History of the Universe's Expansion

FIGURE 16. Different levels of measurement uncertainty in cosmological measurements. It is interesting to note that these are usually dominated by the systematic uncertainty, not the statistical uncertainty.

With a little bit of luck, that kind of measurement can allow us to make what seems like a particularly fundamental experimental measurement, which is a study of the history of the expansion of the universe—the blow-up plot of Figure 17. In this history, we can look for the little differences of the decelerations and the accelerations over the past 15 billion years; the small wiggles and bumps of such decelerations and accelerations are the hallmarks of the different theories of dark energy. That is the task that we have set ourselves with this experiment. I am very optimistic about being able to move into a new level of systematics-controlled cosmology, if we can have someone like Gene on our side.

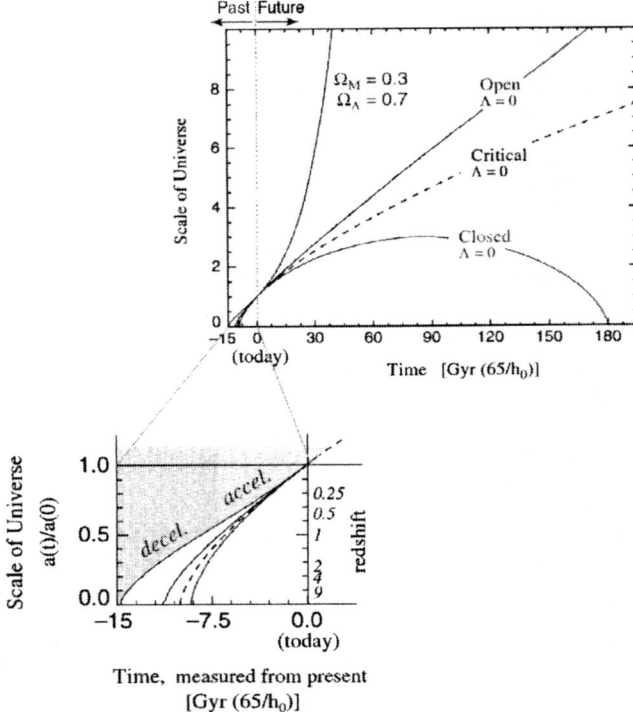

FIGURE 17. The expansion history of the universe. If we can measure the details of deceleration and acceleration in the past we can begin to learn about the identity of the dark energy that now apparently dominates the energy densities of the universe.

I want to finish up with one last philosophical comment. Although I've never actually had the chance to work with Gene, I did have a chance to play chamber music with him during my first year here at Berkeley as a grad student. Figure 18 is not actually the piece that we played, however it is one of my favorite Hayden quartets, and its nickname, the "Quinten" is quite appropriate for the current topic. I want to draw the analogy between the community of science and the work of scientists, and the act of playing chamber music, which struck me as I was preparing for this talk. While you are playing chamber music, one of the real pleasures is being able to listen and pick up on the styles of what you're hearing across the room from your partners in the music. Your music then begins to reflect what they're doing. Similarly, even though I haven't yet had a chance to work with Gene, and in fact many scientists in the field have not directly worked with him, I think most of us have picked up on the music that is being played across the room in Gene's side of the science. In this way, our work is reflective of what he has been doing—and is certainly the better for it! Thank you, Gene.

FIGURE 18. The first few bars of Haydn's "Quinten" string quartet.m

ACKNOWLEDGMENTS

It is a pleasure to thank all my colleagues in both the Supernova Cosmology Project and in the SNAP collaboration for years of exciting teamwork. The work discussed here was supported in part by the Director, Office of Science, of the U.S. Department of Energy under Contract No. DE-AC03-76SF00098; and by NASA through grant number HST-GO-07336 from the Space Telescope Science Institute, which is operated by AURA, Inc., under NASA contract NAS5-26555.

REFERENCES

1. Perlmutter, S., et al., In *Presentation at the January 1998 Meeting of the American Astronomical Society, Washington, D.C.,* Report No. LBL-42230, astro-ph/9812473 (1998)
2. Perlmutter, S., et al., *ApJ,* **517**, 565-586 (1999).
3. Riess, A. G., et al., *AJ,* **116**, 1009-1038 (1998).
4. Jaffe, A. H., et al., *Phys. Rev. Lett.,* **86**, 3475-3479 (2001).
5. Bahcall, N. A., Ostriker, J. P., Perlmutter, and S. Steinhardt, P. J., *Science,* **284**, 1481-1488 (1999).
6. Garnavich, P. M., et al., *ApJ,* **509**, 74-79 (1998).
7. See http://snap.lbl.gov
8. Steinhardt, P. J., Wang, L., and Zlatev, I., *Phys. Rev. D,* **59**, 123504 (1999).
9. Weller, J., and Albrecht, A., *Phys. Rev. Lett.,* **86,** 1939-1942 (2001).
10. Weller, J., and Albrecht, A., Report No. DAMTP-2001-53, astro-ph/0106079 (2001).

Mid-infrared Stellar Interferometry and Diameters of Old Stars

C. H. Townes

University of California at Berkeley

Abstract. Modern technology, including laser interferometry and sensitive high speed detectors, have allowed recent developments in "stellar" interferometry at visible and infrared wavelengths. Three movable 1.65 meter telescopes are now installed on Mt. Wilson for interferometry in the 10 micron wavelength region using heterodyne detection with CO_2 laser local oscillators. They allow imaging of dust surrounding stars by use of phase closure, and high precision measurement of stars surrounded by dust and gas, since mid-IR wavelengths penetrate the dust well and narrow spectral regions can be chosen where there are no spectral lines due to surrounding gas. Results show these older stars are somewhat larger than indicated by previous measurements due to limb darkening effects suffered by measurements at shorter wavelengths and broad spectral bandwidths.

INTRODUCTION

Professor Eugene Commins has been strongly interested not only in physics on a miniscule atomic scale, but also on a large scale, as I learned when I attended an excellent and perceptive course he gave on stellar characteristics and structures. Hence, it is perhaps appropriate for me to discuss my present work on stellar sizes at this conference honoring his many contributions to science and to his colleagues.

Michelson and Pease made the first measurement of the size of a star in 1921, using stellar interferometry on Mt. Wilson. They found α Orionis, or Betelgeuse, to have a diameter of 48 milliarcseconds with a precision of about 10%, using visible light, the 100 inch telescope mount, and the human eye as a detector. Michelson pointed out that limb darkening, due to absorption by material surrounding the star, might make it appear somewhat smaller than its actual size, but did not give an estimate of the magnitude of this effect. Pease continued this type of work, building an interferometer of larger baseline, but the work was technically very difficult and he never published any further journal articles as a result of this effort. The project stopped in the early 1930's.

Modern technology, including new and efficient detectors, lasers which can easily control or determine distances to high precision, and versatile computers have made stellar interferometry increasingly practical and there are now a number of places throughout the world where successful stellar interferometry is being carried out.

Our group at Berkeley has built and is operating a unique type of stellar interferometer, the ISI (Infrared Stellar Interferometer), which operates at wavelengths in the 10-micron region and uses heterodyne detection of the stellar radiation. It involves three movable telescopes of 1.65 meter apertures in trailers so that many different baselines can be obtained, and is located on Mt. Wilson, very close to the Michelson building where Pease and Michelson built their second interferometer. Figure 1 is a picture of two of the telescopes in their trailers on Mt. Wilson with the 100-inch telescope, where Michelson's first measurement of stellar size was made, behind a tree in the background. Figure 2 is a schematic of the system with two telescopes. Heterodyne detectors using CO_2 lasers as local oscillators convert the infrared radiation to the radio region, with a bandwidth of approximately 2.5×10^9 Hz. Both sidebands are used, which doubles the effective frequency range received. The two radio signals are brought together by waveguide transmission, with the waveguide length automatically varied in order to bring signals from two telescopes together at very close to the same total travel time from star to interference. A rather complete description of the system and its performance can be found in Hale et al (2000).

Measurements of Two Red Giants

The diameter of the red giant α Orionis has been now been measured with visible light a number of times with different interferometric systems. Figure 3 shows a plot of the results obtained prior to 1980 (White, 1980).

The various measurements are plotted against a small postulated periodic variation in the luminosity of α Ori, and show perhaps a characteristic precision of about 10%. The average diameter is close to the 48 mas which was Michelson and Pease's value, and these later results did not appreciably improve the precision they obtained. A precision of 1% or better can now be achieved with modern techniques.

The visibility of an interference measurement is defined as the ratio of the interfering part of stellar power received (which makes the intensity of the combined signals go up and down as the relative delay of the two signals change) to the square root of the product of power in both telescopes. Visibility is calibrated to be unity when the source is a point so that the two entire signals interfere. Visibility at various effective baselines can also be simply viewed as the Fourier spatial power spectrum of the object being observed, with the Fourier frequency determined by the effective baseline.

Figure 4 shows the square of the visibility at 11.15 microns wavelength for α Orionis over a certain spatial frequency range as recently measured by the ISI (Weiner et al, 2000). From this, the derived stellar diameter is 55 ± 0.6 mas. This is notable larger than previous measurements, partly due to limb darkening. The best recent measurement of α Ori's apparent diameter at optical wavelengths is 52.7 ± 1.3 mas (Tuthill, 1994). This is both more accurate and larger than earlier measurements shown in Figure 3.

FIGURE 1. Picture of two telescopes in their trailers on Mt. Wilson with the 100-inch telescope, where Michelson's first measurement of stellar size was made, in the background behind a tree. The telescopes are positioned in their shortest baseline, of 4 meters.

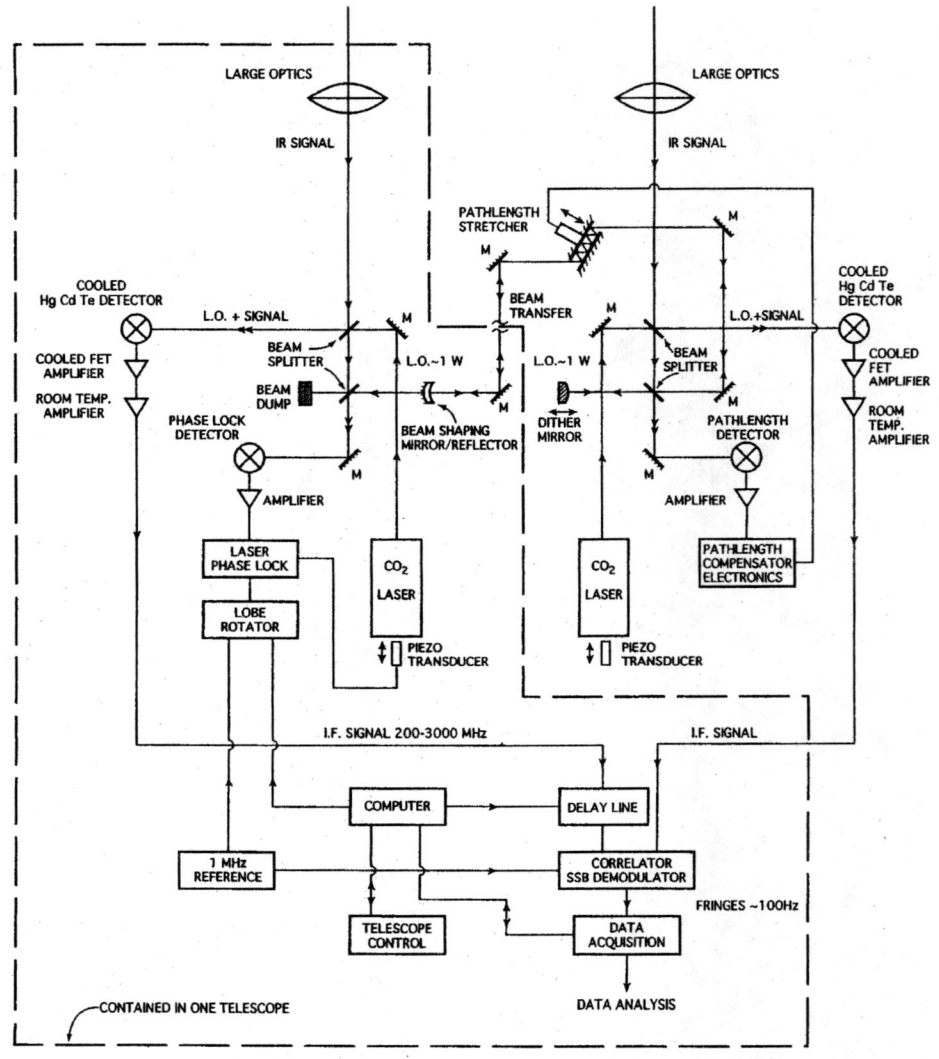

FIGURE 2. Conceptual block diagram of the ISI optics, circuitry, and heterodyne detection system, including laser phase lock and path length compensation systems. One telescope, with associated optics and circuitry, is represented by the equipment enclosed within the dashed line. A second telescope is represented by equipment outside the dashed line.

Theoretical estimates of reduction in apparent size by limb darkening are approximately 10% for visible light and less that 1% for wavelengths of 11 microns. This represents approximately the measured difference between the best recent measurements of α Ori at visible and 11 micron wavelengths, and is the first experimental determination of the actual amount of limb darkening. Allowing for the very small expected limb darkening at 11 microns, the size of α Ori is 56 ± 1 mas.

Another red giant, α Scorpii or Antares, has also been measured recently with the ISI. Its actual size is 48 ± 1 mas, again approximately 10% larger than the best visible light measurements of (White, 1980 and Tuthill, 1994).

The Size of Mira and Surprise

From the ISI measurements, the sizes of α Ori and α Sco are clearly larger than has generally been accepted but do not clash with previous expectations. The size of the Mira variable, o Ceti or Mira itself, as recently measured with the ISI, is on the other hand very surprising. A plot of the square of its visibility at 11.15 microns is shown in Figure 4. This yields a diameter of 46 ± 1 mas, which is much larger than the value near 35 mas which has generally been expected and which matches theoretical derivations of its oscillation period (330 days), assuming the lowest mode of oscillation. This large discrepancy requires careful examination of both our experiment and theoretical analysis of such stars.

Figure 5 provides an example of the very large variations in apparent size of stars with wavelength, which confuse any evaluation of actual size. The measurements were made by Tuthill et al (2000) using aperture masking on the 10 meter Keck telescope to obtain interference from many separate apertures. One of these stars, α Tau, which is not surrounded by much gas and dust and hence can be seen clearly over a wide range of wavelengths, shows a rather constant diameter as measured over a range of wavelengths. However others, and notably o Ceti, show large changes in apparent size with wavelength. This is because of the dust and gas surrounding them and the rather large spectral range used in each of the measurements which likely involve emission and absorption from molecular gas such as TiO and H_2O. The ISI is well adapted to avoiding these difficulties. First, its wavelength is much better transmitted through the dust surrounding stars than is visible or near IR radiation. Secondly, its bandwidth is very narrow, $\sim 5 \times 10^9$ Hz or 0.16 cm^{-1}, and it can use frequencies selected to avoid any known spectral lines. α Ori and o Ceti have H_2O line emission from surrounding gas and o Ceti is heavily surrounded by gases and dust. John Lacy of the University of Texas has measured the spectra of these stars in the mid-IR region, which allows the ISI to choose a frequency to avoid any strong molecular or atomic lines. Hence, it is believed that the ISI sees clearly into the stellar surface and for the first time is measuring the real stellar diameter.

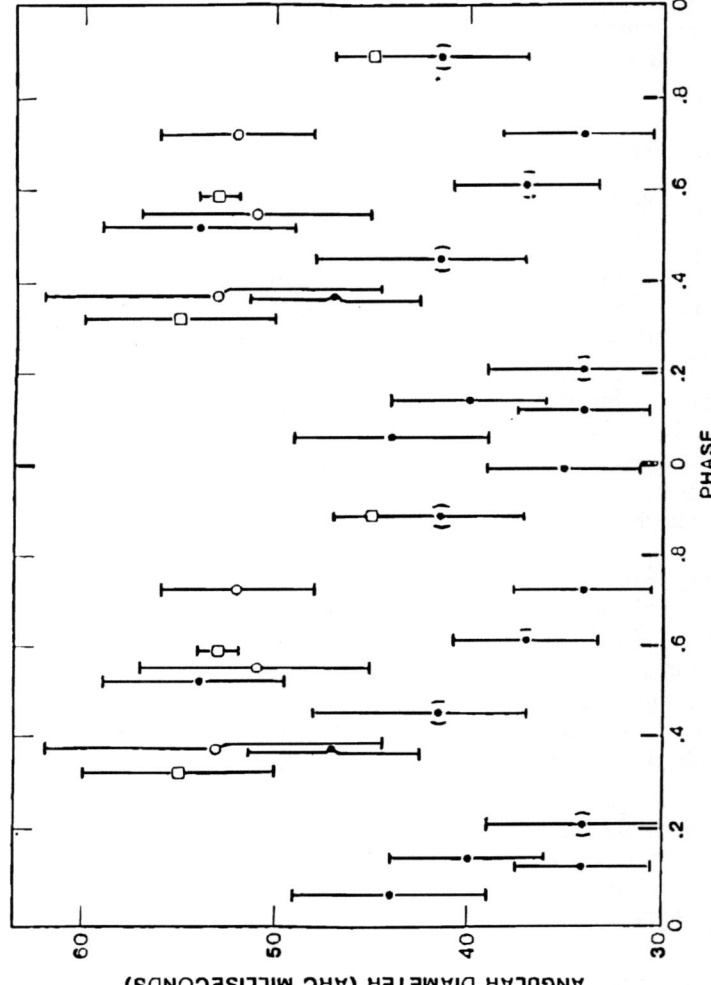

FIGURE 3. Various measurements of the diameter of α Orionis made prior to 1980 by interferometry at visible wavelengths. These have been arranged by White (1980) in accordance with the apparent phase of α Ori's somewhat irregular variations in luminosity. The measured size variation seems substantially larger than the stated probable errors of measurements. However, this now appears to be primarily due to non-uniformity and change of intensity distributions over the stellar surface rather than a simple size change.

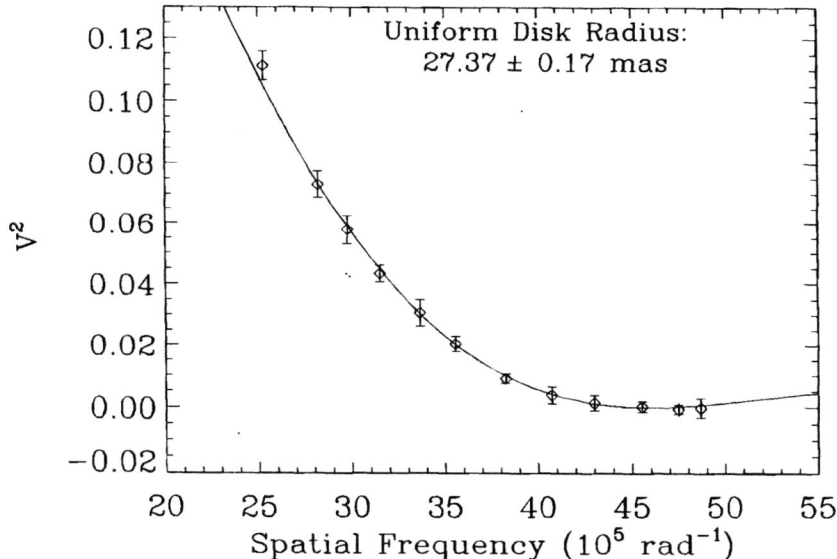

FIGURE 4. Long-baseline visibility data from 1999 for alpha Orionis (*individual points*) and uniform disk model fit (*solid curve*). The theoretical curve gives a visibility of zero at a spatial frequency of 46.0×10^5 rad^{-1}.

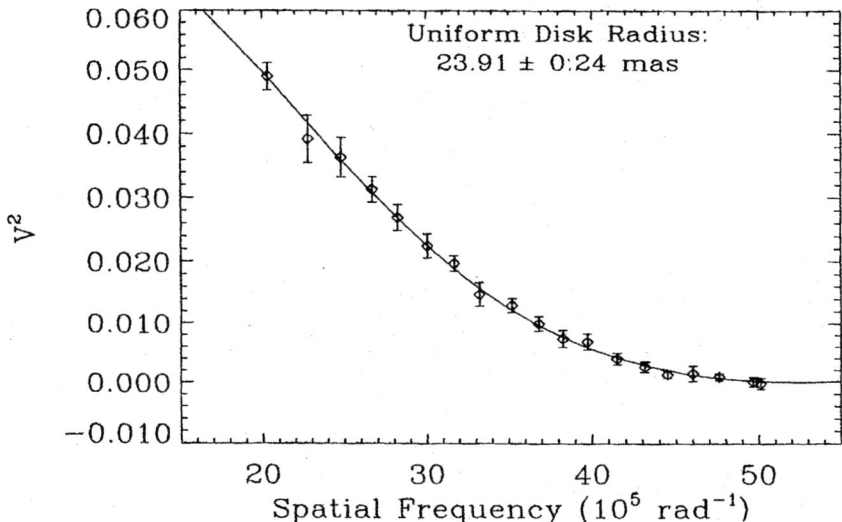

FIGURE 5. Long-baseline visibility data from 1999 for omicron Ceti (*individual points*) and uniform disk model fit (*solid curve*). The theoretical curve gives a visibility of zero at a spatial frequency of 52.7×10^5 rad^{-1}.

Nevertheless, the atmosphere of o Ceti must continue to be studied to be sure there is no reason for misleading results, because the size obtained for it is quite surprising. The apparent size may be distorted, for example by contributions of surrounding dust to the interference signal

Standard theories of the structure and dynamics of o Ceti, including the assumption that it oscillates in its lowest mode, which involves simple radial expansion and contraction, require a stellar size appreciable smaller than that measured at 11 microns. An alternative is that it oscillates in a higher mode, in which case its period of 330 days can be produced by a star of the size measured. There is, however, another promising explanation which involves a refinement and considerable changes in theoretical treatment. Aarons and Tuchman (1998) have concluded that as miras oscillate the strong radial oscillation affects the radial distribution of matter. After about 1000 cycles of oscillation, matter is redistributed in such a way that the oscillation frequency is increased. According to Aarons and Tuchman, who developed this theory involving "second order dynamics", a size for o Ceti comparable with that found by the ISI is then not inconsistent with the lowest mode of oscillation. So far, this second-order dynamical theory has not been generally accepted, but it does appear promising and may lead to an improvement in our understanding of stellar oscillations.

During an oscillation, present theory furthermore predicts a change in size of o Ceti of about 30%. Such changes have never been actually observed, and the ISI has so far measured o Ceti only near its maximum luminosity. Further measurements will be made over the coming year which should easily detect such changes. The relative phases of luminosity and stellar radius oscillations have been predicted by theory, with maximum size occurring slightly later in the cycle than maximum luminosity. We look forward to a careful examination of o Ceti's size change with time and of its phase, which should then provide an excellent check on our understanding of its dynamics.

Conclusion

Modern developments in stellar interferometry have attracted a number of physicists and astronomers to the field. Substantially improved precision has been obtained in the measurement of stellar size, details of stellar emission of material, and of temperature variations over the surface of large stars. A wide range of wavelengths are now exploited in the measurements. Surprises are being found. As a result, we can probably look forward to a very considerable development in our understanding of the characteristics and dynamics of stars.

REFERENCES

1. Ya 'ari, A, and Tuchman, Y., *ApJ*, 514, L35 (1999)
2. Tuthill, P. G., Ph.D. Thesis, Cambridge University (1994)
3. Weiner, J., Danchi, W. C., Hale, D. D. S., McMahon, J., Townes, C. H., Monnier, J. D., and Tuthill, P. G., *ApJ*, 544, 1097 (2000)
4. White, N. M., *ApJ*, 242, 646 (1980)

Author Index

A

Abele, H., 193

B

Bay, F., 72
Bennett, Jr., W. R., 123
Bickman, S., 72
Bucksbaum, P. H., 176
Budker, D., 84, 108

C

Chu, S., 21

D

DeMille, D., 72

E

English, D. S., 84, 108

F

Fiolka, D., 193
Flambaum, V. V., 232
Fortson, E. N., 39, 47
Freedman, S. J., 108

G

Ginges, J. S. M., 232
Griffith, W. C., 47

H

Harris, R. A., 186
Hinds, E. A., 62

Hudson, J. J., 62
Hunter, L., 72

J

Jackson, J. D., 3
Jacobs, J. P., 47

K

Kawall, D., 72
Kimball, D. F., 84, 108
Klein, M., 193
Krause, Jr., D., 72

L

Li, C.-H., 84, 108
Lu, Z.-T., 171

M

Maxwell, S., 72

N

Nguyen, A.-T., 84, 108

P

Perl, M. L., 156
Perlmutter, S., 253

R

Rochester, S. M., 84, 108
Romalis, M. V., 47

S

Sauer, B. E., 62
Schmidt, C. J., 193
Stalnaker, J. E., 108
Sushkov, A. O., 84

T

Tarbutt, M. R., 62
Townes, C. H., 275

U

Ulmer, K., 72

V

Vetter, P. A., 13

W

Walls, J. D., 186
Wichmann, E. H., 201

Y

Yashchuk, V. V., 84, 108

Z

Zee, A., 246
Zolotorev, M., 84, 108